DEVELOPMENTS
IN
APPLIED
SPECTROSCOPY
Volume 4

A Publication of the Chicago Section of the Society for Applied Spectroscopy

DEVELOPMENTS IN APPLIED SPECTROSCOPY

Volume 4

edited by

Elwin N. Davis

Sinclair Research, Inc.
Harvey, Illinois

Proceedings of the
Fifteenth Annual Mid-America Spectroscopy Symposium
Held in Chicago, Illinois
June 2-5, 1964

Distributed by

PLENUM PRESS
NEW YORK
1965

ISBN 978-1-4684-8693-3   ISBN 978-1-4684-8691-9 (eBook)
DOI 10.1007/978-1-4684-8691-9

Library of Congress Catalog Card No. 61-17720

©1965 Chicago Section of the Society for Applied Spectroscopy
Softcover reprint of the hardcover 1st edition 1965

PREFACE

This volume presents some of the papers from the 15th Mid-America Symposium on Spectroscopy held in Chicago on June 2-5, 1964. The Mid-America Symposium is sponsored annually by the Chicago Section of the Society for Applied Spectroscopy, in cooperation with the St. Louis, Niagara Frontier, Cleveland, Detroit, Indianapolis, and Milwaukee sections of the Society and the Chicago Gas Chromatography Discussion Group. Although basically a regional meeting, the Symposium continues to draw enthusiastic attendance from coast to coast and numerous foreign countries.

The present volume contains 45 of the 110 papers presented at the Symposium. It is with some misgiving that we offer this volume as the proceedings of the meeting with less than half of the total papers included. However, it is the opinion of the Symposium Committee that publication of the excellent material available from the Symposium provides a valuable addition to the literature in the field of spectroscopy. Response to previous volumes of this series seems to verify this opinion.

As Chairman of the Symposium and Editor of this volume, I express my sincere appreciation to the authors whose manuscripts make up this volume. I also extend my gratitude to the following members of the Symposium Committee whose time and effort made the 15th Mid-America Symposium a success. Mr. Russell J. Hansen, Exhibits; Mr. Robert L. Smick, Publicity; Mr. E. A. Piotrowski, Dr. L. S. Gray, Infrared; Miss Joan Westermeyer, Dr. Emmett Kaelble, X-Ray; Dr. E. L. Grove, Miss Vivian Biske, Emission; Dr. Herman Szymanski, Mr. Stuart Armstrong, Nuclear Magnetic Resonance; Mr. George Kincaid, Dr. William Baer, Ultraviolet; and Mr. Donald Ford, Mr. Jack O'Neil, Gas Chromatography.

ELWIN N. DAVIS

CONTENTS

X-Ray Spectroscopy

Infrared and Raman Spectroscopy

Ultraviolet and Visible Spectroscopy

Gas Chromatography

X-Ray Spectroscopy

Soft and Very Soft Fluorescence Analysis: Spectrographic and Electronic Modifications for Optimum, Automated Results

A. K. Baird, D. B. McIntyre, and E. E. Welday

Department of Geology, Seaver Laboratory
Pomona College, Claremont, California

The use of soft X-ray sources (designed by B. L. Henke, Pomona College) in a commercial vacuum spectrograph requires modifications in the optic path dependent upon the wavelength region being considered. Changes of targets, tube and counter windows, primary and secondary collimators, and analyzing crystals in a Philips spectrograph will be described; these changes yield optimum results for quantitative analyses for elements oxygen through iron. Additional instrumental and electronic modifications will be discussed which increase analytical efficiency and provide automated output in the form of punch tape. With this equipment it is possible to use computer programs which perform the following functions: detect and correct for instrumental drift, compute and statistically evaluate calibrations, test standards for contamination, and test replicates of unknowns for precision. In these programs flexibility is emphasized so that details of analytical conditions and statistical tests can be altered to fit particular requirements. Plans of instrumental changes and Fortran II program listings are available from the authors.

INTRODUCTION

Our interest in applied spectroscopy lies in obtaining large numbers of silicate chemical analyses, necessary for the determination of relationships between rock bodies and to make conclusions regarding their origins. A major problem at the present time is that, for most large rock bodies, very few chemical analyses have been obtained by conventional techniques. For example, the granitic batholith of Southern California, which extends over more than 10,000 square miles,

3

has had less than 50 chemical analyses published, an inadequate number even for conclusions concerning average composition and its variability. In 1961 we began sampling the batholith under a plan designed to test 400 field localities with over 1600 chemical analyses. To date 350 analyses for 8 elements each have been completed in this program, as well as an additional 1100 silicate analyses for other purposes. Without the rapid X-ray techniques, this research would have been impossible. Along with the running of routine analyses, continual testing of the X-ray method has resulted in increased sensitivity, precision, and efficiency. These points are to be discussed in this paper.

INSTRUMENTAL MODIFICATIONS

The original equipment used was a standard Philips vacuum path spectrograph with tungsten-target FA-60 tube and conventional electronics. Specimens, prepared by fusion [9], were ground and pressed with a bakelite backing and rim at 15 tons [1]. Specimens for oxygen analyses were plain ground powders of rocks and minerals, not fused. Our analytical precisions in 1961 are indicated in Table I. It can be noted that precisions for silicon and aluminum were worse than those of wet chemistry. Magnesium sensitivity was too poor for quantitative

TABLE I

Precisions of Silicate Analyses 1964 Versus 1961

		Standard deviation (weight percent)				
		X-ray				
Element	Weight percent	1961	1964	Wet chemistry	Emission	Flame
O	48.0	—	0.4	0.3		
Na	2.6	—	0.02	0.15	0.11	0.07
Mg	1.8	—	0.01	0.15	0.12	
Al	9.3	0.32	0.04	0.19	0.53	
Si	30.0	0.20	0.08	0.14	1.1	
K	3.7	0.05	0.03	0.21	0.15	0.07
Ca	2.5	0.03	0.01	0.07	0.14	
Fe	3.1	0.04	0.02	0.21	0.14	

analysis and sodium and oxygen were not detectable with the equipment at hand.

To overcome these problems the demountable soft X-ray tubes designed by Burton L. Henke, Pomona College Physics Department, were adapted to the vacuum spectrograph in a cooperative research program [3,4]. These tubes provide sources of high intensity X-rays, the wavelengths of which lie close to the *K* absorption edge of the light element being analyzed. Use of these tubes requires the following modifications through the optic path:

1. The X-Ray Tube. Design of the tubes is indicated in Fig. 1. Four target metals are used according to the wavelength region of interest. Change of target requires change of window material in the ultrasoft region, and these combinations are

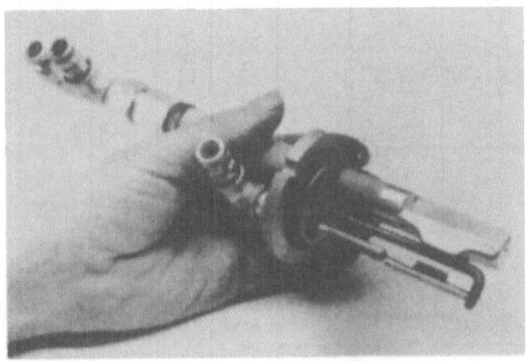

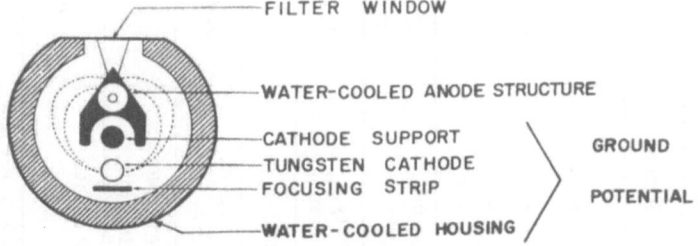

Fig. 1. Demountable source for soft X-ray excitation. Photo shows the anode—cathode structure removed from its housing. Lower drawing is a cross section through the assembled X-ray source. (Figure courtesy of B. L. Henke.)

TABLE II

Analytical Conditions

Element	B	C	N	O	F	Na	Mg	Al	Si	P	S	Cl	K	Ca	Ti	Cr	Mn	Fe
Excitation	CuL				AlK			AgL					Cr K* → CuK W* →					
Power kV / mA	6 / 330				12 / 150+			18 / 200+					20 kV 200 mA (CuK) / 40 + kV 40 + mA (W)					
X-ray tube window	2K Å parlodian				6 μ Al foil			25 μ Be					250 μ Be (Cr) 25 μ Be (Cu) / 1500 μ Be (W)					
Primary collimation	4 x 1 x 0.5 in.					4 x 0.6 x 0.035 in.							4 x 0.6 x 0.015 in.					
Crystal			stearate		kap		gypsum		EDDT or PET				LiF					
Flow counter window	2K Å parlodian					6 μ Al foil			aluminized mylar 6 μ									
Flow counter collimation	0.5 x 1 x 0.022 in.								0.5 x 1 x 0.011 in.									
Flow counter gas	methane								P-10									

*Sealed X-ray tubes.

TABLE III

Reflection Angles and Useful Ranges for Several Crystals

Element with atomic number	Å K-line	LiF 2D = 4.028	PET 8.742	EDDT 8.808	ADP 10.648	Gypsum 15.185	KAP 26.6	Stearate 100.38
				2θ				
26 Fe	1.93	57.5	25.5	25.4				
25 Mn	2.10	62.9	27.8	27.6				
24 Cr	2.29	69.3	30.4	30.2				
23 V	2.50	76.9	33.2	33.0				
22 Ti	2.74	86.1	36.5	36.4	29.9			
21 Sc	3.03	97.7	40.6	40.3	33.1			
20 Ca	3.35	113.1	45.1	44.8	36.8			
19 K	3.73	136.6	50.5	50.3	41.2	28.5		
18 Ar	4.19	—	57.3	56.8	46.4	32.0		
17 Cl	4.73	—	65.5	64.9	52.7	36.2		
16 S	5.37	—	75.8	75.1	60.6	41.4	23.3	
15 P	6.15	—	89.4	88.6	70.6	47.7	26.7	
14 Si	7.13	—	109.3	108.0	84.0	55.9	31.1	
13 Al	8.34	—	145.1	142.0	103.1	66.5	36.5	
12 Mg	9.89	—	—	—	136.5	81.1	43.7	
11 Na	11.9	—	—	—	—	103.0	53.2	
10 Ne	14.6	—	—	—	—	148.0	66.6	
9 F	18.3	—	—	—	—	—	86.9	
8 O	23.6	—	—	—	—	—	125.0	27.3
7 N	31.6	—	—	—	—	—	—	36.7
6 C	44.6	—	—	—	—	—	—	52.8
5 B	67.8	—	—	—	—	—	—	85.1

References: 1. Powers, X-Ray Fluorescent Spectrometer Conversion Tables, Philips Electronic Instruments.
2. Henke, Table I, II, 1963.

TABLE IV

Collimation for Sodium Analysis

(Three percent sodium in complex silicate, gypsum crystal, Al K excitation)

Collimation	Signal (cps)	Signal/noise
None	480	3/1
Holder only	347	15/1
1 Blade and spacers	275	16/1
2 Blades and spacers	234	13/1

indicated in Table II. In order to avoid repeated pumpdowns after the changes, three tubes are kept under high vacuum (10^{-6} mm Hg) continuously using Varian VacIon systems of 8 liter/sec capacity.

2. Collimation and Analyzing Crystal. Use of primary excitation of long wavelength, combined with crystals of large d-spacing (see Table III), allows coarse collimation to be employed without spectral interference and gives high total signal. Coarse collimation is possible only because the primary X-rays are incapable of exciting heavier-element interfering lines in either the specimen or the analyzing crystal. However, optimum collimation must be tested for each particular application; see, for example, the results of collimation tests in sodium analyses of silicate rocks (Table IV). For our applications collimation and crystal selections found to be optimum over the range oxygen through iron are indicated in Table II.

3. Detector. The standard side-window flow proportional counter is used with a P-10 gas mixture flowing at 0.5 cfh. For oxygen analyses, methane is used. Different detector collimations and window materials are selected according to the wavelengths sought and the transmission-filtering properties needed. These are indicated in Table II. It is desirable to have several replacement window-collimator assemblies for rapid changes from one element analysis to another.

4. Vacuum Optical Path. Absorption of both primary and secondary X-rays by air in the optical path becomes more critical as longer wavelength excitation is used and lighter elements are analyzed. Control of the spectrograph vacuum at a level of 100-150 μ is essential for sodium analysis; for oxygen analysis 50 μ is required. Figure 2 shows wt.% varia-

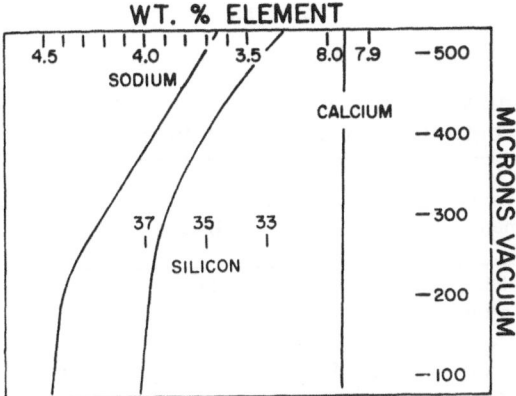

Fig. 2. Optic-path vacuum conditions for sodium, silicon, and calcium: variations in weight percent element caused by pressure fluctuations in the spectrograph. Values indicated are determined by comparison with standards read at constant pressure.

tions for calcium, silicon, and sodium as a function of vacuum in microns. Liquid air trapping of the pumpline, regulated by a valve, allows pumpdowns to 100 μ in less than 30 sec. In order to further reduce the time spent pumping, an 8-position lid was built for the spectrograph [2]. Duplicate sets of holders, machined and numbered for particular positions, allow specimen loading during analysis. Height of the specimen above the X-ray tube has been found to be critical. For example, a range of 0.008 in. in height between eight holders results in a range of 0.04% potassium in a specimen containing 2.30% potassium.

RESULTING IMPROVEMENTS IN ANALYTICAL CONDITIONS

The instrumental modifications listed above have resulted in significant improvements in our analytical conditions over those obtained in 1961. These are summarized in Table V, which shows that detectability in silicate analyses has been extended to oxygen and to near-trace quantities of phosphorus and sulfur. Additionally, a 30-fold increase in signal has been achieved for silicon and aluminum accompanied by a significant improvement in the signal-to-noise ratios.

Examples of the results obtainable for the light elements oxygen, sodium, and magnesium are given in Fig. 3. Oxygen in standard W-1 and apparently pure SiO_2 shows that good

TABLE V

Silicate Analysis 1961 vs. 1964

(Improvement by soft X-ray excitation and
optimum optic path conditions, etc.)

Element	Weight percent	1964		1961	
		Signal cps	Signal/noise	Signal cps	Signal/noise
O	53.0	9400	20	—	—
Na	3.0	450	18	—	—
Mg	0.4	190	9	—	—
Al	3.0	2000	17	70	12
Si	14.0	10000	185	290	32
P	0.02	20	4	—	—
S	0.01	45	4	—	—

counting statistics can now be obtained for this very light element (Fig. 3a). Further improvements have been described by Henke [5]. Figure 3b shows sodium and magnesium pulse amplitude distributions and line profiles for the U.S.G.S. rock standards.

Continuous testing of analytical precision in our routine silicate analyses (see descriptions below) has resulted in good estimates of the improvement over our 1961 figures. These are listed in Table I, column 4, where they may be compared with our earlier values, with wet chemistry, optical emission spectrography, and flame photometry.

ELECTRONIC MODIFICATIONS FOR AUTOMATED OUTPUT AND COMPUTER ANALYSIS

Because spectrographic conditions must be changed element by element, we have found it desirable to run unknowns in large batches. To provide calibration by comparison with standards five major problems were encountered:

1. Direct and immediate ratioing of unknowns to standards meant that as much time was spent in repeated readings of a few standards as was spent in reading all of the unknowns and was thus an inefficient operation.
2. The few good silicate standards available to us proved to yield slightly, but significantly, quadratic calibration

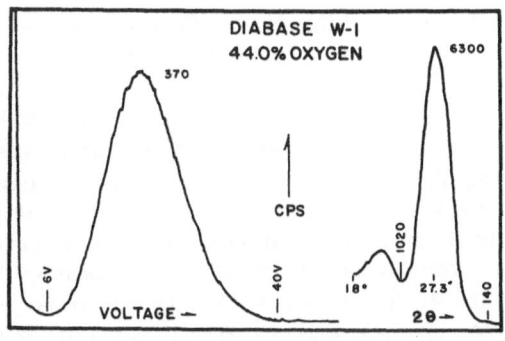

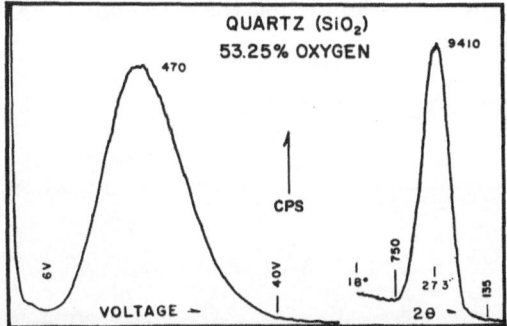

Fig. 3a. Oxygen in U.S. Geological Survey standard W-1 and quartz: pulse amplitude distributions and line profiles. Conditions of analysis are given in Table II. Pulse distribution obtained with a 1-V window; line profile with a 34-V window at settings indicated.

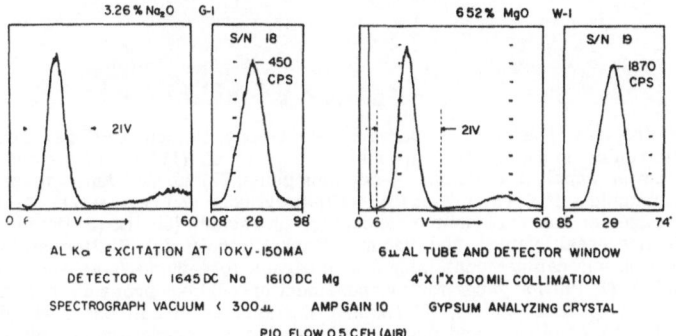

Fig. 3b. Sodium and magnesium in U.S. Geological Survey standard rocks: pulse amplitude distributions and line profiles. S/N is the signal to noise ratio.

curves for some elements. This suggested that, if more standards were used (even if less well-known in composition) and least-squares curve fitting was employed, bias could be minimized, and each unknown would be related to the pooled values from many wet analyses.

3. Counting rate drift with time was discovered, the magnitude of which varied inversely with composition [3]. This drift is in part attributable to electronics and in part to contamination of the demountable X-ray tubes. Fortunately, it can be shown to be linear with time (at least to 12 hr) and in the direction of lower rates.

4. In long elemental runs, operator errors in misreading timers and misrecording the data added greatly to overall inaccuracies.

Fig. 4. Soft X-ray laboratory, Department of Geology, Pomona College. Equipment in semicircular grouping (left to right) consists of: (1) Specimen trays on cart holding power supply for VacIon X-ray tube pump; (2) 20kV—300mA power supply for demountable soft X-ray tubes; (3) P-10 gas cylinder and regulators; (4) thermocouple gauge for measurement of spectrograph vacuum (shelf); (5) liquid nitrogen trap, bellows-seal valve, and vacuum flasks (stool); (6) Philips vacuum-path spectrograph on cabinet containing spectrograph fore-pump (note water lines to X-ray tube); (7) Philips gauge for measurement of roughing-pressure in X-ray tube during start-up (rear shelf); (8) Hamner Electronics circuit panel with detector supply, linear amplifier and pulse-height analyzer, decade scaler, two crystal-controlled timers, data scanner, rate-meter and recorder; and (9) IBM output writer and Friden 8-channel paper-tape punch to record information from the Hamner data scanner in the form of total count, total counting time to hundredths of a second, and elapsed time (from start of run) in tens of seconds.

5. Computations of weight percent from the raw spectrographic data proved to be especially tedious, and subject
to additional introduction of error when done by hand with
a mechanical calculator. It became obvious that hand
calculation seriously limited both the speed and precision
of a presumed rapid method.

In order to reduce these problems, studies of automation
techniques and computer utilization were undertaken. The
resulting operations now in use involve relatively simple
automated devices on the spectrograph and relatively elaborate
Fortran II programs designed explicitly for teleprocessing
(IBM 1013) to an IBM 7094 computer (Western Data Processing
Center, University of California, Los Angeles). Spectrographic output equipment, shown in Fig. 4, consists of elapsed
time register, tape punch, and typewriter (the latter for operator
checks of equipment error and for operator notations), in addition to conventional electronics. The computer programs are
adaptable to other systems and the reader is referred to
McIntyre et al. [8] and McIntyre [7] for flow charts, listings,
and detailed explanations and examples.

In a routine analytical run, which may include several
hundred unknowns, the following procedure is followed:

1. All standards (usually eight) are read with predetermined
counting precision in either fixed-time or fixed-count
mode (see below).
2. All unknowns, including replicates, are read; peak heights
are measured. Unusually high or low counting rates on
individual unknowns may require individual background
counting rate determinations; otherwise a standard background reading is selected. (In our work an attempt is
made to organize unknowns into groups suitable for use
with certain ranges of standards; thus uniform background
counting rates are to be expected and are used. The
validity of this assumption depends on specimens being
analyzed and on specimen preparation techniques used.)
3. All standards are reread. The form of spectrograph
output for each standard and unknown in sequence on tape-
punch and typewriter is: (a) total count (a constant for
fixed count operation); (b) total time in seconds to
hundredths of second (a constant for fixed time operation);
and (c) elapsed time from an arbitrary zero time in tens

of seconds, taken at the end of an (a) and (b) reading.

The punch-tape generated during the run is passed through an IBM Document Writer to produce cards for all subsequent computations. The typewriter output with specimen identifications and other operator notes of analytical conditions is filed with the tape as a permanent record of the analyses for possible future reference. No human recording errors can be present for the spectrograph output.

Punch cards with data for standards and unknowns are handled as follows by the computer:

1. Drift equations for the standards are computed.
2. Statistics of drift, such as magnitude and significance, are compared standard by standard with the counting precision.
3. Calibration curves for two pre-set elapsed times are computed and tested for linear or quadratic character, and tables for manual plotting are computed if desired.
4. Standards with significant departures from the calibration curve or line are marked. If marked, it suggests contaminations, damage, or misplacement in the spectrograph.
5. Determinations of unknowns: (a) For a single unknown, all standards are dedrifted to the time of running of the unknown according to (1) above; (b) a least-squares calibration curve is generated; (c) using (b), a weight percent is computed; (d) extrapolations beyond standards' range is noted; and (e) steps (a) — (d) for all unknowns and replicates are repeated.
6. Replicate unknowns are matched and differences tabulated in weight percent.
7. Standard deviation of differences in (6) is computed; any replicate pair exceeding a pre-set multiple of this value is marked for future consideration and possible re-analysis because of a gross error in preparation or handling.
8. Mean differences between replicates are tested for zero character.
9. A sign-test for + and - differences is conducted.
10. Runs of + and - differences are tested.
11. An analysis of variance is performed yielding a "within unknown replicates" standard deviation, which is the measure of analytical precision.

Output on Standards

Examples of the computer output for typical analytical runs are shown in Tables VI and VII. The following remarks explain the form of presentation in print-out (note that computations may not be performed in the same order as output).

Lines 1 and 2 contain explanatory statements of identification and conditions. Line 3 is the calibration at 60 min elapsed time, with F-test for curvature. Line 4 represents the

TABLE VI

Example of Computer Output: Detection of Gross Errors and Extrapolations Beyond the Range of Standards

---

```
SAN BERNARDINO BLOCK.   POTASSIUM RUN 92 2/12/64

COUNTS  100000. PEAK     1000. BACKGROUND
AT   60 MINS QUADRATIC NOT SIGNIFICANT   F 0.26 AGAINST 5.99
ST   4  SD IN WT 0/0   POINT 0.063  LINE 0.012  COUNTING 0.006
```

COMP	NR	MEAN	A	CPS/HR	F	CP	CL	C	DIFF	
3.63	8	8982.3	9249.4	−71.09	24.51	3.17	1.52	0.32	−0.06	C
3.25	8	8396.0	8648.2	−67.05	126.12	2.93	0.67	0.32	0.09	
2.99	8	7578.3	7811.2	−61.84	843.89	2.94	0.27	0.32	0.03	
1.92	8	5000.9	5167.8	−44.23	180.71	3.22	0.62	0.32	0.07	
1.61	8	3882.5	4002.3	−31.70	13.46	3.52	2.11	0.32	−0.07	C
1.25	8	3231.8	3339.5	−28.42	237.24	3.19	0.54	0.32	0.03	
0.81	8	2147.1	2218.6	−18.77	72.87	3.26	0.97	0.32	0.04	
0.36	8	890.9	922.0	− 8.12	26.34	3.62	1.69	0.33	−0.02	C

```
LINEAR SY/X 0.063

CALIB TABLE      −0 −0.01    5000  1.93  10000  3.87
AT 120 MINS QUADRATIC NOT SIGNIFICANT F 0.25 AGAINST 5.99
CALIB TABLE      −0 −0.01    5000  1.94  10000  3.90

ST DEVIATION OF DIFFS BTW 2ND AND 1ST READINGS 0.44    DF 185
ANY DIFFERENCE EXCEEDING 5.00*ST DEV IS MARKED X IN COLUMN 40
MEAN DIF 0.079 T 0.97 TABLE 1.97 SIGNS −53 LOW CRIT VAL 73
    42 RUNS 121 −53 LOW

A.O.V.   F 11.33 ST DEVS BTW 0.722 DF 185. WITHIN 0.318 DF 186
```

SPECIMEN	DIFF	MEAN
2−1J2324 8A	0.43	4.03H
2−1J2324 7A	−0.08	3.32
2−1J2324 7B	−0.08	3.41
BA−1	0.	0.18L
W−1	3.50X	2.26
G−1	−3.47X	2.25
.	.	.
.	.	.
.	.	.

TABLE VII

Example of Computer Output: Run Without Gross Error

```
SAN JACINTO BLOCK PART 1    RUN 111/12 9/10/64 MAGNESIUM

FIXED TIME    50.00 PEAK    20.00 BACKGROUND
AT  60 MINS QUADRATIC NOT SIGNIFICANT  F  0.11 AGAINST  5.99
ST  4  SD IN WT O/O   POINT 0.030   LINE 0.018   COUNTING 0.020
```

COMP	NR	WT	MEAN	A	DPH	F	CP	CL	C	DIFF
8.12	8	1	944.0	936.2	3.	13.	0.96	0.58	0.47	-0.02
4.29	8	1	503.5	497.2	2.	11.	1.48	0.95	0.66	-0.00
3.23	8	1	377.6	373.3	1.	35.	1.17	0.48	0.77	-0.02
1.87	8	1	219.9	216.8	1.	14.	1.59	0.95	1.04	-0.02
0.89	8	1	105.6	105.6	0.	1.	1.35	1.33	1.56	-0.01
0.60	8	1	72.7	72.7	0.	1.	3.40	3.34	1.93	0.00
0.34	8	1	42.7	42.7	0.	1.	2.62	2.63	2.66	0.00
0.04	8	1	8.8	8.8	0.	0.	8.46	8.83	5.60	0.00

```
LINEAR SY/X 0.015

CALIB TABLE      -0 -0.03    500  4.31    1000  8.65
AT 120 MINS QUADRATIC NOT SIGNIFICANT   F  0.09 AGAINST  5.99
CALIB TABLE      -0 -0.03    500  4.30    1000  8.63

ST DEVIATION OF DIFFS BTW 2ND AND 1ST READINGS 0.03    DF   87
ANY DIFFERENCE EXCEEDING  3.00*ST DEV IS MARKED X IN COLUMN 40
MEAN DIF-0.006 T -0.04 TABLE  1.99  SIGNS 26  OK TOTAL 67
   31  RUNS    26   -41   OK

A.O.V.  F5697.11 ST DEVS BTW  0.969 DF   87. WITHIN 0.018 DF   88
```

	SPECIMEN	DIFF	MEAN
SJ 1A		-0.02	.56
SJ 1B		-0.02	.58
SJ 2A		-0.01	.77
SJ 2B		0.	.78
SJ 3A		0.	.78
SJ 3B		0.01	.76
SJ 4A		-0.02	.64
SJ 4B		0.01	.68
.		.	.
.		.	.
.		.	.

behavior of the middle standard (coefficient of variation in wt.%) in comparison with counting error, with no drift correction (point) and with drift correction (line). Lines 5–13 are a table which summarizes the computations on the standards. Comp. = wt.% element; NR = number of reading on standards; mean = net counting rate for eight readings; A and CPS/HR = drift equations; DIFF = difference between predicted and accepted values for standards; C, CP, and CL = percentage variance due to counting statistics, total variance ignoring drift, and total variance correcting for drift, respectively (excessive CL or CP values are marked in last column as C); F = statistic for evaluation of significance of drift. Line 14 is the standard deviation of departures of standards from either linear or quadratic calibrations (as the case may be). Line 15 is the calibration table (at 60 min elapsed time) for manual plotting to evaluate amount of curvature, if quadratic. Line 16 is a repeat of line 3 for 120 min elapsed time. Line 17 is a repeat of line 15 for 120 min elapsed time, in order that drift magnitude in wt.% element may be evaluated.

Output on Unknowns

In the first example, 186 rocks (372 briquettes) were analyzed for potassium. Columns headed SPECIMEN, DIFF, and MEAN begin the unknowns' output; only a few are included for this example. Each line of output presents the results of a replicate pair of analyses, each of which has been carried throughout the specimen preparation procedure. MEAN composition of the replicate pair is presented to hundredths of a wt.%. DIFF is the difference between replicates.

Lines 18-22 show a statistical analysis of the determined weight percents for the unknowns. A changeable comment indicates that differences between replicate readings exceeding five times the standard deviation in line 18 will be marked with an X. As an example, in the output, specimens labeled W-1 and G-1 have this X. Consideration of the means and differences for these two specimens shows that the source of this gross error was a spectrograph operator error in which replicates for the two specimens had been interchanged in the running order.

Specimens labeled 2-1J2324 8A and BA-1 have an H and L, respectively, marked after their mean compositions. These are high and low extrapolations beyond the range of standards used. If deemed necessary, they may be rerun with appropriate standards.

The mean difference is computed (line 20) and tested to see whether it is significantly different from zero. Observed and tabled values of the t statistics are given. The results of sign and run tests are printed on lines 20 and 21.

Line 22 presents the results of an analysis of variance on the output, showing values of standard deviation in wt.% element for the total precision, for the "between" specimen variance and for the "within" specimen (between replicates) variance. The latter is the best estimate of analytical precision. In this example, it is poor because of the gross errors in specimens W-1 and G-1. Figures quoted in Table I above, however, are taken from analyses of variance in runs without gross error, such as the second example (Table VII) for magnesium analyses. In this sample, the "within" value of precision on line 22 is 0.018 wt.% Mg, which compares favorably with the counting error expressed in wt.% Mg of 0.020 (line 4). Because our analytical results are reported as the mean of two replicate preparations carried independently though the procedures, we may reduce the value for precision to 0.013% (i.e., $s/\sqrt{n}$, where n is the number of replicates, in this case, two).

The "between" specimen component of variance may or may not be useful depending on the analytical job to be done. Obviously, if the specimens analyzed are from widely varying sources, then high variance is known to be present before analysis. If, however, all specimens have been selected from a single population (e.g., one stockpile or one rock body), then this value is the best estimate of the population variability. The F-value and DF (degrees of freedom) allow tests of significance of population variability over analytical error.

IMPORTANCE OF RAPID AND AUTOMATED DATA HANDLING

In addition to the factors of speed, reduction of human error, and facility for precision tests, automation of the form described above yields permanent records of analyses which can then be used directly in various types of evaluations of the meanings of the analyses. For example, in assessing the variability on several field sampling scales, output data cards of analyses may be reordered and used as input data for further computations. Table VIII shows a typical application of this type. Aluminum, for instance, had a theoretical counting error 0.04 wt.%; this was measured to be 0.06%, estimating the contribution above the theoretical value of instrument readings. Between replicate preparations yielded a measured 0.07%. Thus, overall analytical error can be quoted as 0.05 wt.% $(0.07/\sqrt{2})$ because two analyses

TABLE VIII

Chemical Variability by Sampling Scale

Element	Weight	Standard deviation (weight percent)					
		Theory counting	Briquettes			qz monzonite	
			within	between	analytic	20*20	1000*1000
Na	2.0	0.02	0.05	0.06	0.04	0.40	0.48
Mg	0.6	0.01	0.01	0.01	0.007	0.09	0.12
Al	8.0	0.04	0.06	0.07	0.05	0.37	0.54
Si	30.0	0.09	0.04	0.28	0.20	0.85	1.20
Ca	1.5	0.01	0.01	0.02	0.010	0.26	0.26·
K	2.0	0.01	0.02	0.02	0.01	0.39	0.41
Ti	0.4	0.004	0.004	0.05	0.03	0.04	0.05
Fe	2.5	0.01	0.01	0.06	0.04	0.36	0.34

Two briquettes are included in standard analytical procedure.

per sample were made. Thirty samples taken over an area 20 × 20 ft show a variability of 0.37 wt.% Al, and over an area 1000 × 1000 ft this is increased to 0.56 wt.%. The latter values can be used to determine how many specimens must be collected

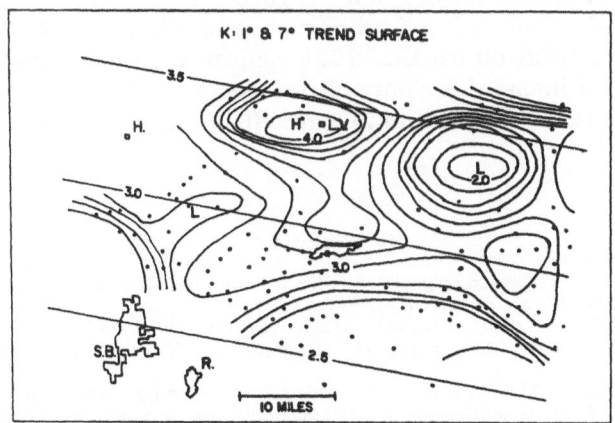

Fig. 5. Trend surface map for potassium over the San Bernardino Mountains, Southern California. Values are in weight percent element. Surfaces shown (first and seventh degree) are computer least-squares polynomial fits to the data obtained from field localities indicated as dots. At each map dot, a 600 by 600-ft square, two groups of nine rocks each were collected. Bulked samples of each group were analyzed in replicate providing measures of both analytical error and collection locality variability. For geographic reference: S. B. is San Bernardino city; R is Redlands, L. V. is Lucerne Valley, and H. is Hesperia. Big Bear Lake is indicated in outline.

in the field to yield compositional estimates with a given confidence. The importance of applications of these studies to ore body evaluation, for example, is obvious.

Other direct uses of output data cards for further computations include compositional trend analysis [6]. Figure 5 shows the distribution of potassium over 2000 square miles of granitic rocks in the San Bernardino Mountains of Southern California. The results of 350 analyses of 1500 samples from 85 localities were used in a computer program to generate least-square fits to surfaces. Both the first-degree, or planar, fit and the seventh-degree are indicated. The planar trend shows a regional gradient of 1% potassium per 30 miles. The seventh degree surface explains 80% of the total sums of squares; highs and lows in the surface can be studied for relationships to other geologic phenomena.

ACKNOWLEDGMENTS

We are indebted to Burton L. Henke for the application of his X-ray tubes to our problems of silicate analysis. His cooperation in developing our methods of analysis is greatly appreciated. Western Data Processing Center (UCLA) has contributed time on its IBM 7094 system for the computations reported. Financial support of our research was given by the National Science Foundation, Grant GP-1336.

REFERENCES

1. A. K. Baird, A pressed-specimen die for the Norelco vacuum-path X-ray spectrograph, Norelco Reptr. 8, no. 6 (1961), p. 108.
2. A. K. Baird, An eight-specimen holder for the Philips vacuum-path spectrograph, Technical Report no. 11, Department of Geology, Pomona College (1963).
3. A. K. Baird, D. B. McIntyre, and E. E. Welday, Sodium and magnesium fluorescence analysis — Part II: application to silicates, Advances in X-Ray Analysis, Vol. 6, Plenum Press, New York (1963), pp. 377-388.
4. B. L. Henke, Sodium and magnesium fluorencence analysis - Part I: method. Advances in X-Ray Analysis, Vol. 6, Plenum Press, New York (1963), pp. 361-376.
5. B. L. Henke, X-ray fluorescence analysis for sodium, oxygen, nitrogen, carbon and boron, Advances in X-Ray Analysis, Vol. 7, Plenum Press, New York (1964), pp. 460-488.
6. D. B. McIntyre, Program for computation of trend surfaces and residuals of degree 1 through 8, Technical Report no. 4, Department of Geology, Pomona College (1963).

7. D. B. McIntyre, Fortran II programs for X-ray spectrography, Technical Report no. 13, Department of Geology, Pomona College (1964).
8. D. B. McIntyre, A. K. Baird, and E. E. Welday, The use of a high-speed digital computer in light-element analysis by X-ray fluorescence, Technical Report no. 8, Department of Geology, Pomona College (1963).
9. E. E. Welday, A. K. Baird, D. B. McIntyre, and K. W. Madlem, Silicate sample preparation for light-element analyses by X-ray spectrography, Am. Mineralogist 49, 889-903 (1964).

The Soft X-Ray Emission Band Spectra of Metals and Alloys

Dr. Brian J. Thompson and Paul F. Kellen

Technical Operations Research
Burlington, Massachusetts

Soft X-ray emission bands provide information concerning the structure of the occupied levels in the valence bands of metals. The emission takes place upon transition of electrons in the levels of the valence band to vacancies in inner core levels. The density of states in the band is dependent on the particular atom and upon the lattice structure. To carry these ideas over to alloys is not, however, self-evident, especially if the alloy has a different structure than either of the constituents. In this paper, recent experimental results for both the pure metals and the alloys are reviewed with particular reference to those measurements made since the excellent review by Tomboulian in 1957 [4]. An examination is made of the effects of lattice symmetry on the structure of the conduction band.

INTRODUCTION

The structure of the occupied levels in the valence band of metals and alloys has been investigated by experimentally determining the soft X-ray emission band spectra. This emission takes place on transition of electrons in the valence band to vacancies in inner core levels. These studies have been carried out with varying degrees of success since the excellent and, in certain circumstances, unsurpassed work of Skinner and his co-workers [1,2,3]. An excellent review of the experimental methods and a discussion of the results are to be found in an article by Tomboulian [4]. Mention must also be made of the very useful annotated bibliography on soft X-ray spectroscopy compiled by Yakowitz and Cuthill [5]. A number of other reviews also exist; one of the most useful is by Parratt [6] and contains a large number of references.

It is not our intention in this particular paper to give a

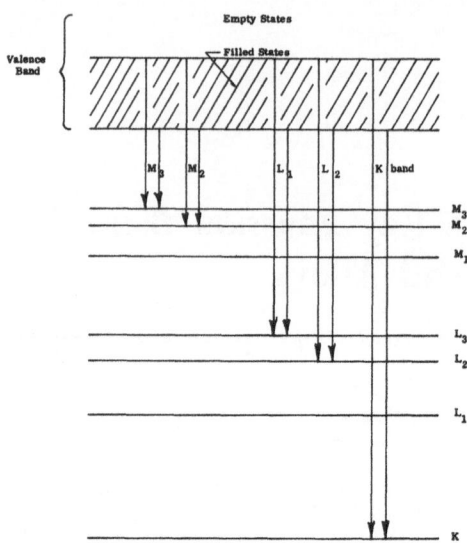

Fig. 1. Illustration of the electron transitions in emission (not to scale).

cataloged review of recent measurements, but rather to take a critical look at the subject and to attempt to determine the main problem areas. Of particular importance is the extent to which the model of soft X-ray emission in a pure metal may be extended to alloys.

The emission spectra then should provide direct information concerning the outer occupied levels. This point is usually

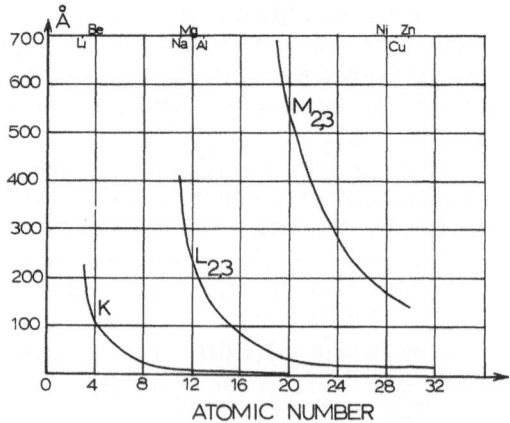

Fig. 2. The spectral position of the K, L, and M bands for various elements.

illustrated by the so-called "one-electron-jump" diagram as shown schematically in Fig. 1. In this diagram, the vertical arrows indicate single electron transitions from occupied levels in the valence band to inner core vacancies; the allowable transitions produce the characteristic K, L, and M spectra. As Parratt points out, this type of diagram is unsatisfactory for real interpretation; however, it still retains its illustrative value. The spectral range in which the soft X-ray spectra occur is 5–500 A. The disposition of the various spectra is indicated approximately in Fig. 2. The individual spectra give information about the various types of electron states in the valence band; i.e., K spectra give a measure of the density of states of p symmetry, and L spectra, the density of states of $s + d$ symmetry.

EXPERIMENTAL METHODS

The experimental determination of soft X-ray emission spectra is accomplished with the use of a concave-grating, grazing-incidence, vacuum spectrograph as shown diagrammatically in Fig. 3. The specimen is the anode of an X-ray tube, and the emitted radiation is incident at grazing incidence onto a concave grating whose radius of curvature is usually of the order of 1 m. The entrance slit and the grating are both placed on the Rowland circle, and the spectrum is formed and recorded on this same circle. The grating normally has about 1000 lines/mm and is ruled on either glass or aluminum. The spectrum may be recorded in a number of ways. Photographic

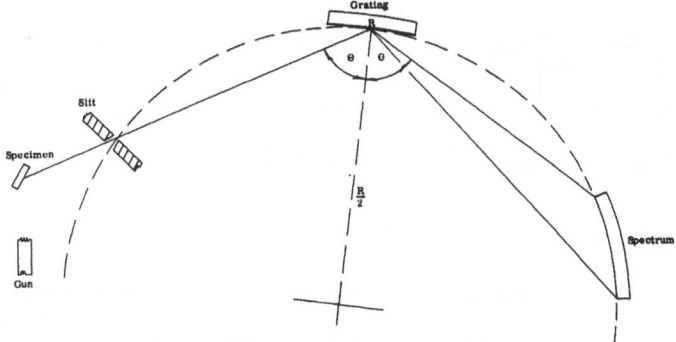

Fig. 3. Schematic diagram of the spectrograph.

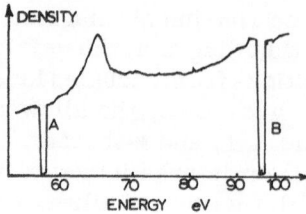

Fig. 4. Typical microdensitometer trace from a photographic record of a Ni *M* spectrum; A and B are fiducial marks for calibration purposes [7].

recording has been the most widely used, but photoelectric and Geiger-counter methods are also becoming of considerable use. However, all the types of recording have their own difficulties, and, while at present photographic methods are being super-seded, new emulsions and emulsionless films as well as improved microdensitometry may still give the photographic method an advantage. Figure 4 shows a typical micro-densitometer trace taken from a photographic record of a Ni $M_{2,3}$ spectrum. A and B are fiducial marks for calibration of the observed spectrum. The main problem in converting this curve to an intensity vs. energy curve for the band alone is that of removing the background. This problem is, of course, common to all procedures and is overcome by taking a record from a target that has no features in the area of interest.

SOME FEATURES OF EXPERIMENTAL RESULTS

Figure 5 shows a selection of typical results. Figure 5a is the L_3 emission spectrum of magnesium as determined by Cady and Tomboulian [8]. The main peak in this curve indicated by

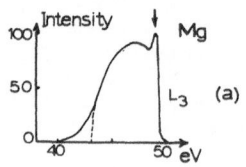

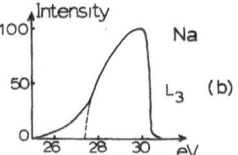

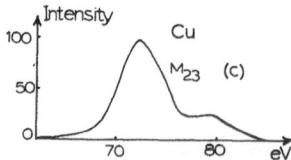

Fig. 5. Some typical band spectra: (a) L_3 emission spectrum of magnesium [8], (b) L_3 emission spectrum of sodium [8], and (c) $M_{2,3}$ emission spectrum of copper [7].

the arrow is almost coincident with the Fermi energy. This and other features of the curve are explained from the Brillouin-zone model; the first maximum occurs when the first zone is crossed, and then the intensity decreases until the position of minimum energy of the second zone is reached. The contribution from the second zone starts to increase, and the maximum of the curve is produced by the overlap of the first and second zones. To produce this curve, the L_2 emission band has been subtracted. In Fig. 5b, the L_3 emission spectrum of sodium [8] is shown. Here, the L_2 band has not been subtracted since it is fifty times less intense than the L_3 band and therefore can be neglected. The dotted line shows the low-energy $E^{1/2}$ variation. Once again, the position of the high-energy edge agrees well with the Fermi energy. Unfortunately, only very few metals exhibit these sharp edges in their emission spectra. More commonly, a spectrum similar to that shown in Fig. 5c is observed. This figure shows the $M_{2,3}$ emission bands of copper [7]. Here the ratio of intensities of the two bands is only 2:1 and, hence, they are more difficult to separate. The extra peak at about 80 eV is a satellite band arising from transitions involving atoms which are doubly-ionized in the inner shells; the double ionization is caused either directly or by an Auger process.

THE PROBLEM OF INTERPRETATION

The examples chosen above illustrate some typical spectra and also show some of the difficulties of interpretation. To determine the actual shape of the bands, it is necessary to estimate the effect of the low-energy tail, the origin of which is accounted for by either nonradiative transitions of electrons within the valence band or by an Auger effect. At the high-energy end of the spectrum, a satellite band may have to be subtracted; this is particularly troublesome when the satellite is not widely separated from the main band. When this is done, a reliable value for the width of the band can be obtained which characterizes the energy spread of the occupied levels. The remaining spectrum still has to be interpreted with care, with full allowance being made for the resolution limit set by the limiting aperture of the system, as well as by the effective width and shape of the inner level involved in the transition. The instrument effect (spectral window) can be reliably estimated.

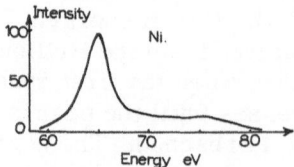

Fig. 6. $M_{2,3}$ emission spectrum of nickel [7].

The width of the inner level can sometimes by estimated from the appropriate line spectrum.

RESULTS

The main effort in recent years has been in determining the effect on the spectrum of alloying a metal. These and other studies on the pure materials themselves have in some instances yielded added information about the pure metal spectra, although most have only confirmed earlier results. As an example, let us mention the recent report [7] that the Ni $M_{2,3}$ band has a much longer high-energy projection forward from the main band. This result is illustrated in Fig. 6. Previous determinations had shown a cutoff in the region of 70 eV.

The simple idea of a common valence band shared by both constituents of an alloy is not generally supported by the experimental evidence. The general result seems to be that the separate identities of the bands are retained on alloying, even if some minor (and perhaps not always real) differences are observed. Let us illustrate these conclusions by a few examples.

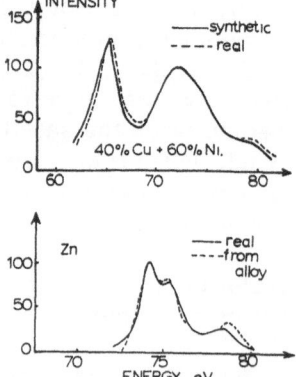

Fig. 7. (a) Comparison of 40%Cu—60%Ni alloy spectrum with synthetic curve derived from pure metal spectra [7]. (b) Comparison between pure zinc spectrum and zinc spectrum obtained from alloy [9].

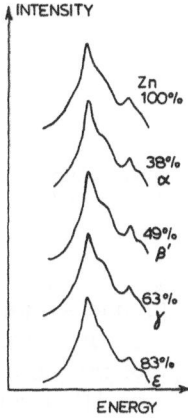

Fig. 8. The L spectra of zinc in the alloys of Cu—Zn [10].

The $M_{2,3}$ emission bands for the whole range of Cu—Ni alloys has been extensively investigated by Clift, Curry, and Thompson [7]. The main result can be summarized by Fig. 7a. A synthetic curve obtained by superposition of pure copper and pure nickel spectra in appropriate amounts is compared with the real spectrum for a 40%Cu—60%Ni alloy; obviously, no distinction can be made between the curves. It is to be noted, however, that the copper and nickel atoms are quite similar and that the face-centered structure persists through the whole range of the alloys. Figure 7b shows the zinc spectrum obtained from an α-brass (30%Zn—70%Cu) [9]. Again, the two spectra are indistinguishable even though a change in structure is involved, since α-brass is face-centered and pure zinc is hexagonal. This result is also confirmed by the L spectra determined by Rumyantsev and Korsunskii [10], illustrated in Fig. 8, which covers the whole range of phases. These workers claim that the differences, though very slight, are significant. Other results on different alloy systems also support the results of the copper—nickel experiments. Catterall and Trotter [11] found that the M band of copper apparently remained the same when alloyed with gold in both the ordered and disordered states.

There are, however, a number of results that do not entirely agree with the results discussed. It is perhaps significant that all these results concern the alloys of lithium, magnesium, and aluminum. Farineau [12] worked with the ordered compounds Mg_2Al_3 and Mg_3Al_2, and reported that the Al K and Mg K bands were alike in any one alloy and that the breadth increased

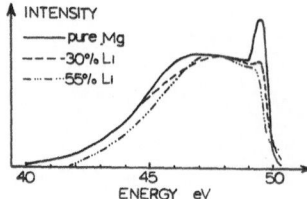

Fig. 9. The L spectra of magnesium in lithium alloys [13].

progressively from magnesium through Mg_3Al_2 and Mg_2Al_3 to pure aluminum. Catterall and Trotter [13] found evidence of electron transfer from magnesium to lithium in their work on the K emission bands of lithium and L bands of magnesium in evaporated alloys. Some of their results are illustrated in Fig. 9. These results are confirmed by Crisp and Williams [14] using solid specimens as well as evaporated alloys. It is perhaps not without meaning that the effect was more pronounced in the evaporated alloys.

DISCUSSION

If the emitting atom contains two vacancies in its inner shells, then a shifted satellite spectrum appears. This is due to a change in the energy of the final state. One might then be led to expect a change in the general features or location of the emission band if the initial states are perturbed, such as in alloying. However, in the copper—nickel and copper—zinc systems, the results can be obtained by adding the spectra of the constituents in proper proportions. This is especially disturbing in the copper—zinc case since the copper is found in a lattice of symmetry different from that of the pure metal.

It has been suggested that a single energy band would be formed in the case of an alloy and that the emission band for each constituent would change to some intermediate shape. This shape would be identical for each constituent save for the relative intensity and position on the energy scale. This is clearly not the case. This apparent indifference of the density of states $N(E)$ to their surroundings may be due to a change in the transition probability $T(E)$, which just compensates for the changes since the product $N(E)\,T(E)$ is what is measured, but this is unlikely.

A similar problem appears in the study of the cohesive

energy of alloys and metals.   There does not seem to be any significant contribution from the coordination number or the lattice symmetry.   Rather, the critical factor seems to be atomic volume.   The heat of solution is markedly dependent on the ratio of the specific volumes of the solvent and solute (it is small when this ratio is nearly one) and is independent of charge and symmetry consideration.   Mott [15] suggests that electron correlations explain this volume dependence in the case of alloys of monovalent and divalent metals.

This correlation can be introduced by taking states of properly symmetrized pairs of wave functions.   Then, in the alloy, one of the electrons of the divalent constituent appears in a localized state.   This state is not a physically separate bound state and so does not contribute to the paramagnetism. It can be thought of as arising from the states above the Fermi surface, and so can have an s-orbital character.   In the case of nearly-completed d-orbitals, strong correlation forces may act within the atomic polyhedra to give a solid-core aspect to these states.   This would explain why nickel has a higher cohesion energy than copper, though less of its electron density is in the bonding $4s$ state than in the antibonding $3d$ states.

At the edges of the Brillouin zone, the energy of the one-electron states is depressed relative to the free-electron values. Differences in crystal structure will make small contributions to the free energy by shifting the Fermi level in this depressed region.   However, strong correlation forces will dominate this effect and the location of the Fermi level will be more sensitive to these forces.

It is suggested here that the effect of alloying metals appears mainly at the band edge which is masked by overlapping bands in the copper systems.   The changes in the band edges of magnesium and lithium in alloy form have already been mentioned.   There still is, however, a need to explain the fact that the result appears more markedly in the cases where the alloy was formed by evaporation of the two metals than in spectra obtained from more conventionally formed alloys.   In any case, it is in systems which have a sharp emission edge, such as magnesium, that one would expect to find the most pronounced effects.

Perhaps the greatest need at the present time is to further investigate alloys in which one or both of the constituents exhibit sharp emission edges.

REFERENCES

1. H. W. B. Skinner and J. E. Johnson, Proc. Roy. Soc. (London) A161, 420 (1937).
2. H. W. B. Skinner, Phil. Trans. Roy. Soc. (London) A239, 95 (1940).
3. H. W. B. Skinner, T. G. Bullen, and J. E. Johnson, Phil. Mag. 45, 1070 (1954).
4. D. H. Tomboulian, Handb. Phys. 30, 246 (1957).
5. H. Yakowitz and J. R. Cuthill, N. B. S. Monograph No. 52 (1962).
6. L. G. Parratt, Rev. Mod. Phys. 31, 616 (1959).
7. J. Clift, C. Curry, B. J. Thompson, Phil. Mag. 8, 593 (1963).
8. W. M. Cady and D. H. Tomboulian, Phys. Rev. 59, 381 (1941).
9. J. Clift, C. Curry, and B. J. Thompson, Phil Mag. 8, 639, (1963).
10. I. A. Rumyantsev and M. I. Korsunskii, Opt. Spectr. 7, 498 (1959).
11. J. A. Catterall and J. Trotter, Proc. Phys. Soc. (London) 79, 691 (1962).
12. J. Farineau, Ann phys. (Paris) 10, 20 (1938).
13. J. A. Catterall and J. Trotter, Phil. Mag. 4, 1164 (1959).
14. R. S. Crisp and S. E. Williams, Phil. Mag. 5, 1205 (1960).
15. N. F. Mott, Rept. Progr. Phys. 25, 218 (1962).

Demountable X-Ray Tube for Light Element Fluorescence Analysis

J. A. Dunne* and W. R. Muller

Philips Electronic Instruments
Mt. Vernon, New York

A primary X-ray source suitable for the fluorescence analysis of elements ranging from chlorine to boron ($Z = 17 - 5$, respectively) is described. The tube, based on the concept of Prof. B. L. Henke, Pomona College, Claremont, California, has been designed to allow utilization in the general X-ray laboratory. Targets can be changed and filament replaced. A vacuum isolation gate is provided to permit the convenient utilization of extremely thin windows, which can be changed without disturbing tube vacuum. Separate water-cooling systems are provided for the anode itself and the outer shell. Detailed data on tube operating parameters are given, and the X-ray generator designed to power this tube is described.

INTRODUCTION

This paper describes a commercial execution of the Henke demountable ultrasoft X-ray fluorescence analysis tube. The original tube has been described by Henke [2]. The principal objective of the design described here was the evolution of a practical, safe (shockproof), and easily operable device for the generation of the primary radiation required for the efficient excitation of the fluorescence spectra of the elements ranging from chlorine to boron. This design program has resulted in the development of an integrated ultrasoft fluorescence analysis system.

THE X-RAY TUBE

A schematic diagram of the demountable tube is given in Fig. 1. Principal features include a positive accelerating voltage

*Present address: Jet Propulsion Laboratory, Pasadena, California.

applied to the water-cooled anode, grounding of all other
structures, and placement of the filament in a position "hidden"
from the electron impact area. The first characteristic avoids
high-speed, elastically-scattered, electron impact in the window, and the second, tungsten contamination of the anode.
These effects would preclude the use of thin windows and high-
beam currents, respectively, in conventional X-ray tube geometry.   In addition, the field configuration characteristic of
the Henke geometry results in space change limitation at quite
low voltages, a feature necessary for the high-current, low-
voltage operation required for the efficient production of high
fluxes of low-energy primary radiation. The electrical characteristics of the tube operated full-wave, single-phase and
full-wave, three-phase (virtually constant potential) are given
in Figs. 2 and 3.   Figure 2 plots anode current as a function
of anode voltage, with filament current as the parameter. The
sharper knee in the full-wave, three-phase mode of operation
should be noted. Figure 3 shows anode current as a function of

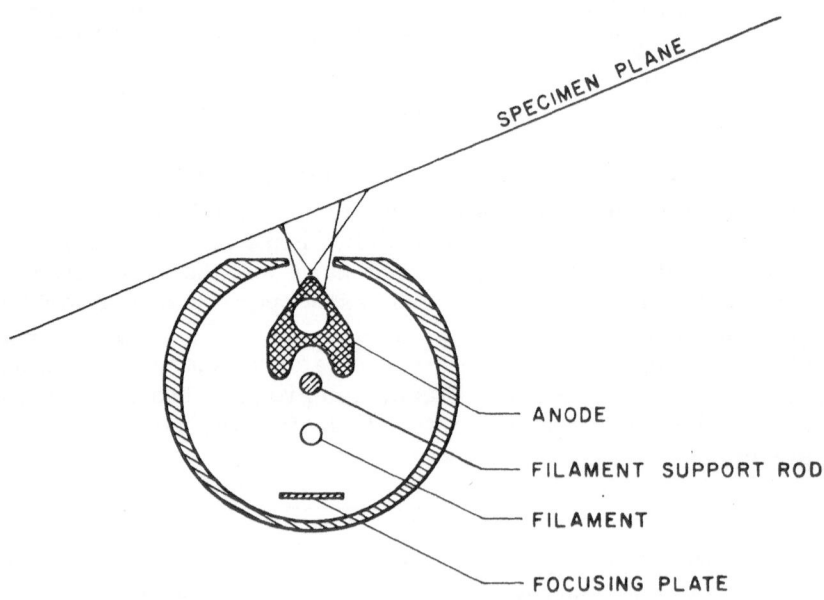

Fig. 1. Schematic diagram of the demountable tube.

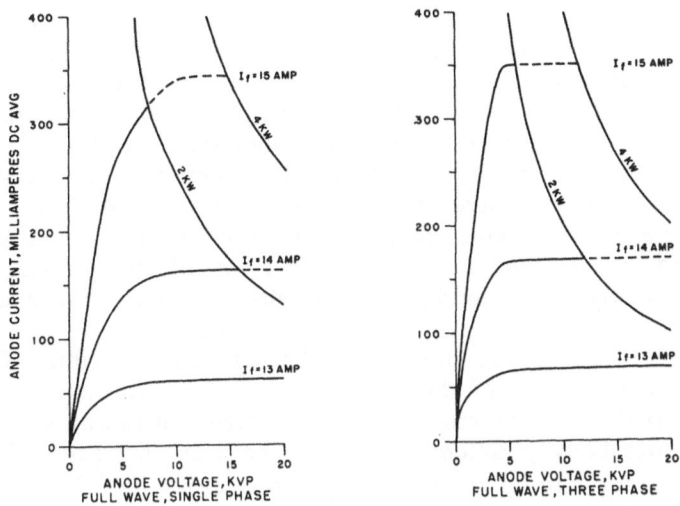

Fig. 2. Anode voltage vs. anode current for full-wave, single-phase and full-wave, three-phase operation.

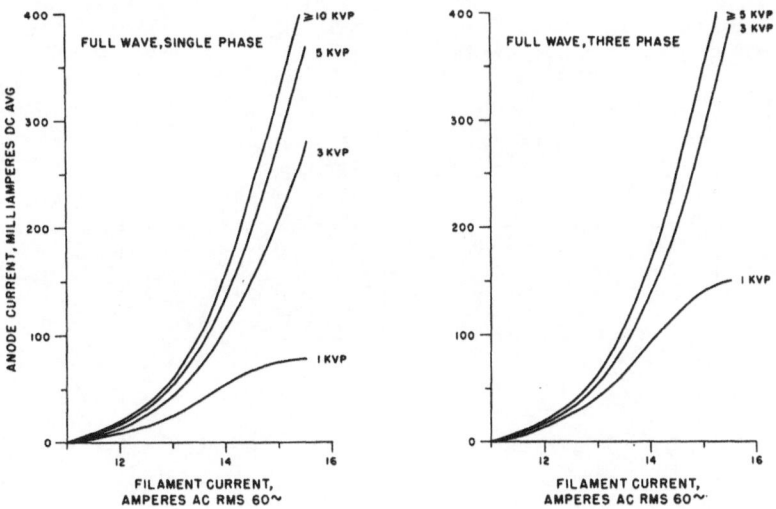

Fig. 3. Filament current vs. anode current for full-wave, single-phase and full-wave, three-phase operation.

filament current, with anode voltage the parameter. Since a filament current of 17 A is allowable, it can be said that the region of space change limitation of anode current extends only up to 3 KVP full-wave single-phase. The load lines shown in Fig. 2 indicate that anode loading limits prevail once the space change limited region is exceeded. It should be noted here that the anode loading limits of 2 kW for aluminum anodes and 4 kW for copper anodes have been arbitrarily determined. No tests have been conducted to fix the true limits, which are known to exceed those selected in light of many hours of operation at the 2 kW and 4 kW points with no evidence of anode damage.

Figure 4 shows the internal construction of the tube. The main vacuum seal can be seen on the forward edge of the ceramic body. An O-ring between the base of the anode and the ceramic body provides a vacuum seal in that area. The filament leads can be seen entering into the vacuum line. The locating pin just above the base of the anode assures accurate

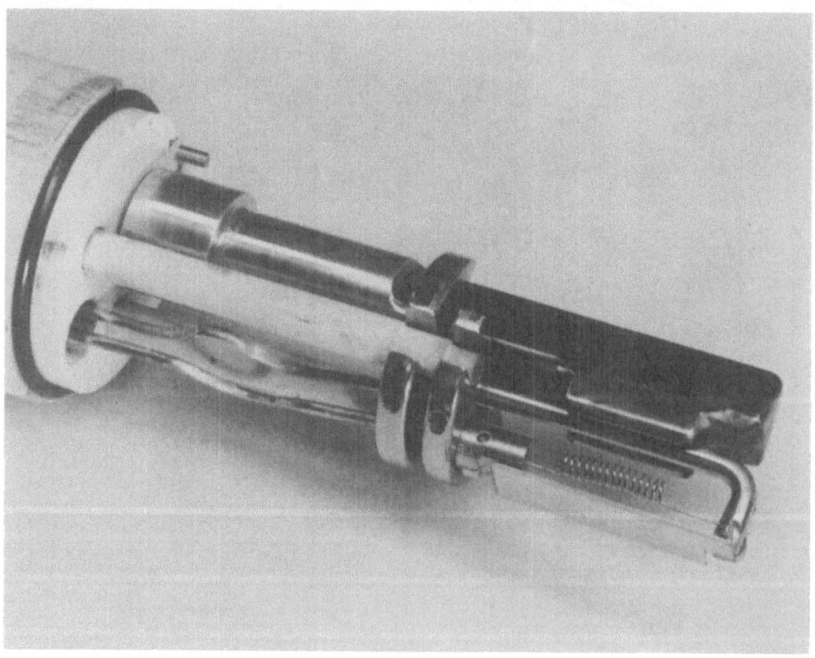

Fig. 4. Internal tube construction.

positioning of the assembly within the tube jacket. The rather massive filament structure support plates provide both structural support and optical shielding, minimizing problems of electrical creepage along the ceramic surface from the anode to the grounded tube jacket, which would result from the buildup of evaporation deposits in this area. The ease of filament replacement is evident. Anode removal is readily accomplished from the rear by the use of a standard socket wrench extension. An exploded view of the entire tube assembly is given in Fig. 5. The flexible vacuum bellows terminated with a quick connect fitting allows convenient alignment of the tube in the spectrograph. The exterior water lines are for cooling the tube jacket. Anode cooling water is brought in through the high-voltage cable.

Since the usefulness of the tube in practical fluorescence analysis requires windows of appropriate X-ray transmission, special arrangements must be made to allow the use of extremely thin and fragile windows. The vacuum isolation arrangement which has been adopted is shown in Fig. 6. An operating tube is shown as viewed through a lucite test plate on the vacuum spectrograph sample chamber. When the spectrograph is at operating vacuum, the window sees a pressure differential of the order of 10^{-4} bars. When air must be admitted into the chamber, the sliding mechanism is moved into the blanked off position. It follows that windows can be readily interchanged without disturbing tube vacuum. Figure 7 is a photograph of the window plates used. The plate on the left is shown upside down to illustrate the position of the O-ring groove. The window material shown is the $1-\mu$ polypropylene film used on copper anode tubes. The window will last for about 6–8 hr of operation, and so must be replaced daily. Window replacement, however, takes only a few minutes.

The vacuum system supplied with the tube consists of a 15-liter/sec ion pump, two high-vacuum valves, and appropriate accessory fittings. Ion pumps enjoy several advantages in this application, not the least of which is relative freedom from hydrocarbon backstreaming contamination of the X-ray tube anode. This effect is illustrated in Fig. 8, which is a plot of measured fluorine and boron fluorescence emission as a function of running time for oil diffusion and ion pumping systems. The loss in fluorine intensity results from hydrocarbon contamination of the anode surface, which lowers

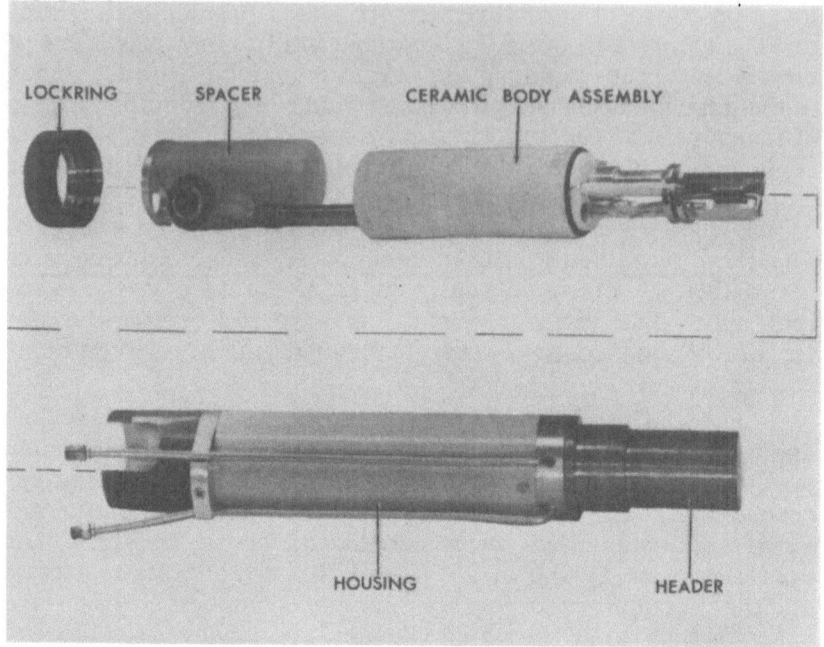

Fig. 5. Exploded view of the entire tube assembly.

Cu L_a primary emission due to incident electron energy loss in the contaminating film. A corresponding increase in B K_a intensity is observed, reflecting the greater efficiency of C K_a(44 A) as opposed to Cu L_a(13 A) in the excitation of boron ($K_{ab} \cong 64$ A). It should be noted that only two boron readings are shown for each sequence, and the dotted lines connecting these points have no experimental significance. In order to encourage the hydrocarbon buildup, the anode was run overcooled (2 kW), but the indicated relative performance of the two pumping systems is nevertheless valid. From the data shown in Fig. 8, it can be said that the ion pump system is far less likely to present severe contamination problems, although more efficient cryogenic trapping arrangements on the oil diffusion system could be envisioned (a single-stage liquid-nitrogen trap located in the throat of the pump was used). An additional advantage of the ion pump lies in its freedom from mechanical pump backing, which results in more reliable and quiet operation. Unexpected power failures will not result in downtime for system cleaning. One of the valves supplied with the system is provided so that the ion pump can be isolated from the tube, and thus can be run continuously.

X-RAY GENERATOR

The high-voltage generator for powering the tube described above is designed to deliver the required heavy currents with low voltage drop, to withstand the occasional arc-over that may be anticipated in the continuously pumped tube, and to protect the tube in the event of arc-over, overload, or vacuum failure. Low voltage drop is obtained through the use of a full-wave bridge silicon rectifier and a low-impedance transformer. The rectifier bridge consists of series-connected, individual diodes properly compensated and protected by a transient voltage clipper. The low transformer impedance is

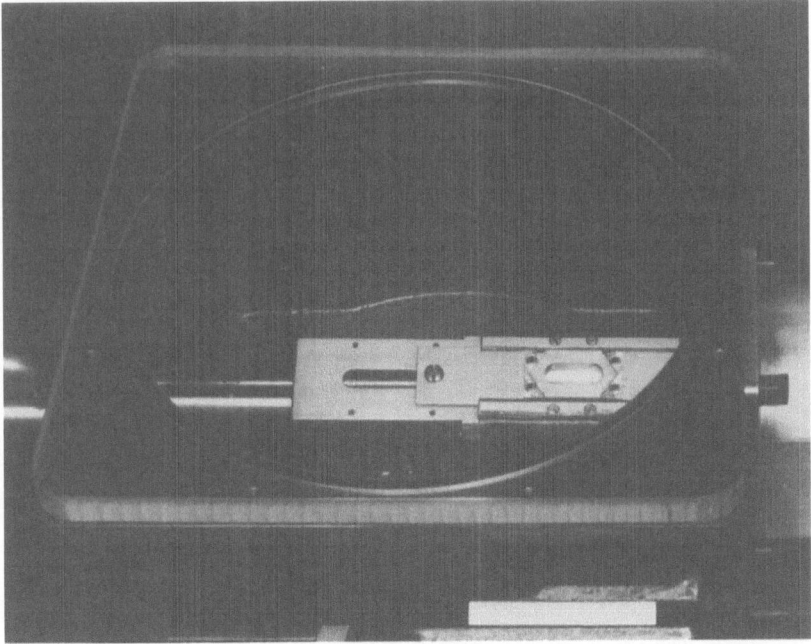

Fig. 6. The vacuum isolation arrangement; the operating tube can be seen through the lucite test plate on the vacuum spectrograph sample chamber.

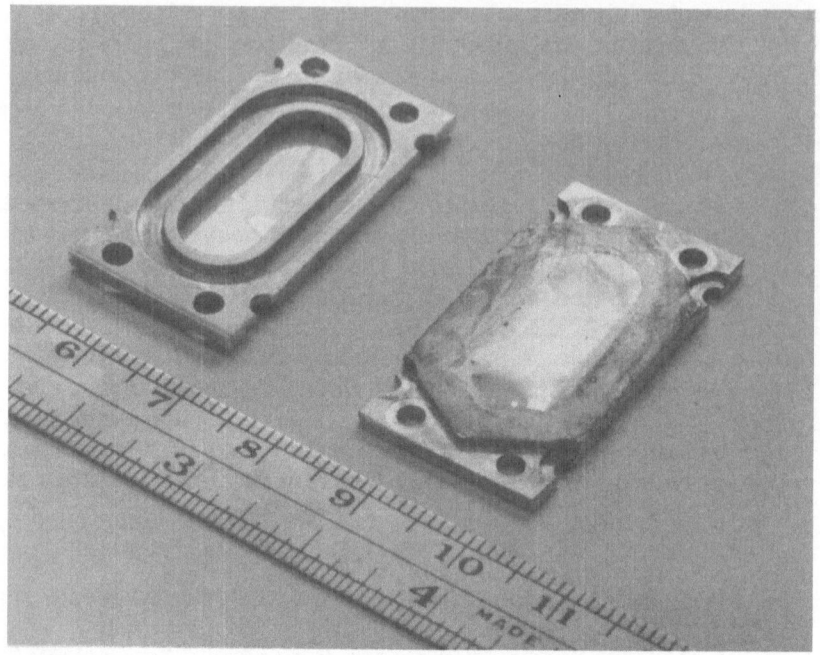

Fig. 7. Window plates made of 1-μ polypropylene film. The left plate (upside down) shows the O-ring groove.

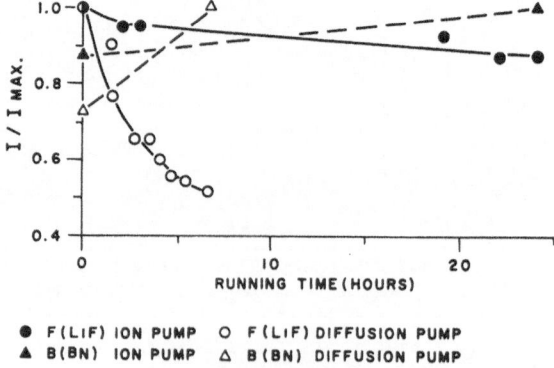

● F(LiF) ION PUMP   O F(LiF) DIFFUSION PUMP
▲ B(BN) ION PUMP   △ B(BN) DIFFUSION PUMP

Fig. 8. Fluorine and boron emission as a function of running time for oil diffusion and ion pumps.

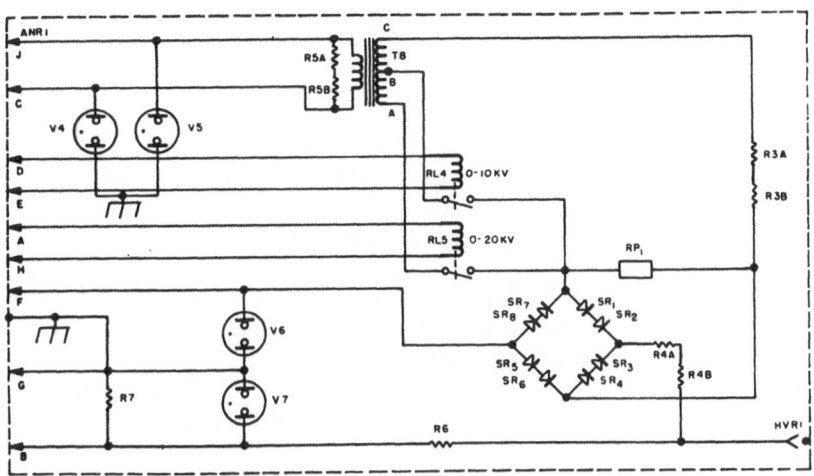

Fig. 9. Schematic diagram of the high-voltage tank.

further reduced in the low-voltage range where the heaviest currents are delivered by using only half of the secondary winding. On the schematic of the high-voltage tank in Fig. 9, note the externally activated relays RL-4 and RL-5, which switch the rectifier input between center tap and full secondary winding. The efficiency of the transformer—rectifier system has been found to exceed 95% at 4 KWP. Arc-over protection for the power supply is afforded by high-speed circuit breakers and by gaseous and solid state transient clippers. Since there is no filter in the higher-voltage supply and consequently no significant energy storage, the energy delivered to the tube in the event of arc-over is insufficient to cause damage. Further protection for the tube is effected by a high-speed circuit breaker, set at 500 mA, which senses the tube anode current and removes power in event of overload. In addition, provision is made for de-energizing the tube in the event of vacuum failure or failure of coolant water flow.

Figure 10 is a diagrammatic representation of the integrated ultrasoft fluorescence analysis system, consisting of a vacuum spectrometer, the demountable tube and its vacuum system, the X-ray generator, and analyzing electronics. The water separator is an oil-insulated tank which separates high voltage and anode cooling water, grounding the latter through two, parallel, ten-foot lengths of tygon tubing. Typical losses in the

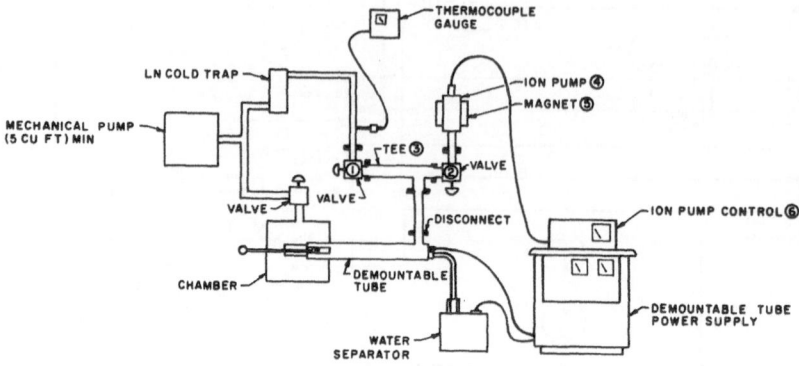

Fig. 10. The integrated ultrasoft fluorescence analysis system.

TABLE I

Minimum Detectable Limits of Some Light Elements Using Ultrasoft X-Ray Fluorescence Analysis

Element	Wt. % analyzed	Matrix	3σ Minimum detectable limits (wt. %)
Sulfur	0.01	Silicate rock	< 0.001*
Phosphorus	0.02	Silicate rock	< 0.001*
Silicon	13.6	Silicate rock	0.003*
Aluminum	3.0	Silicate rock	0.006*
Magnesium	0.4	Silicate rock	0.003*
Sodium	3.0	Silicate rock	0.013*
Oxygen	2.18	Fluorspar	0.08†
Carbon	0.11	Steel	0.03†

*Data from Baird.
† Data from Henke.

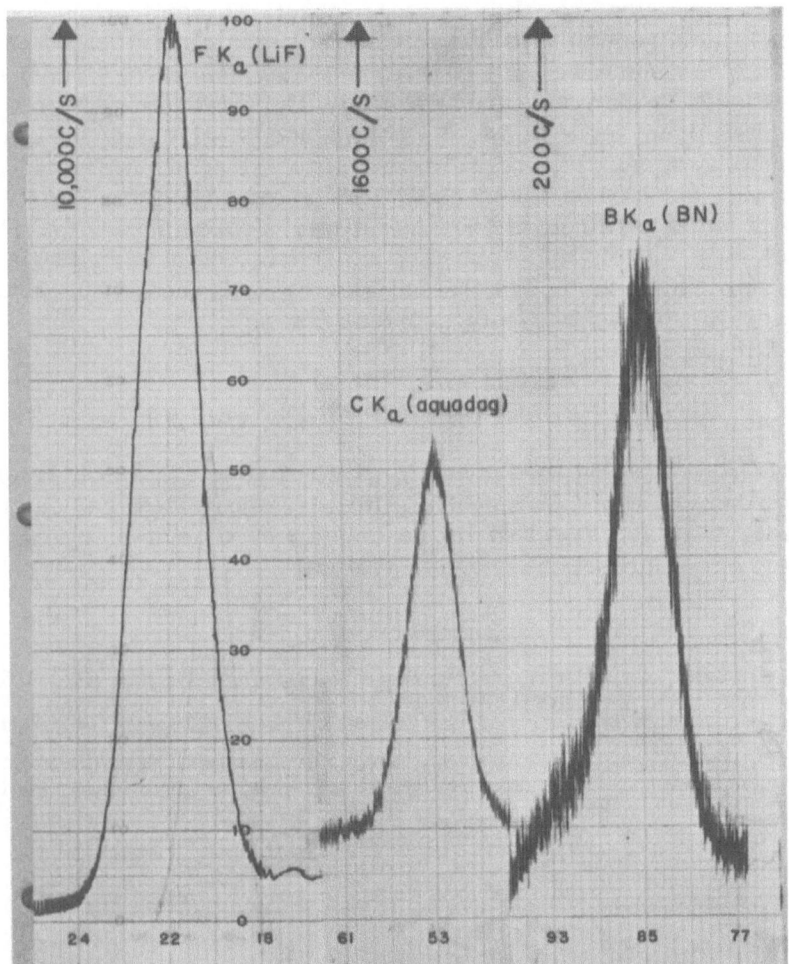

Fig. 11. Spectral scans of fluorine (F K_α—LiF), boron (B K_α—BN), and carbon (C K_α—aqua dag) using the integrated ultrasoft fluorescence analysis system (Fig. 10).

water path amount to about 1—2 mA at 20 kV using Mt. Vernon city water.

APPLICATION

Results obtained from systems such as the one described here have been given by Henke [2], Baird [1], and Volborth [4]. Henke's developments have brought ultrasoft X-ray fluorescence analysis to the point of practical application with reasonably good sensitivity. Table I gives minimum detectable limits for a number of light elements calculated from results published by Baird [1] and Henke [3]. The minimum detectable limits given refer to the 3 σ confidence level using 100-sec counting time. Figure 11 shows spectral scans of fluorine, boron, and carbon taken using the system shown in Fig. 10. Instrument parameters were as follows: Copper anode, 6 KVP, 200 mA; collimation 4 by 0.5−in. source, 0.75 by 0.020−in. detector; analyzing crystal, lead stearate−decanoate; tube and detector window, 1−μ polypropylene film.

ACKNOWLEDGMENT

Grateful acknowledgment is given to Prof. B. L. Henke of Pomona College, Claremont, California, who guided the design program which resulted in the development of an integrated ultrasoft fluorescence analysis system.

REFERENCES

1. A. K. Baird, D. B. McIntyre, and E. D. Welday, Advances in X-Ray Analysis, Vol. 6, Plenum Press, New York (1962), pp. 377–388.
2. B. L. Henke, Advances in X-Ray Analysis, Vol. 5, Plenum Press, New York (1961), pp. 288–305.
3. B. L. Henke, Advances in X-Ray Analysis, Vol. 7, Plenum Press, New York (1963).
4. A. Volborth, Materials Science and Technology for Advanced Applications, Vol. II (1964), pp. 117–142.

Relationship Between X-Ray Tube Target Materials and X-Ray Emission Intensities

F. Bernstein

General Electric Co.
Milwaukee, Wisconsin

Factors affecting fluorescent intensity from a sample in an X-ray spectrograph are reviewed. Excitation efficiencies of monoenergetic X-ray sources are discussed and data on low atomic number elements are given. For polychromatic X-ray spectra, effective wavelengths and excitation efficiencies are measured as a function of atomic number. A new dual-target X-ray tube is shown to be an effective answer to the problem of covering a broad range of elements with a single X-ray tube.

INTRODUCTION

The fluorescent intensity from a sample in an X-ray spectrograph is a complex function of the interactions between the primary X-ray photons and the sample. These interactions are affected by a number of factors: the energy or wavelength distribution of the primary X-ray spectrum, the geometry of the X-ray spectrograph, the fluorescent yield, and the relative magnitude of photoelectric absorption and absorption leading to scattering. Of these factors, the only one that can be varied by the operator of a modern commercial spectrograph is the first one, namely, the wavelength distribution of the primary X-ray tube. This can be accomplished by changing the X-ray tube excitation, or by the use of different X-ray tube targets. This paper deals primarily with the latter, being a discussion of the effect on intensity of a number of different commercially available X-ray tube targets.

Before considering the more complex case of the polychromatic X-ray generator, it is desirable to review the simpler case of monoenergetic excitation of the sample, since the

45

polyenergetic case can be considered a summation of the effects of the many energies comprising the primary radiation from an X-ray tube. For monoenergetic excitation, the fluorescent excitation efficiency can be defined as the ratio of the total number of photons of characteristic radiation leaving the sample over the total number of exciting photons which are incident on the sample. The excitation efficiency varies with the wavelength of the exciting radiation and the atomic number of the material being excited. It is well known that the excitation efficiency increases as the atomic number of the material increases and that the most effective wavelength for exciting an element is just shorter than the absorption edge of the element. Birks [1,2] has reported on the excitation efficiency of elements from silver through aluminum as a function of primary wavelengths. He measured values of excitation efficiencies of 76% maximum for silver and 1% maximum for aluminum, showing the sharp decrease in excitation efficiency as the atomic number is reduced. In addition, he reported that the calculated values of excitation efficiencies for the lower atomic numbers were higher than his observed values.

PROCEDURE AND RESULTS

In view of the fact that this study was aimed primarily at the improvement of excitation efficiencies for low atomic number elements below titanium, it was felt desirable to measure these excitation efficiencies experimentally. To accomplish this, an experimental setup similar to that utilized by Birks in his investigation was used. A schematic diagram of the setup is shown in Fig. 1. Normal geometry of the vacuum X-ray spectrograph was used in these tests, with the exception that the collimator in front of the counter tube was split in the center so that the sample could be inserted at an angle of 45° to the collimated X-ray beam. An additional counter tube was placed alongside the split collimator so that radiation from the sample could be detected nondispersively. To measure the incident exciting radiation, the sample was removed and the counter tube, in the normal position, was utilized to scale the number of photons which were incident on the sample. The wavelength of the primary radiation was varied by using different pure materials in the normal sample-holder position

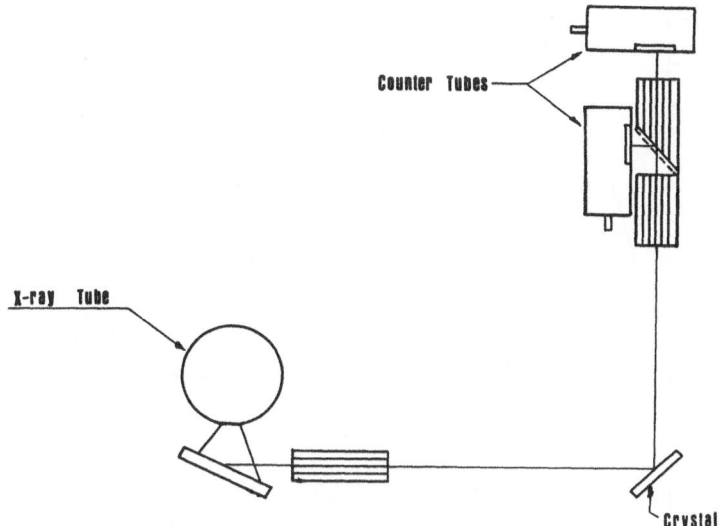

Fig. 1. Schematic diagram of setup for measuring excitation efficiencies.

and by using a suitable crystal and goniometer setting to receive this radiation. Since the crystal will also reflect $\lambda/2$, $\lambda/3$, etc., components in the primary radiation, pulse height discrimination was utilized in measuring the intensity of primary exciting radiation to eliminate these unwanted components. The fluorescent intensities from the samples in the collimator were also measured using pulse height discrimination. In addition, wavelengths slightly longer than the K edge of the elements being measured in the collimator were utilized in order to correct for higher order excitation and scatter effects from the samples.

Utilizing these procedures, relative excitation efficiencies were measured for copper, calcium, chlorine, silicon, and aluminum. The measurements for the latter four elements were made with the system in vacuum. Calcium carbonate and sodium chloride were the materials used in addition to pure samples of copper, silicon, and aluminum. Figure 2 shows the relative excitation efficiencies for the five elements as a function of the wavelength of the incident radiation. It should be emphasized that the values shown in the graphs do not relate to photon energy of the incident photons. They are merely a ratio of the number of photons incident to the number of photons excited. Obviously, the measured intensities used in the calculations had

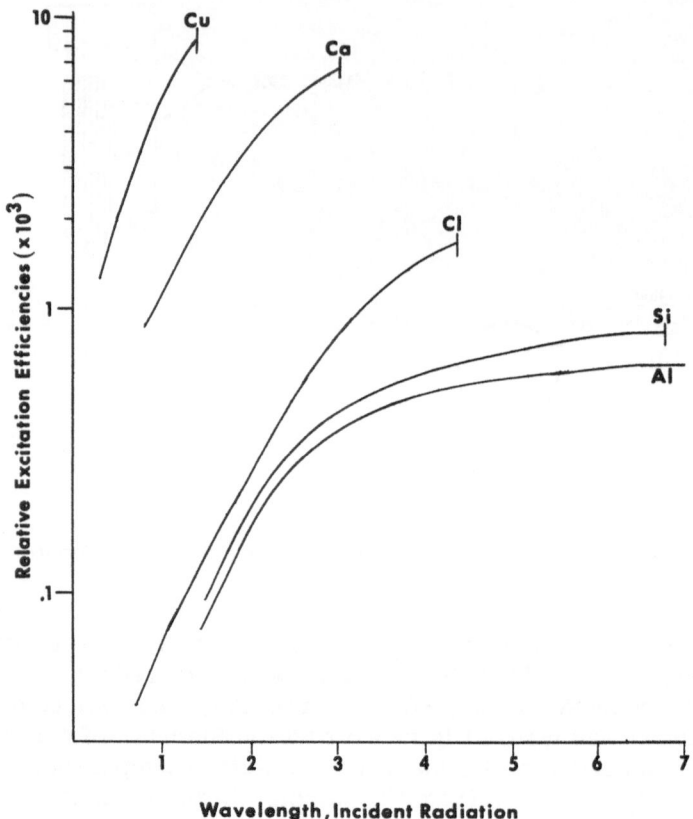

Wavelength, Incident Radiation

Fig. 2. Relative excitation efficiencies for light elements.

to be corrected for counter tube efficiencies. In order to convert the values shown in Fig. 2 to absolute excitation efficiencies, it is necessary to multiply the values shown by the ratio of 4Π to the solid angle actually utilized in the experiment. This assumes that there is uniform fluorescent radiation over the full 4Π solid angle; the specimen is assumed to be a point source so that the relative solid angle intercepted by the detector can be more easily calculated. Thus, in this experiment, the solid angle was calculated to be about 4% so that it would be necessary to multiply the values shown by 25. The chlorine and calcium curves would also have to be corrected for their respective weight percents in the salts used. Based upon published values of absorption jump ratio and fluorescent yield for copper, an excitation efficiency of 25% was calculated for copper using excitation which is just below the K absorption edge in wavelength. This is in fairly good agreement with the observed value of approximately 21% in this experiment; this is felt to be good particularly in view of the difficulty in estimating the solid angle accurately. It should be noted that Fig. 2 shows the K excitation efficiency and not the K_α excitation efficiency, since the experimental setup did not discriminate between K_α and K_β radiation.

It is evident from Fig. 2 that ratios of excitation efficiencies of 10 or more are more readily obtainable for copper as compared to silicon or aluminum. In addition, it should be noted that the excitation efficiencies of both silicon and aluminum fall off rather slowly as the wavelength becomes farther and farther removed from the K edge, until the $3-4$ A excitation range is reached. After this, it can be seen that the excitation efficiency decreases rapidly with decreasing wavelength. For example, the excitation efficiency of CrK_α at 2.29 A is approximately three times that of WL_α at 1.48 A for silicon and aluminum.

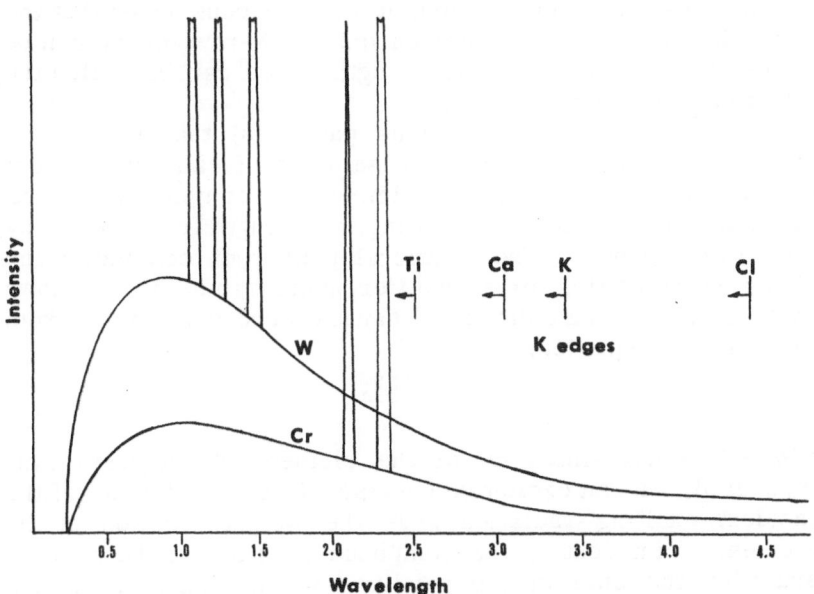

Fig. 3. Typical X-ray spectra of chromium and tungsten target X-ray tubes.

Typical X-ray spectra, which are shown in Fig. 3, are intensity—wavelength plots of chromium and tungsten target tubes.
These spectra were taken at 50 KVP and show the K characteristic lines for chromium and the L series for tungsten.
These lines are superimposed upon the continuous background
which would be a few times higher for tungsten than chromium,
roughly in the ratio of atomic numbers of tungsten to chromium.
In order to determine the relative excitation efficiencies of two
different target materials from a theoretical standpoint, it
would be necessary to know the distribution of the number of
photons at each wavelength in each of the tube spectra. This
is a difficult matter to ascertain because of the problem of
dispersing the spectra, the nonuniform spatial distribution of
X-ray energies, corrections for counter tube efficiencies,
multiple orders of reflections which can occur, and, finally,
variations in crystal efficiencies with angle of reflection. If
such spectral distribution information were known, it would then
be possible to measure relative excitation efficiencies by
multiplying the ordinates of Fig. 3 by the excitation efficiencies
for each wavelength as determined in Fig. 2 and making a
summation of all the values. Obviously, it is a far simpler task
to perform this comparison experimentally using two tubes of
similar makeup.   In addition, it may be considered that the
sum of the effects of a polychromatic X-ray source can be
represented by a single wavelength which can be called the
effective wavelength.
It was shown in a previous paper [3] that the relative
intensities of a series of thick samples of salts of different
compositions containing one element in common could be
predicted if the effective exciting wavelength from the X-ray
tube were known.   The relationship between calculated and
observed intensities for eight different potassium salts is shown
in Fig. 4.   The calculated intensities were determined from
the following equation:

$$I_a = \frac{kI_0 W_a}{A_a}$$

where I_a is the intensity of the element of interest being
measured, k is an excitation constant, I_0 is the intensity of the
incident exciting radiation, W_a is the weight fraction of the
element of interest in the compound constituting the sample,
and A_a is the sum of the products of the mass absorption
coefficients and the cosecants of incident and take-off angles

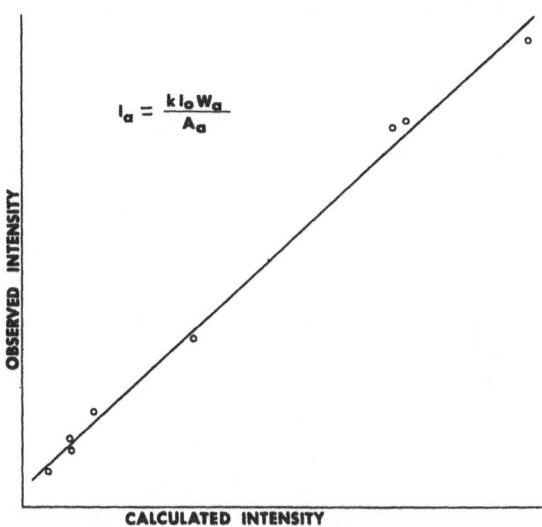

Fig. 4. Relationship between calculated and observed intensities for thick films of potassium salts.

for the incident and characteristic radiations, respectively. Chromium excitation was used to measure the observed intensities, and the effective exciting wavelength was chosen to be that of $\mathrm{Cr}\,K_{a}$, which is 2.29 A. Agreement between calculated and observed points is very good.

Based upon the equation above, it is possible to calculate relative intensities for these different potassium salts as a function of effective exciting wavelength. This was done for wavelengths from 1.2 — 3.4 A and is shown in Fig. 5. By measuring the relative potassium intensities of four of the salts with chromium and tungsten radiation at 50 KVP excitation, it is possible to determine the effective exciting wavelength of each of these tubes. This was done by measuring the relative intensities of the salts and fitting the observed values to the curves in Fig. 5. The experimental points for tungsten coincide with the curves at an effective wavelength of 2.1 A while those for chromium coincide at approximately 2.6 A. It is interesting to note that the effective wavelength for tungsten excitation of potassium is considerably removed from the L series characteristic lines of tungsten. Obviously, the effective wavelength would be a function of the particular element for which it is being studied.

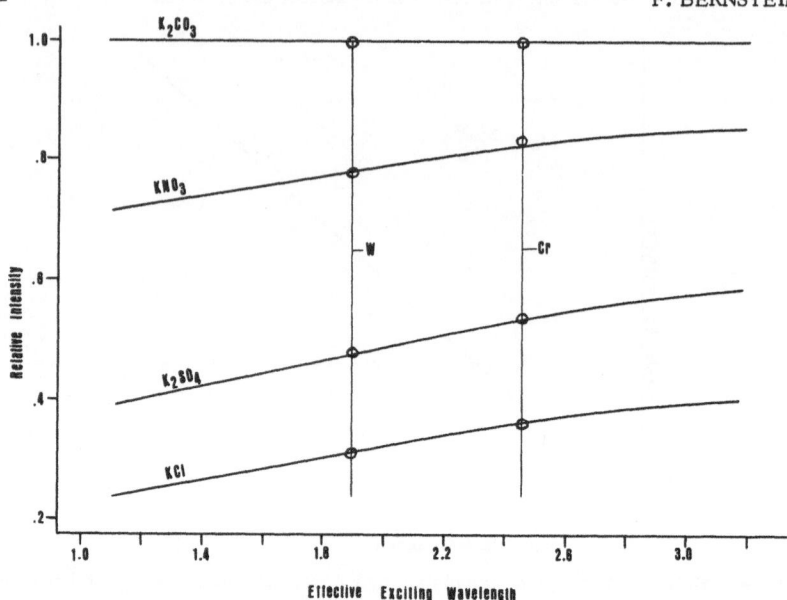

Fig. 5. Relationship between relative intensity of potassium salts and effective exciting wavelength.

As was noted above, the problem of predicting the excitation efficiency of a polychromatic spectrum which emerges from a tube of a given target material is extremely complex, if not well nigh impossible. For this reason, the excitation efficiencies for five different target materials were compared experimentally using 50 KVP and 50 mA, and the results are shown in Fig. 6.

Relative intensities for chromium, silver, tungsten, titanium, and platinum target tubes are plotted against wavelength and element. The intensity from tungsten or platinum target tubes with 30-mil beryllium windows was taken as a base of one. The results obtained from both platinum and tungsten were very close, so that for simplicity these are plotted as a single curve for thin and thick window targets. The thin window targets refer to 10-mil beryllium X-ray tube windows. It is evident from these curves that the most efficient excitation of elements of atomic number 22 and below can be obtained using chromium or titanium radiation, while above atomic number 22, the most efficient excitation is either tungsten or platinum target tubes. A thin window silver target tube can be seen to be almost as effective as chromium for atomic number 17

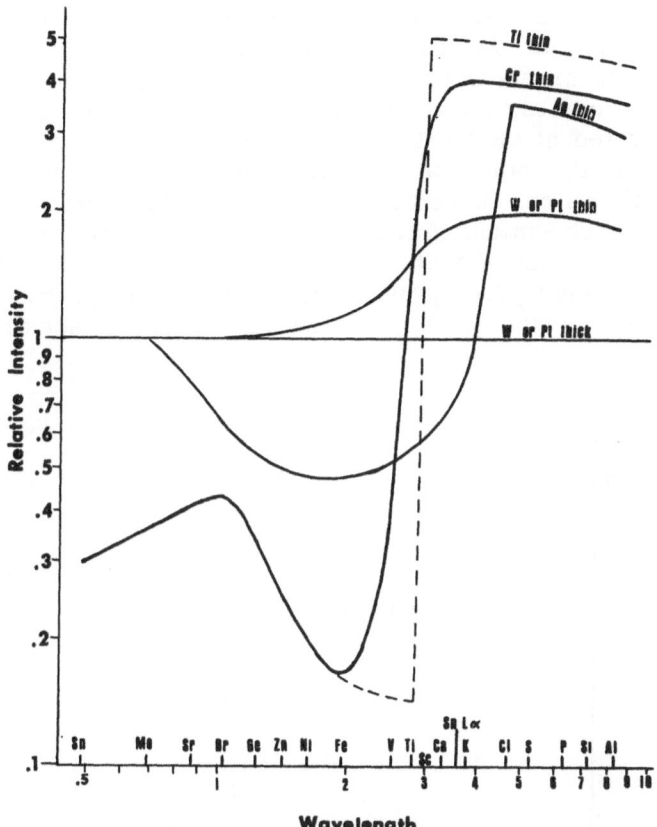

Fig. 6. Relative excitation efficiencies of various X-ray tubes.

(chlorine) down, due to the emission of a strong Ag L line at 4.15 A. In general, for atomic number 22 and below, it can be seen that chromium gives essentially a 2/1 gain over an equivalent tungsten tube, while titanium is approximately 2.5/1 over the equivalent tungsten tube. These results are in agreement with the curves of Fig. 2, which predict significant improvement in excitation efficiencies as the wavelength of the exciting radiation is increased to about 3 A.

If one were to design an X-ray tube for optimum excitation based upon the results shown in Fig. 6, this tube would have the characteristics of either chromium or titanium targets for atomic numbers below 22 and would have the characteristics of thin window tungsten or platinum targets above atomic

number 22.   It was these considerations which motivated the
development of the dual-target X-ray tube.   Tungsten and
chromium targets were chosen for the initial design, but other
combinations are currently being developed.   Both targets can
be operated at 75 KVP, and it is possible to switch from one
target to the other in a matter of seconds by means of an
external switch.   In addition, the tube has been designed so as
to give a chromium spectrum which is enhanced in intensity
over the normal EA-75 X-ray tube.   Thus, coupled with the
ease of changing X-ray tube targets, this dual-target tube offers
the added advantage of increased intensities and sensitivity in

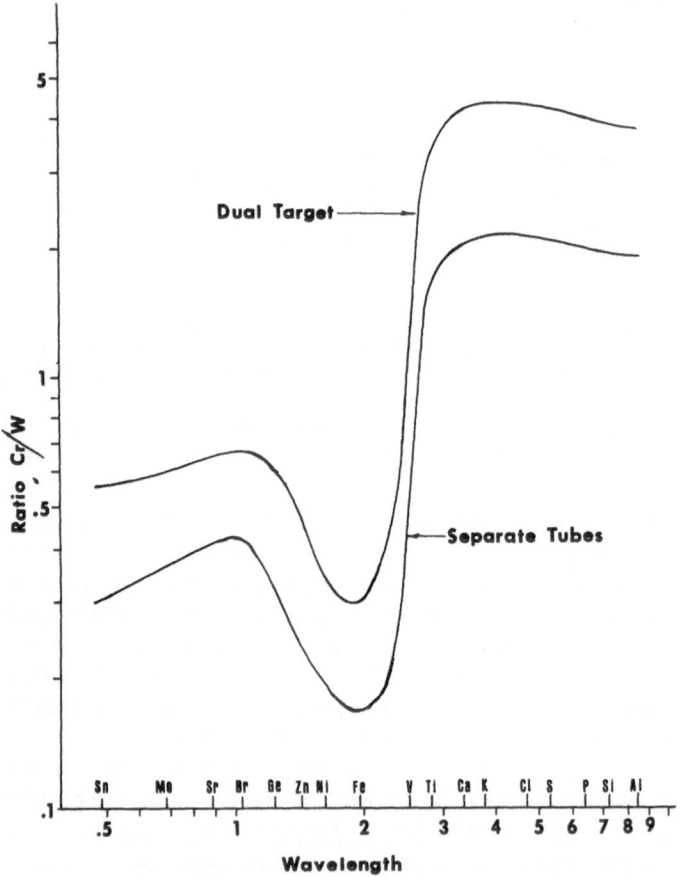

Fig. 7. Relative excitation efficiencies of chromium and tungsten target X-ray tubes.

the lower atomic number regions (below atomic number 22). This is demonstrated in Fig. 7, which shows a plot of the relative intensities of chromium-to-tungsten excitation as the function of atomic number for thin window tubes and for a dual-target chromium-tungsten tube.   Since the tungsten output of the dual-target tube is only slightly reduced as compared to a single-target EA-75, the gain in intensity of the chromium spectrum is principally attributable to the increased chromium output.

SUMMARY

1. Low fluorescent intensities which are obtained from elements having atomic numbers below 22 are due primarily to their low excitation efficiencies.

2. It is shown that the use of X-ray target materials which have a significant percentage of their emission spectra in longer wavelength regions, such as chromium and titanium, greatly enhances the intensities which can be achieved in measurement of light elements.

3. A method is presented for determining the effective wavelength when using a polychromatic spectrum for excitation of a sample.

REFERENCES

1. L. S. Birks, J. Appl. Phys. 32, 387 (1961).
2. L. S. Birks, Spectrochim. Acta 17, 148 (1961).
3. F. Bernstein, Twelfth Annual Conference on Applications of X-Ray Analysis, Denver, Colorado (1963).

Applications of Chemical Precipitation Methods for Improving Sensitivity in X-Ray Fluorescent Analysis

Joseph S. Rudolph, Owen H. Kriege, and Robert J. Nadalin

Westinghouse Research Laboratories
Pittsburgh, Pennsylvania

X-ray fluorescent analysis has limited application to the determination of traces in metals because of insufficient sensitivity when absorption by the matrix is significant. By applying chemical precipitation methods to concentrate the elements to be determined, orders of magnitude increase in sensitivity can be obtained. In addition to gaining sensitivity, absorption and enhancement effects become negligible since the precipitate ·is exposed to the X-ray beam as a thin film, producing a linear relationship between intensity and concentration. Because the elements to be determined are separated from the matrix, a single set of standards can often be used for the analysis of samples with varying composition. Application of chemical precipitation methods to the determination of parts-per-million concentrations of zirconium in cobalt, iron, and nickel alloys and to the determination of chlorides in high-purity titanium metal has been accomplished.

INTRODUCTION

X-ray fluorescent methods are used routinely for the analysis of major and minor constituents. The accuracy of these X-ray methods at concentrations of 0.1% or greater is, in many cases, equal to or greater than that of classical chemical procedures. However, low fluorescent yield and poor precision have imposed limitations when X-ray analysis has been applied to trace analysis or concentration levels less than 0.1%. There are several reasons for this problem: (1) Most obvious, the fluorescent yield is low because of dilution by the matrix of the element to be determined. (2) The matrix may be a strong absorber of the characteristic X-ray wave-length. (3) The major constituents may coincide or overlap the characteristic wavelength. These limitations in X-ray analysis can be circumvented by applying chemical separation techniques to concentrate the elements to be determined from the matrix. In this manner, matrix dilution, absorption, and interference are removed.

Because of simplicity, rapidity, and other advantages, which will be described later, most separations have been made by chemical precipitation. The chemical precipitation methods described in this paper are inaccurate as gravimetric methods for the determination of trace concentrations because of co-precipitation and interferences. However, substitution of the X-ray spectrometer for the analytical balance extends these methods for determining microgram quantities. In addition to a large increase in sensitivity, chemical precipitation methods have other advantages as compared to analysis of solid samples. A more representative sample is analyzed by this technique since one or more grams of drillings or millings can be dissolved prior to precipitation. Standards are easily synthesized by precipitation of the elements to be determined from solutions prepared in the same manner as the samples. One set of standards is sufficient for the determination of an element in a number of matrices, since the matrix is removed and often the stability of the precipitate provides a permanent standard. Since the precipitates are present as thin films, absorption and enhancement effects are negligible, though several elements may be present, and a linear relationship exists between intensity and concentration.

To illustrate the advantages of chemical precipitation methods for increasing sensitivity in X-ray fluorescence analysis, two methods to which we have applied this technique will be described. The first method, which is the determination of microgram quantities of zirconium in steel, cobalt—nickel alloys, beryllium, and copper, is an example of how one set of standards can be used for analyzing several different alloys, since zirconium is precipitated from the various matrices. By precipitation, zirconium can be quantitatively determined at the 20-ppm level, which is an order of magnitude increase in sensitivity as compared to analyzing a solid sample. The second method, which is used for the determination of chloride in high-purity titanium metal, is an example of an increase in sensitivity by concentrating the chloride from a strongly absorbing matrix, titanium. By analyzing the precipitate, we can determine chloride at the 15-ppm level, as compared to 0.1% in a solid sample.

EXPERIMENTAL DETAILS

X-Ray Procedure

The X-ray procedure for analyzing precipitates is as follows: After dissolving the sample, the elements to be determined are precipitated, filtered onto a Millipore filter pad, and prepared for X-ray analysis as shown in Fig. 1. The Millipore filter pad and precipitate are sandwiched between two sheets of 0.00025-in. Mylar film. A flat surface is obtained by stretching the Mylar film over a 30-mm plastic ring which has a 25-mm aperture. An O-ring is placed over the outside surface of the plastic ring to maintain tautness. The characteristic X-ray spectra are obtained by placing the precipitate side of the Millipore filter in an inverted position to the path of an X-ray beam emitted from a tungsten target X-ray tube

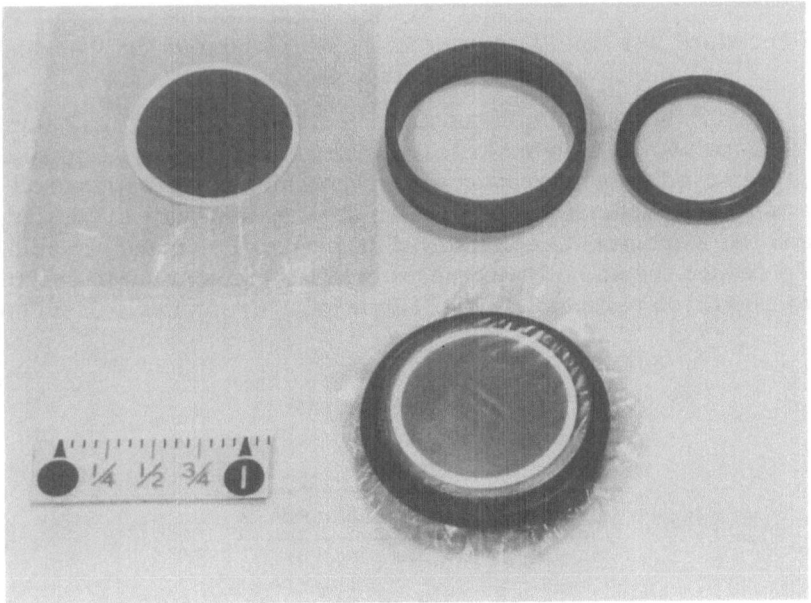

Fig. 1. Plastic ring and O-ring used to obtain a flat surface on the Millipore filter pad and precipitate.

operated at 50 kV and 45 mA. The sample holder of a Norelco Bulk Spectrograph is apertured to 1 in., which is the diameter of the Millipore filter pad. The optical system consists of a 20 mil by 4 in. source collimator, lithium fluoride crystal, and a 5 mil by 1 in. detector collimator. The pulses from a scintillation counter are registered on a fixed-count scaler system. Each sample or standard is exposed twice. Exposure time for a fixed number of counts is recorded at one sample position, and then 90° from the original position to decrease the effect of any nonuniform distribution of the precipitate across the surface of the Millipore pad. Background counts/sec are subtracted from the peak counts/sec. The net counts/sec for both sample positions (0° and 90°) are averaged, and the amount of the zirconium or chloride is determined by reference to an analytical curve. The synthetic standards used to plot the analytical curve are prepared by precipitating the zirconium or chloride from solutions containing added quantities of the respective elements over the concentration range to be analyzed. The precipitation is made in the same manner as for the samples.

Procedure for the Determination of Zirconium in Iron and Nickel—Cobalt Alloys

A $0.1 - 1$ g sample containing $20 - 700 \mu$ g of zirconium is dissolved in hydrochloric and nitric acids. The nitric acid is removed by evaporation with hydrochloric acid. Hydroxylamine hydrochloride and $1 - 2$ mg of paper pulp are added, and the zirconium precipitated with para-bromomandelic acid. To reduce the solubility of the precipitate, the solution is cooled to room temperature before filtration. Precipitates from an

TABLE I

The Effect of Aging the
Precipitate on the Recovery
of Zirconium

Aging time	Zirconium found, %
5 min	0.018, 0.017
1 hr	0.016, 0.017
4 hr	0.018, 0.018
24 hr	0.017, 0.016

TABLE II

The Analysis of Standard Samples from the National Bureau of Standards

Sample	Zirconium content, %	
	Certified value	Determined by X-ray
Low alloy steel 462	0.063	0.062 0.061
Low alloy steel 464	0.010	0.0099 0.0097
Synthetic stainless steel*	0.029	0.029 0.029

*Composed of 42% NBS steel 464 and 58% NBS nichrome 169.

iron-base alloy were aged for periods ranging from 5 min to 24 hr before filtration to determine the effect of aging on the recovery of zirconium. As shown in Table I, this had no effect on the zirconium recoveries. The standard deviation of these results is 0.0008%.

An evaluation of the accuracy of this method for the separation and determination of zirconium was made by analyzing standard samples from the National Bureau of Standards. Results obtained for the determination of zirconium in these materials are presented in Table II.

The results were fitted by standard statistical procedures to the model $y = bx$, where y is the amount found, x is the certified NBS value (assumed to be without error for this calculation), and b is the slope. The calculated estimate of b is 0.98 with a 95% confidence interval of $0.97 - 1.00$. The estimated standard deviation for a single determination is 0.0006% for the concentration range shown in Table II. The results appear to be slightly low, but when it is considered that the certified NBS values also contain some error, there is justification for considering the method to yield quantitative values over the comparison range. It is of interest to note that the results of Table I yield an estimated standard deviation of 0.0008% which agrees quite favorably with the 0.0006% calculated from the results of Table II.

Several samples of iron-base and cobalt-base alloys were analyzed in order to ascertain the precision of the X-ray fluorescence method for zirconium in concentration levels other than those reported in Tables I and II. In most cases, replicate determinations were made on two or more days. Results are presented in Table III. From these results, it

TABLE III

The Determination of Zirconium in Iron- and Cobalt-Base Alloys

Material	Heat no.	Zirconium content, %
Kromarc ® — 58 steel*	1	0.0028, 0.0031, 0.0029, 0.0029
Kromarc ® — 58 steel*	2	0.064, 0.060
Nivco ® — 10 alloy†	1	0.21, 0.21
Nivco ® — 10 alloy†	2	0.55, 0.56

*Kromarc ® —58 steel contains 50% iron, 22% nickel, 15% chromium, 10% manganese, and 2% molybdenum.
†Nivco ® –10 alloy contains 75% cobalt, 22% nickel, and 2% titanium.

appears that a satisfactory precision is obtained for these other zirconium levels. This procedure has been applied to the determination of zirconium in beryllium and copper with comparable precision and accuracy.

Procedure for the Determination of Chloride in High-Purity Titanium

This procedure is used to analyze high-purity titanium metal containing 15 — 2000 ppm of residual chloride. A 0.5 —3 g sample is dissolved in hydrofluoric and nitric acids and the chloride precipitated with $AgNO_3$ as $AgCl$. Because the chloride is precipitated as the stoichiometric compound of silver chloride, the Ag K_α spectra can be used as a measurement of the chloride present. This results in added sensitivity, since the AgK_α fluorescent yield is greater than ClK_α, and the Ag K_α radiation is less absorbed by the protective Mylar which encloses the Millipore filter pad.

A measure of the accuracy and precision of the chloride method was obtained by analyzing two series of samples. The first series consisted of aqueous solutions of hydrochloric acid and two samples of dissolved titanium chips to which are ladded known amounts of hydrochloric acid. The results of these analyses are summarized in Table IV. The second series of samples analyzed to determine the accuracy and precision of the X-ray procedure were samples previously analyzed by potentiometric titration. These data are summarized in Table V. Samples K-1, K-2, and K-3 are titanium powder samples. Samples K-4, K-5, and K-6 are sample K-3 mixed with low-

TABLE IV

Accuracy Data—X-Ray Fluorescent Determination of Chloride in Titanium

Samples	Amount of chloride	
	Added, mg	Found, mg
Aqueous solution of HCl #1	1.06	1.06
Aqueous solution of HCl #2	1.06	1.06
Aqueous solution of HCl #3	1.06	1.08
Aqueous solution of HCl #4	0.090	0.090
Titanium chips* + 0.53 mg Cl⁻	0.56	0.59
Titanium chips + 1.06 mg Cl⁻	1.09	1.08

*Titanium chips contained 0.03 mg Cl⁻.

chloride-content titanium chips. The data in Tables IV and V were fitted by standard statistical procedures to the model $y = bx$, where y is the amount found by the X-ray method, x is the amount added or found chemically, and b is the slope. Standard least-squares techniques were used to calculate b. Slopes of 0.994 and 0.987 for the data in Tables IV and V, respectively, show the X-ray determinations are in excellent agreement with the chloride added or determined chemically. The estimated standard deviations for a single determination for the concentration ranges shown in Tables IV and V are 17 μg and 32 μg, respectively.

TABLE V

Accuracy Data—Comparison of Chemical and X-Ray Fluorescent Determinations of Chloride in Titanium

Samples	Milligrams of chloride found	
	Chemical	X-ray
K-1, 60-140 Mesh	1.37	1.42
K-1, > 140 Mesh	1.23	1.24
k-2, 60-140 Mesh	1.31	1.32
K-3	1.37, 1.37	1.34, 1.34
K-4	0.89	0.90
K-5	0.51	0.52
K-6	0.22, 0.22	0.24, 0.24

Solubility of the precipitate and possibly other chemical variables during sample preparation limit the method to a lower chloride content of 45 μg. Chloride concentrations of 15 ppm can be determined by analyzing a 3-g sample. The length of time required to dissolve larger samples, possible salt effects on the solubility of the AgCl, and higher reagent blanks preclude using larger samples. The X-ray method can be extended to determine chloride concentrations higher than 0.2% by decreasing the amount of sample analyzed. Chloride can be determined in any material provided the silver chloride alone can be precipitated quantitatively and be isolated from the matrix. This procedure can also be applied to the quantitative determination of bromide and iodide by calibration with appropriate standards.

SUMMARY

Chemical precipitation methods extend the applicability of X-ray fluorescence analysis to the determination of microgram quantities. The data we have obtained have shown these methods to be rapid and reliable. In addition to increasing sensitivity, precipitation methods require only one set of standards for analyzing various matrices, standards are easily synthesized as solutions, and a more representative sample can be analyzed as compared to solid samples.

X-Ray Spectrochemical Determination of Niobium and Tantalum in High-Alloy and Stainless Steel

Roger W. Taylor

A. O. Smith Corporation
Milwaukee, Wisconsin

In alloys for nuclear applications, tantalum must be maintained at low levels to avoid neutron loss and induced radioactivity. Since varying amounts of tantalum generally will be present in alloys containing niobium, the chemist needs a method of determining tantalum in the presence of niobium, or else he must perform the tedious job of quantitatively separating them. An X-ray spectrochemical method was developed for the determination of niobium and tantalum in a variety of high-temperature alloys and steels. A curved crystal spectrometer with a tungsten target tube and lithium fluoride crystals is employed. The earth acids are separated by dissolving the alloy in aqua regia, evaporating with perchloric acid, boiling with sulfurous acid, and then filtering. After ignition the oxides are blended with cellulose and iron oxide. A pellet is briquetted for the X-ray analysis. The iron oxide serves to reduce the background and scattered radiation from the Compton effect. Niobium and tantalum are run simultaneously on the scanners with the lithium fluoride crystals. Silicon could not be determined accurately on the pellets because of the varying absorption effects from different amounts of niobium and tantalum. Silicon can be determined by the conventional wet chemical method—volatilizing with hydrofluoric acid. The niobium and tantalum can then be reprecipitated and determined by X-ray. Since the niobium and tantalum are separated from most of the alloy matrix, it is possible to analyze these elements from one curve for nearly all types of alloys.

INTRODUCTION

The chemical properties of niobium and tantalum are very similar, and therefore their separation by chemical methods is quite difficult. Niobium is seldom encountered without varying amounts of tantalum present, and often determinations in steels and alloys are actually reported as total niobium plus tantalum. Usually tantalum does not exceed 10% of the niobium content,

65

and this method of reporting may be acceptable when the analysis is performed by the hydrolysis of the earth acids. However, in nuclear reactors niobium is a good structural material because of its corrosion resistance, high-temperature strength, and low neutron absorption, while tantalum exhibits a high neutron absorption. Therefore, tantalum must be maintained at low concentration levels [1].

It has been stated that there is no other class of minerals so beset with analytical difficulties [2]. A method developed by Marignac [3] in 1866 was the only method used commercially up to 1945. It is based on the difference in solubility of the two double potassium salts. A method by Schoeller uses the difference in hydrolysis of the niobium and tantalum oxalates [4]. A gravimetric precipitation method for tantalum with N-benzoyl-N-phenylhydroxylamine is described by Moshier and Schwarberg [5]. Two to three precipitations and close pH control are required for the quantitative separation from niobium. Anion exchange methods have been fairly successful, and separations 99% pure have been effected according to Huffman [6]. Tantalum is preferentially absorbed from mixed HF−HCl acids and niobium from HCl solutions [6−8]. Martin and Magee [9] described a partition chromatography method for the separation and determination of niobium, tantalum, and titanium. Since 1952 a considerable number of papers dealing with liquid−liquid extraction methods have appeared [10]. Often these methods are inconvenient for analytical work because the degree and rate of extraction may not be favorable. A recent paper by Kim and Meinke [11] describes a thermal neutron activation method; however, few laboratories have access to this type of equipment.

The literature describes a number of X-ray methods for determining niobium and tantalum. Luke [12] describes a borax fusion of the separated oxides of niobium, tantalum, molybdenum, and tungsten in high alloys. Barium oxide was added as a heavy absorber to reduce interelement effects and a platinum target tube employed. Mitchell [13] determined niobium and tantalum oxides in the presence of other refractory oxides at levels from 0.1−99%. Interelement absorption was corrected by an arithmetic factor method. Campbell and Carl [14] have determined tantalum in niobium oxide using a molybdenum target tube and a quartz analyzing crystal to give a more efficient excitation and better resolution of tantalum due to a

TABLE I

Operating Conditions for X-Ray Analysis Determination of Niobium and Tantalum

Operating conditions	Nb	Ta
Analytical line	K_α 0.747 A	L_α 1.522 A
Crystal (curved)	11 in. LiF	4 in. LiF
Slits	0.010 − 0.030 in.	0.020 − 0.030 in.
Detector voltage	1500	2050
Tube	Tungsten, Machlett OEG 60	
Voltage	50 kV	
Amperage	35 mA	
Integration time	2 min	
Sample mask	$3/4$ in.	

lower reflectivity of second order Nb K_α radiation. Tompkins, Borun, and Fahlbusch [15] have described a method for tantalum, niobium, tungsten, and zirconium in high-temperature alloys by an X-ray fluorescent solution technique. A platinum target tube was used and line intensities were integrated against scattered radiation at 0.57 A. Burton, Jacobs, and Valecko [16] determined tantalum in niobium at 0.05−0.70% with a tungsten target tube. A curved 11-in. LiF crystal and 0.015−0.020-in. primary and secondary collimation slits were employed. The oxides were blended with cellulose and pressed into a pellet. Hanson [17] used an oxide sample and a gold target tube to analyze for tantalum in the X-ray analysis of F-48 alloy.

This paper describes a method for determining niobium and tantalum in stainless steels and many high-temperature alloys. Niobium is covered in the 0−3% range and tantalum from 0−0.5%. In the ARL vacuum X-ray quantometer, a tungsten target tube and curved lithium fluoride analyzing crystals are employed. While it is true that other tubes and crystals may offer advantages in terms of intensity and resolution for the determination of tantalum, it is usually not convenient to make a change-over—especially in the larger instruments that may be programmed for a variety of elements. The tungsten tube and lithium fluoride crystal are probably the most versatile combination for X-ray analysis. The instrumental conditions are shown in Table I.

Various types of samples were employed in this study,

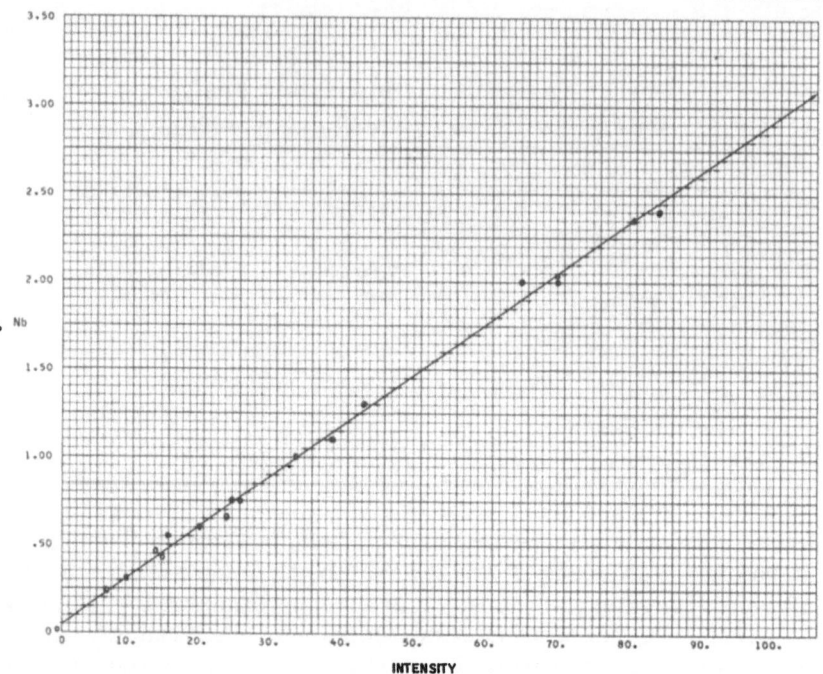

Fig. 1. Calibration curve for niobium.

which included chemical and spectrographic standards, secondary standards, and some A. O. Smith samples carefully analyzed by other chemical methods. The tantalum and most of the niobium values are not certified. This may be partially responsible for some scattering of the plotted data. The calibration curves were plotted by the method of least squares using the IBM 7070 computer to calculate the equations of the lines. An alternative equation for the determination of niobium was calculated on the 7070 running a multiple regression, taking into account the variables tantalum, tungsten, and silicon. The equation is

$$\% \, Nb = 0.04427 + 0.27522 \, (\%Ta) + 0.12701 \, (\%W) - 0.05164 \, (\%Si)$$
$$+ \, 0.02871 \, (\text{intensity Nb})$$

Although the calibrations were made using noncertified values for tantalum and, in many cases, for niobium, the deviations from the plotted line appeared to be small enough to justify the method. Figure 1 shows the calibration data for niobium and

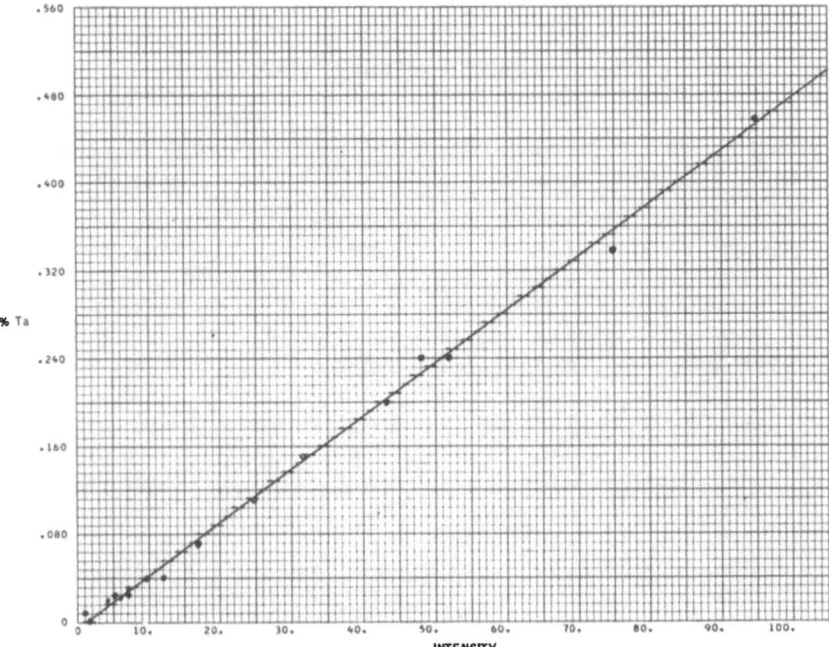

Fig. 2. Calibration curve for tantalum.

Fig. 2 that for tantalum.* The lines were calculated on the computer and then plotted with a printer.

SAMPLE PREPARATION

Two grams of the alloy is weighed in a 400-ml beaker and dissolved in 40 ml HCl and 15 ml HNO₃ using gentle heat. After the sample is completely dissolved, 30 ml of 70% perchloric acid is added; the sample is covered with a watch glass and evaporated to fumes; finally, the HClO₄ is boiled for a few minutes. After removal from the heat and cooling,

*Author's Note: The analytical values for niobium and tantalum on Inconel alloys X-750, HT 8209X, and 8211X have been corrected by the International Nickel Company. HT 8209X originally reported to contain 0.55% Nb and 0.24% Ta was changed to 0.52% and 0.23%, respectively. HT 8211X originally reported to contain 0.46% Nb and 0.24% Ta was changed to 0.45% and 0.22%. These new values fit the calibration curves better, especially in the case of niobium. The corrected values are reported in ASTM Special Technical Publication No. 58-E, Report on Available Standard Samples and High-Purity Materials for Spectrochemical Analysis, 1963, p. 104, table 61.

TABLE II

X-Ray Determination of Niobium

Sample	Reported value	% Nb obtained	
		Linear equation*	Multiple regression equation†
BCS 261	(0.67)	0.73	0.71
NBS 123	0.435	0.458	0.437
123a	0.75	0.77	0.75
123b	0.75	0.74	0.78
345	0.23	0.23	0.19
NBS D 846	0.60	0.61	0.56
D 849	0.31	0.31	0.30
1184	0.49	0.45‡	0.57
167	3.15	2.49‡	3.03
349	0.00	0.02	0.00
1188	1.11	1.14	1.13
1189	0.00	0.02	0.03
1203	1.00	0.99	1.03
1204	1.31	1.27	1.36
1205	1.95	1.76‡	1.90
Inco 8209X¶	0.55	0.47	0.52
8211X¶	0.46	0.43	0.47
AOS W 2160	2.00	1.89	1.90
AOS A 4123-2	2.00	2.04	2.04
AOS Y 179	2.03	2.04	2.06
AOS W 1479	2.35	2.35	2.34
AOS W 1041	2.40	2.45	2.43

*%Nb = 0.0515 + 0.0288 (int.).
†%Nb = 0.0443 + 0.2752 (Ta) + 0.1270 (W) −0.0516 (Si) + 0.0287 (int.).
‡These samples were not used in computing the linear equation line because of very high tungsten or tantalum content.
¶See Author's Note.

150 ml warm water and 50 ml sulfurous acid (6%) are added. It is stirred, placed on the hot plate, and boiled gently for 15—20 min. It is removed to a cooler section of the hot plate, and some ashless filter paper pulp is added. The hydrolyzed earth acids are allowed to digest for at least 15 min. Filtration is accomplished through a S+S No. 589 blue ribbon or equivalent paper. The beaker and paper are washed with hot 2% HCl. The sample is ignited at 950—1000°C in a muffle furnace for 1 hr. The ignited oxides are transferred to a plastic vial containing two plastic balls, 1.0 g Whatman CF11 cellulose powder, and

TABLE III

X-Ray Determination of Tantalum

Sample	Reported value	% Ta obtained *	Deviation
BCS 261	0.04	0.053	+ 0.01
NBS 123	0.027	0.029	+ 0.002
123a	0.02	0.015	
123b	0.20	0.203	
345	0.002	0.002	
D846	0.030	0.029	−0.001
D849	0.021	0.019	−0.002
1184	0.022	0.023	+ 0.001
167	0.08	0.082	
349	< 0.01	0.001	
1188	0.11	0.114	
1203	0.34	0.355	+ 0.02
1204	0.46	0.449	−0.01
Inco 8209X†	0.24	0.243	
8211X†	0.24	0.226	−0.01
AOS A 4123-2	0.07	0.074	
AOS Y 179	0.15	0.147	

*%Ta = -0.0056 + 0.0048 (int.).
†See Author's Note.

0.5 g analytical grade ferric oxide. The contents are blended for 4—5 min in a mixer mill. This mixture is briquetted into a $7/8$-in.-diameter pellet with a hand-operated hydraulic press at 30 tsi. (It is convenient to press the pellet on a cellulose backing to give it strength and facilitate identification.) The pellet is placed in the rotating sample holder and irradiated for 2 min at 50 kV and 35 mA. The niobium and tantalum intensity ratio values are read out from their respective channels on the strip chart recorder. Niobium is calibrated from 0—3% and tantalum from 0—0.5%. Tables II and III show results obtained for niobium and tantalum.

If a tantalum determination is not required, a shorter method may be used for niobium. Briefly, it is as follows: Dissolve approximately 1.0 g of the alloy in aqua regia, add 15 ml perchloric acid, and evaporate the sample to dryness on the hot plate. Scrape the charred residue from the beaker into a porcelain crucible and ignite at 500°C for 1 hr. Blend 1.00 g of

TABLE IV

X-Ray Determination of Niobium
from Ignited Oxides of the Alloy

	% Nb	Found	Deviation
Stainless steel:			
NBS 123a	0.75	0.76	+ 0.01
123b	0.75	0.76	+ 0.01
1184	0.49	0.50	+ 0.01
345	0.23	0.21	−0.02
D846	0.60	0.62	+ 0.02
High-temperature alloys (nickel base):			
NBS 1188	1.11	1.08	−0.03
1203	1.00	0.96	−0.04
1204	1.31	1.32	+ 0.01
1205	1.95	1.98	+ 0.03
Inco 8209X*	0.55	0.51	−0.04
8211X*	0.46	0.42	−0.04

*See Author's Note.

the oxides with 0.50 g cellulose and press into a $7/8$-in. pellet
This sample preparation technique was described by Goldblatt
and Friedlander [18]. High nickel interferes with the determination of tantalum by this method. Stainless and nickel-base
alloys should be plotted on separate curves.   Manganese,
molybdenum, copper, cobalt, and titanium may also be determined from these pellets.   Table IV shows data obtained on
stainless and nickel-base alloys.

DISCUSSION OF METHOD AND INTERFERENCES

A serious problem encountered in the determination of
tantalum by X-ray fluorescence is that of adequate resolution.
This is especially true when using a tungsten tube and lithium
fluoride crystal and when exciting the L line of tantalum. Approximately 70 kV is required to excite the K line of tantalum.
Lines that may cause interference, depending upon the resolution
attainable, are: Nb K_{α} (second order) 1.496 A; Ni K_{β}, 1.500 A,
Compton scattering at approximately 1.50—1.53 A; and Cu K_{α},
1.541 A.   Sufficient resolution was obtained from the second

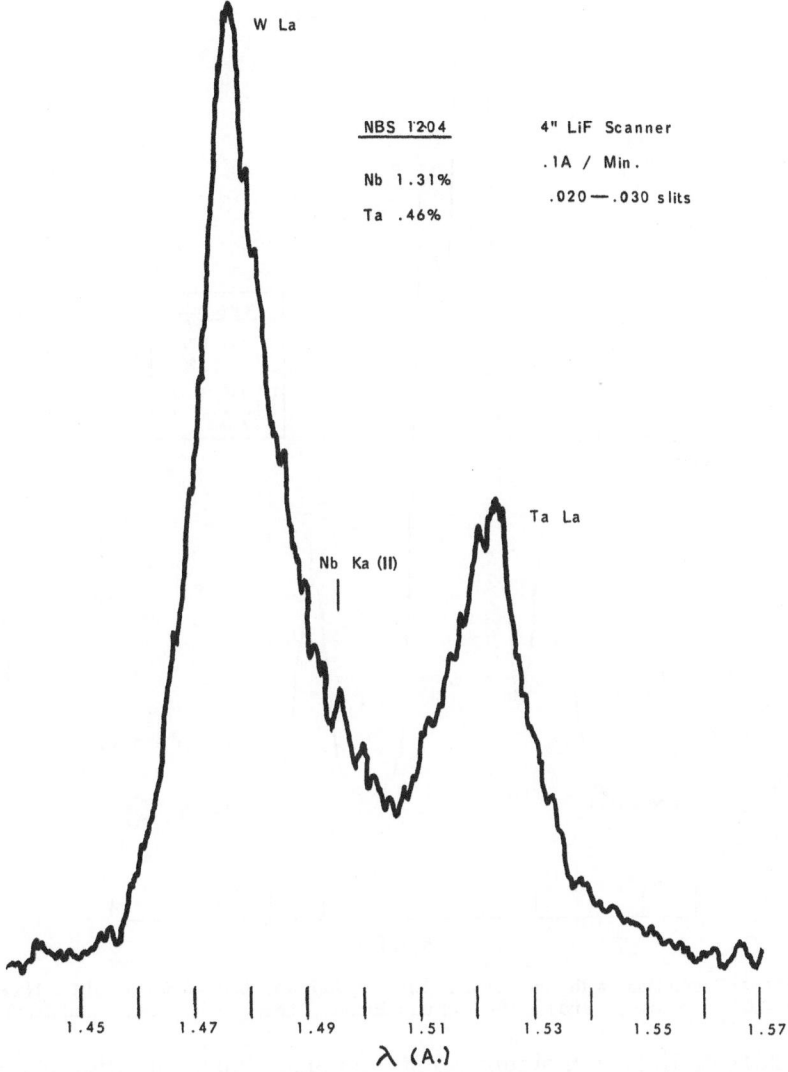

Fig. 3. Resolution of tantalum in NBS 1204.

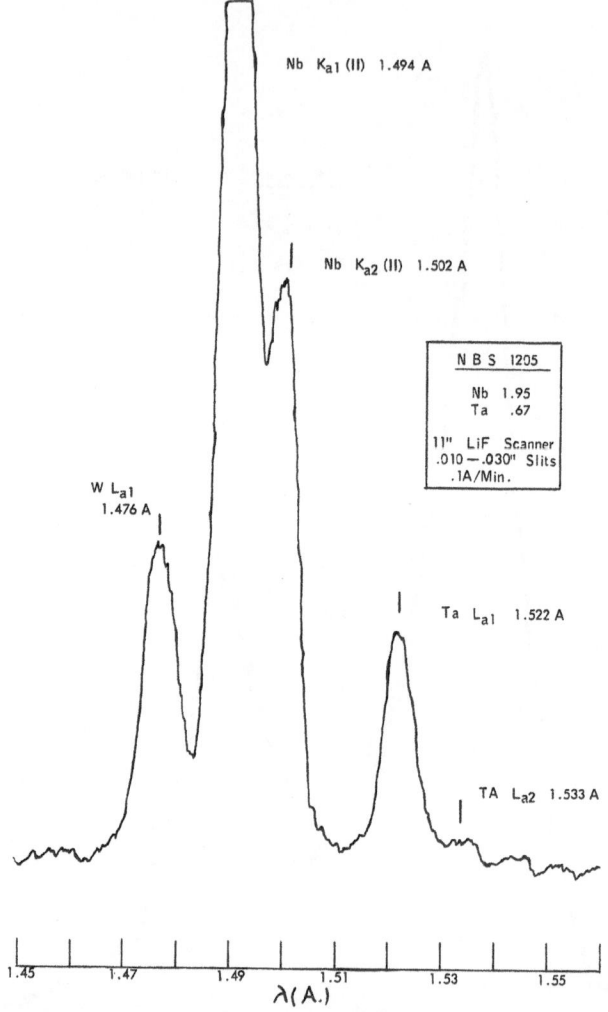

Fig. 4. Resolution with the 11-in. LiF crystal and 0.10—0.30-in. slits. Note resolution of the L_{α_1} and L_{α_2} lines of tantalum and the K_{α_1} and K_{α_2} lines of niobium.

order K_α line of niobium using the 0.020—0.030-in. slits and a 4-in. curved lithium fluoride crystal. The resolution of tantalum in NBS 1204 is shown in Fig. 3. With the 11-in. LiF crystal and 0.010—0.030-in. slits, the L_{α_2} line (1.533 A) càn be resolved from the L_{α_1} at 1.522 A. Also, the second order Nb K_{α_1} and K_{α_2} lines are separated (Fig. 4). Although the resolution was not as good with the 0.020—0.030-in. slit combination

TABLE V

Separation of Elements with
Perchloric and Sulfurous Acids

Precipitated elements	Elements remaining in solution
Nb	Ni
Ta	Cr
W	Mn
Si	Cu
Mo*	Fe
Ti*	Co
	Al

*Partial precipitation only.

and 4-in. LiF crystal, this scanner was used for tantalum because it provided greater intensity and also permitted the niobium and tantalum to be determined simultaneously on a single integration.

Because of very high nickel content in Inconel and stainless-type alloys, there is considerable interference from the K_β line of nickel at 1.500 A. Therefore, it is necessary to separate the tantalum from the nickel by chemical means. The evaporation with perchloric acid followed by boiling with sulfurous acid effects the separation of niobium and tantalum from most of the other matrix elements except silicon, tungsten, and varying amounts of titanium and molybdenum. Table V shows the elements that are precipitated or left in solution by this treatment.

The ignited oxides cannot be directly blended with cellulose because of severe Compton scattering. The Compton effect occurs when X-rays from the target are scattered by low atomic number elements, such as carbon, hydrogen, and oxygen. Some of the X-ray photons lose energy in colliding with electrons with a resulting increase in wavelength. The broad scattered peak seriously interferes with the Ta L_α line [19]. If ferric oxide is blended with the cellulose in a 1-to-2 ratio, most of the Compton scattering will be eliminated. The main disadvantage of iron oxide is its high absorption of tantalum radiation. Other materials were not investigated. Figure 5 shows the elimination of the Compton effect with the ferric oxide—cellulose mixture.

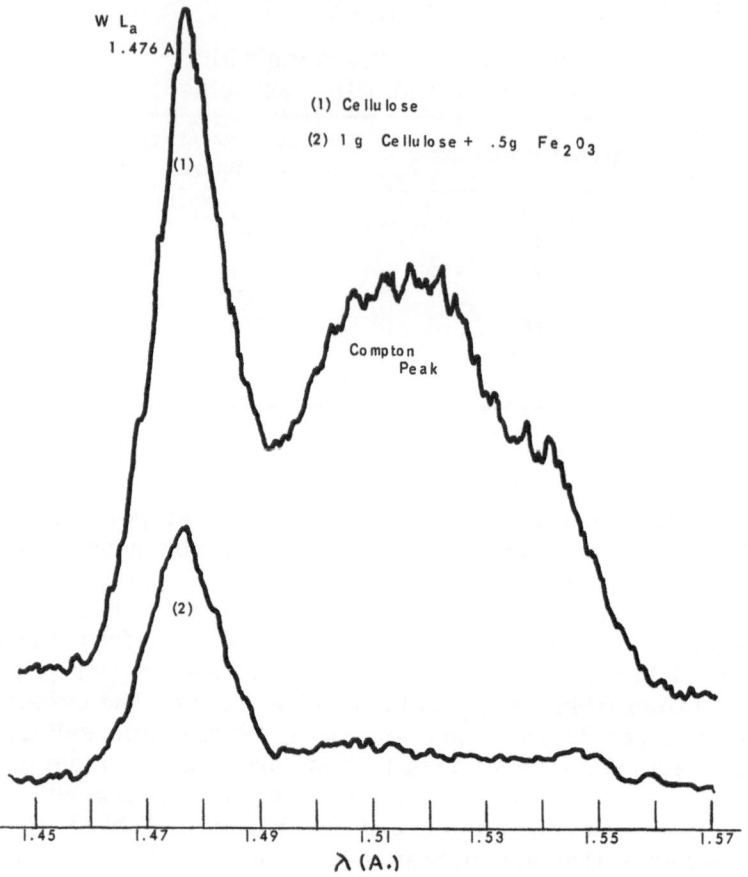

Fig. 5. Elimination of the Compton effect with the ferric oxide—cellulose mixture.

During the perchloric acid evaporation, tungsten and varying amounts of titanium and molybdenum may separate along with the earth acids. The precipitation of tungsten is quantitative if niobium and tantalum are present [12]. The precipitation of molybdenum and titanium is only partial. Table VI shows the approximate percentages of molybdenum that separated along with the niobium, tantalum, and silicon on several samples.

Upon examination of the mass absorption coefficients (Table VII) for niobium, tantalum, tungsten, molybdenum, iron, titanium, and silicon at 0.75 A and 1.52 A, it is possible to predict which elements will have an absorbing or enhancing

TABLE VI

Partial Precipitation of Molybdenum Along with the Earth Acids

	Nb	Ta	Mo	% Mo precipitated	% of total
NBS 1203	1.00	0.34	3.01	1.06—1.29	35—43
NBS 1204	1.31	0.46	4.28	1.78—2.07	42—48
NBS 1205	1.95	0.67	5.75	3.15—3.22	55—56

effect for niobium and tantalum. Tungsten and tantalum will exhibit strong absorption for niobium, while molybdenum and titanium will have little effect. Silicon results in an enhancement effect. Tantalum is strongly absorbed by the iron blended with the cellulose, and this probably overshadows any inter-element effects from other elements that may be present.

During the ignition at 950—1000°C molybdenum and tungsten may slowly volatilize as oxides. Molybdenum begins to volatilize above 550°C and tungsten above 750°C. If desired, about 80% of the precipitated molybdenum may be driven off by increasing the ignition temperature to 1150°C for 1 hr. This step does not appear necessary, however, since tests indicated that the molybdenum had little effect on the niobium intensity and only a very slight enhancement effect for tantalum.

The loss of molybdenum and tungsten will cause errors in the determination of silicon if the determination is by the classical loss-in-weight method with hydrofluoric acid. The loss in weight between the first and second ignition is calculated

TABLE VII

Mass Absorption Coefficients*

For Nb K_α(0.748 A)		For Ta L_α(1.522 A)	
Si	8	Si	58.5
Nb	21.5	Nb	144
Mo	23	Mo	155
Ti	27.5	Ta	166
Fe	42.5	W	173
Ta	115	Ti	208
W	119	Fe	320

*$u = \lambda^n$, J. Leroux, Advances in X-Ray Analysis, Vol. 5 (1961), pp. 153—160.

as silicon, when actually the loss may also be due to the loss of molybdenum and some tungsten during the reprecipitation and subsequent ignition.   Molybdenum does not precipitate during the boiling with sulfurous acid as it does with perchloric acid. Therefore, the molybdenum put back into solution during the HF treatment of the silica does not precipitate again. If appreciable molybdenum is present, most of it could be removed by setting the initial ignition at 1150°C or higher for at least 1 hr.   Then, the second ignition after reprecipitation of the earth acids could be carried out at about 600°C to prevent the rapid loss of small amounts of molybdenum remaining. Tungsten will not be volatilized at this low temperature.

The presence of tungsten will cause serious absorption of niobium.   This was verified from results obtained on NBS 1184 (1.39% W) and NBS 167 (4.50% W) if the niobium was read from the calibration line.   The multiple regression equation will give better results; however, there is a degree of uncertainty where tungsten is involved. This is because the tungsten content of the sample at the time it is being analyzed will be somewhat less than before the ignition due to the volatility of the oxide.   A recent article by Carey, Raby, and Banks [20] indicated that the loss of WO_3 at 1000°C should not exceed $1\frac{1}{2}\%$ of the amount present (Fig. 6). At this point it should be mentioned that the multiple regression equation was calculated assuming the actual tungsten content in the alloy and not taking into account any volatility of the oxide.   Tungsten may be removed from the niobium and tantalum by treatment with NaOH as described in the ASTM method [21].

If significant titanium is present, it will coprecipitate along with the earth acids as mentioned earlier. The amount carried down appears to be directly related to the concentration of earth acids. If the oxides are redissolved by treatment with sulfuric, perchloric, and hydrofluoric acids, the titanium will not reprecipitate during the following treatment with sulfurous acid. The presence of various amounts of titanium and molybdenum did not appear to exhibit any serious interelement effects in this method.

Since the precipitation of silicon by this method is quantitative, its determination along with the earth acids would be most desirable; however, usable intensities could not be obtained on a 2-min integration.   Four minutes provided sufficient intensity, but the varying absorption effects from the other

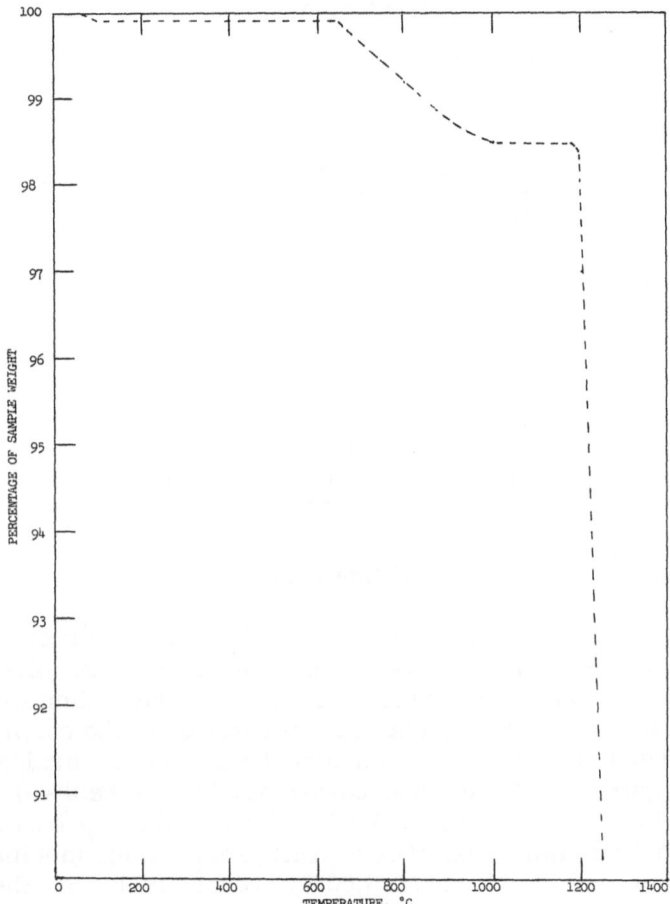

Fig. 6. Thermogravimetric behavior of WO_3 (Carey, Raby, and Banks [20]).

elements did not permit the construction of a suitable calibration
line. Further investigation of the determination of silicon was
not continued because the usual wet chemical method provided
satisfactory results on most of our alloys. Precautions must
be exercised, however, if tungsten or appreciable quantities of
molybdenum and titanium are present in the alloy. Ignition
temperatures and times must be carefully controlled to prevent
errors in the silicon determination.

TABLE VIII

Demonstration of Reproducibility
(Separate samples carried
through entire procedure)

NBS -123a	Nb 0.75%	Ta 0.02%
(1)	0.75	0.012
(2)	0.75	0.013
(3)	0.77	0.013
(4)	0.76	0.015
NBS 1188	Nb 1.11%	Ta 0.11%
(1)	1.14	0.117
(2)	1.14	0.112
(3)	1.11	0.110
(4)	1.13	0.111

SUMMARY

A reliable X-ray spectrochemical method has been developed for niobium and tantalum in high-temperature alloys and stainless steel using a tungsten target tube and lithium fluoride analyzing crystals. A chemical separation of the earth acids eliminated interference from nickel and copper. Fusions are not required. The method covers niobium in the 0—3% range and tantalum from 0—0.5%. Table VIII shows the reproducibility on two NBS standards. The overall precision for niobium and tantalum is ±0.02% and ±0.005%, respectively, at the 95% confidence level.

ACKNOWLEDGMENT

The author wishes to thank Neil Butler for the chemical preparation of many of the samples and Jerome Soboleski of the Data Systems Department for the computer programming of the analytical data.

REFERENCES

1. E. L. Koerner, M. Smutz, and H. A. Wilhelm, Separation of niobium and tantalum by liquid extraction, U.S. Atomic Energy Commission Report ISC-802 (1956).

2. G. E. F. Lundell, J. I. Hoffman, W. F. Hillebrand, and H. A. Bright, Applied Inorganic Analysis, 2nd edition, John Wiley and Sons, New York (1953), p. 589.
3. E. L. Koerner and M. Smutz, Separation of niobium and tantalum - a literature survey, U.S. Atomic Energy Commission Report ISC-793 (1956).
4. Ibid., p. 7.
5. R. W. Moshier and J. E. Schwarberg, Tantalum determination in presence of niobium by precipitation with N-benzoyl-N-phenylhydroxylamine, Anal. Chem. 29, 947-951 (1957).
6. E. H. Huffman, G. M. Iddings, and R. C. Lilly, J. Am. Chem. Soc. 73, 4474 (1951).
7. K. A. Kraus and G. E. Moore, J. Am. Chem. Soc. 71, 3855 (1949).
8. K. A. Kraus, J. Am. Chem. Soc. 73, 2900 (1951).
9. I. Martin and R. J. Magee, Chromatographic separation and determination of niobium, tantalum, and titanium, Talanta 10, 119-1123 (1963).
10. E. L. Koerner and M. Smutz, Separation of niobium and tantalum - a literature survey, U. S. Atomic Energy Commission Report ISC-793 (1956), p. 10.
11. C. K. Kim and W. W. Meinke, Simultaneous determination of niobium and tantalum by neutron activation using niobium -94m and tantalum -182m and rapid radiochemical separations, Anal. Chem. 35, 2135-2138 (1963).
12. C. L. Luke, Determination of refractory metals in ferrous alloys and high-alloy steel by the borax disk X-ray spectrochemical method, Anal. Chem. 35, 56-58 (1963).
13. B. J. Mitchell, X-ray spectrochemical determination of zirconium, tungsten, and vanadium oxides, Anal. Chem. 32, 1652-1656 (1960).
14. W. J. Campbell and H. F. Carl, Fluorescent X-ray spectrographic determination of tantalum in commercial niobium oxides, Anal. Chem. 28, 960-962 (1956).
15. M. L. Tompkins, G. A. Borun, and W. A. Fahlbusch, Quantitative determination of tantalum, tungsten, niobium, and zirconium in high-temperature alloys by X-ray fluorescence solution method, Anal. Chem. 34, 1260-1263 (1962).
16. R. Burton, R. M. Jacobs, and E. R. Valecko, The determination of tantalum in niobium with the applied research laboratory's X-ray fluorescent monochromator, Office of Technical Services, Dept. of Commerce, Washington 24, D.C., WAPD-CTA (GLA)-617-1, July 10, 1958.
17. E. E. Hanson, The analysis of 15-W, 5-Mo, 1-Zr, Cb alloy by X-ray fluorescence spectroscopy, in: Developments in Applied Spectroscopy, Vol. 1, Plenum Press, New York (1962), p. 108.
18. S. Friedlander and A. Goldblatt, A comparison of precision for solid, liquid, and powder sampling techniques in the X-ray spectrochemical analysis of high-temperature alloys, Pittsburgh Conference on Analytical Chemistry and Applied Spectroscopy, Pittsburgh, Pa., March (1959).
19. L. S. Birks, X-ray Spectrochemical Analysis, Interscience, New York (1959), pp. 9-10.
20. M. A. Carey, B. A. Raby, and C. V. Banks, Preparation, high-temperature properties, and analytical uses of tungsten oxide, Anal. Chem. 36, 1166-67 (1964).
21. ASTM Methods Chemical Analysis of Metals (1960), pp. 151-154.

Influence of the Origin of Raw Materials on the X-Ray Analysis of Cements

H. T. Dryer and H. Renton

Applied Research Laboratories
Detroit, Michigan

The analysis of nonmetallics by X-ray fluorescence methods has been widely accepted as a standard method for routine control. As new materials are studied and appropriate methods developed, these are added to the list of control procedures. Each of these materials presents different types of problems to the analyst, and must be investigated with this in mind. One problem common to almost all nonmetallics is the mineralogical differences and origin of raw materials. The materials of the cement industry have been studied primarily to determine the influence of the raw materials and their origin. The results of this study and their relation to analytical performance will be presented.

X-ray spectrochemical methods of analysis have been widely accepted as a means for routine control. However, before the analytical method for each type of material can be accepted for control purposes, the effects of possible variables must be considered and evaluated. Obviously, acceptable methods are possible only when these effects are small, when they can be controlled, or when corrections can be made to compensate for them.

The analytical programs for nonmetallic materials represent some of the most difficult problems and contain more variables than those of other fields. Each of these materials presents a new set of problems to the analyst and must be studied with this in mind. Mineralogical differences and origin of the raw materials are typical problems common to almost all nonmetallics.

Because of the increasing demand on analytical performance, i.e., high speed and large volume of samples at the re-

quired precision and accuracy level, many of the nonmetallic
industries are turning to X-ray analyses for their programs. In
order to study these applications and to provide a general
method for nonmetallics, an investigation was made covering the
analytical problems of the cement industry. Raw mix, a
preliminary material of this industry, was chosen to provide a
wide range of variables. This material is a physical blend of
the required raw materials proportioned to the proper specifications prior to introduction into the cement kilns. Clinker, which
is formed from a semifusion of the raw mix, is further processed to produce finished cement. By control of raw-mix
composition and kiln condition, the required end-product can
be produced more easily and economically.

The number of raw materials can vary widely from plant
to plant and may change drastically in one plant, depending
upon the composition of the raw material type of cement, etc.
For example, raw mix from one plant may be composed of
limestone, shale, sand, iron ore, bauxite, and slag, while another plant may use only the first four raw materials listed;
still another plant may use only limestone and shale to make
raw mix. In many instances, the variations in feed stock at a
single plant may be relatively minor over short periods of
time and major over long periods of time. Limestone, the
basic raw material of the cement industry, is deposited in the
quarry area as shown in Fig. 1. Because of the wide variations
in these deposits, composition changes in limestone require
changes in the blended material to produce a uniform feed.

Sampling of the materials is quite difficult in order to
obtain a proper representative material and to provide feed

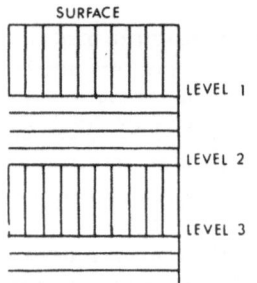

Fig. 1. Limestone distribution in quarry.
Composition and structure may vary widely
in each level.

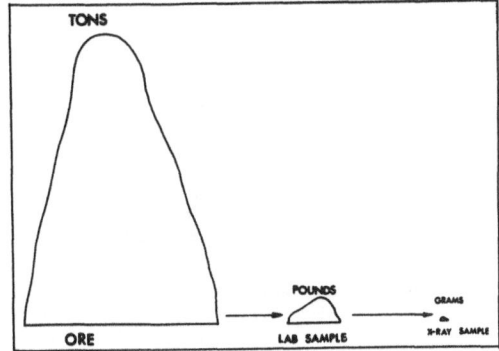

Fig. 2. Sampling factor from original sample to material analyzed (may represent a factor of 1,000,000:1).

control. As illustrated in Fig. 2, the sampling factor is extremely high. Although the sampling problem does not reduce the accuracy and precision of the analytical method, considerable study should be made to ensure that the analytical results are representative of the material and consequently useful for process control.

Several methods of sample preparation were considered for the nonmetallic analyses. A fusion method such as described by Rose, Adler, and Flanagan [1] of the U. S. Geological Survey would be ideal to eliminate many of the variables and to provide analytical curves covering a wide range of material. This method uses a flux to reduce the materials to a common

TABLE I

Sample Preparation of 35 Nonmetallics

	Methods		
	General nonmetallics	Borax flux	Lithium tetraborate flux and lanthanum oxide
Sample weight, g	10 – 50	1.00	0.125
Flux, g	-	2.00	1.000
Buffer, g	-	-	0.125
Temperature, °C	-	1050	950
Time, min	-	5	8
Grinding, min	1.5	1.5	1.5

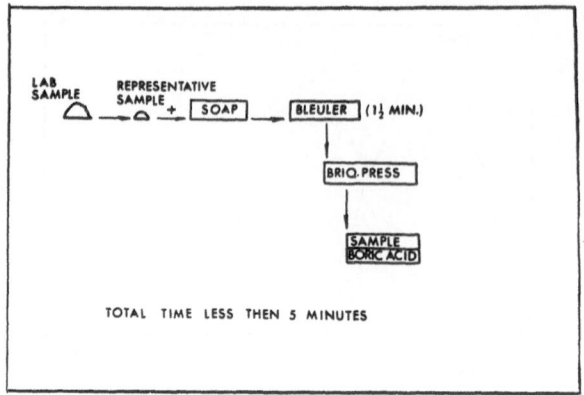

Fig. 3. General nonmetallics method of sample preparation.

structure and a buffer to reduce the material to a common matrix. Although this method provides many of the desirable features of an analytical procedure, it does not provide the high speed of analysis or large volume of samples required by a control laboratory.

The sample preparation used for this investigation has been in use for some time by the Applied Research Laboratories (see Table I, General nonmetallics). Lithium tetraborate flux and lanthanum oxide is the method of the U.S.G.S. described above. The method of general nonmetallics is shown graphically in Fig. 3, indicating the steps and time required for this method. The Bleuler grinder, which is used for the grinding step in sample preparation, is shown in Fig. 4.

Approximately 250 samples from 25 cement companies were used for these studies. These samples covered wide ranges of concentration with systems of two to six components, and raw material sources were from a wide range of geographical formations. Raw material sources are indicated in Fig. 5.

These studies were performed using an ARL VP XQ[2] which provides simultaneous analytical data for the programmed elements from a given sample. The analytical data for the elements are provided as ratios to the intensity of a control channel. One set of samples covering the analytical ranges was used as a control standard throughout this study.

The analytical data for all samples and companies were combined for the evaluation of the results of this study. The samples were analyzed for CaO, SiO_2, Al_2O_3, MgO, Fe_2O_3, MnO, TiO_2, phosphorus, and sulfur.

Fig. 4. Bleuler rotary mill.

Correlation of the analytical data was excellent and indicated that little difficulty would be experienced in the analysis of raw mix for the cement industry. Figures 6–8, which are the analytical curves for MgO, CaO, and Fe_2O_3, respectively, illustrate the correlation of data. The control standards are represented by the solid curves and the extremes of the other samples by the dotted curves.

Most of the analytical curves for the individual companies were coincident upon those of the control standards; several were displaced slightly, and only a few samples were displaced as far as the extremes shown.

Unfortunately, X-ray spectrochemical analysis is a comparative method and is dependent upon chemical or assumed values. Routine methods are based upon the analytical curves constructed from the data of reliable standards. During this

Fig. 5. Distribution of raw material sources.

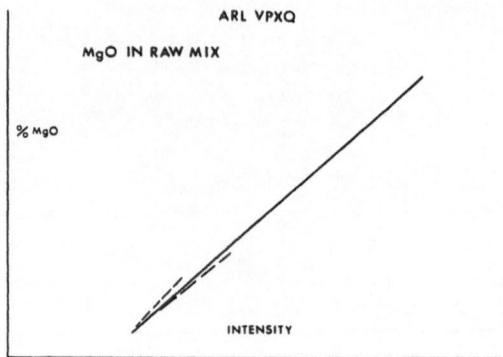

Fig. 6. MgO in raw mix, average curve and maximum spread.

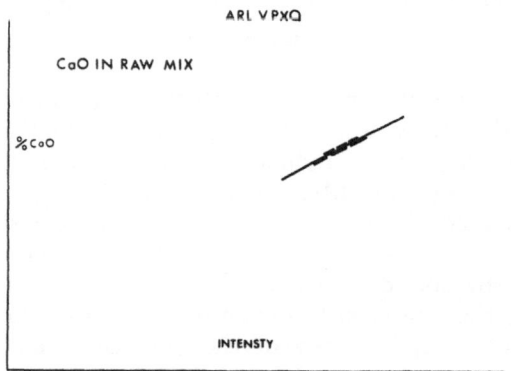

Fig. 7. CaO in raw mix, average curve and maximum spread.

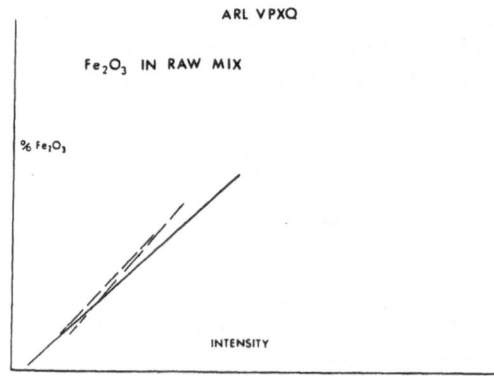

Fig. 8. Fe_2O_3 in raw mix, average curve and maximum spread.

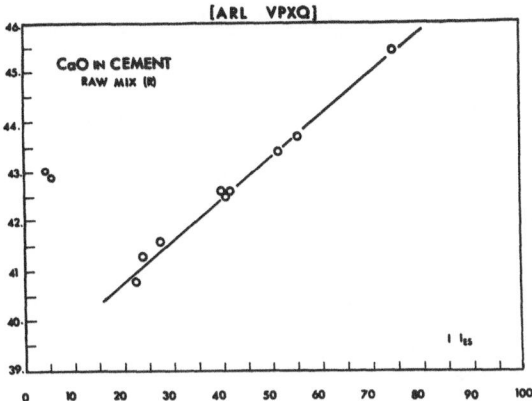

Fig. 9. CaO in raw mix—method of general nonmetallics.

investigation, it became apparent that two sets of samples did not show proper correlation with the other samples. Since these samples were similar in makeup to other sets of samples on both location and raw materials blend, it was apparent that some other variable or variables were affecting the analytical data. Chemical data proved to be the limiting factor for these samples. The chemical methods were not in error, but the manner of calculation was not consistent with those used for the other samples. This procedure would not detract from the use of X-ray analyses for control purposes, but would hinder interlaboratory analyses.

As a further check on this analytical procedure, a portion of these samples was prepared by the fusion technique described by Andermann [3] for cement analyses. The analytical curves for CaO in raw mix, as prepared by the general nonmetallics and boron flux methods, are shown in Figs. 9 and 10, respectively. As demonstrated, no apparent improvement is achieved through the use of the fusion method.

As described previously, the fusion method of Rose et al. of the U.S.G.S. might be unsuited for routine control. This method, however, can be extremely valuable as a means of establishing reliable analytical curves through the preparation of synthetic standards and certification of standard samples.

Since the analytical surface presented to the X-ray equipment for analysis must represent the material to be analyzed, the sampling and sample preparation procedures are of utmost importance [4].

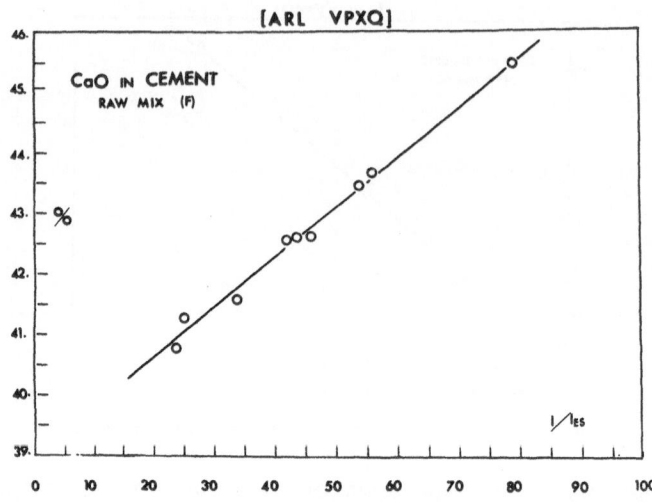

Fig. 10. CaO in raw mix—method of boron flux.

In summary, an evaluation of the data from this study indicates that little, if any, difficulty would be experienced with this analytical program. Obviously, as new raw materials are used, additional data must be evaluated to study their effects upon these analyses.

REFERENCES

1. H. J. Rose, I. Adler, and F. J. Flanagan, Appl. Spectroscopy 17:81 (1963).
2. E. Davidson, A. W. Gilkerson, and H. Neuhaus, "Direct-reading X-ray poly-chromators for research and production control," Pittsburgh Conference (1960).
3. G. Andermann, Anal. Chem. 33:1689 (1961).
4. H. T. Dryer, Advances in X-Ray Analysis, Vol. 6, Plenum Press, New York (1962), p. 447.

Scanning Electron-Probe Techniques for Diverse Nonmetallurgical, Industrial Applications

T. E. Reichard and W. S. Coakley

Monsanto Company
St. Louis, Missouri

Techniques and procedures are described which have proven especially valuable for rapid qualitative and semiquantitative electron-probe microanalysis of nonmetallographic specimens, wherein rough or irregular surface structures must often be analyzed as they occur, and samples may have low physical stability, coarse, porous texture, fragile structures, high and varying electrical resistivity, and/or surface conditions which prevent deposition of a conducting film. The techniques include: (1) the use of various conducting epoxy resins and glues for specimen mounting, (2) a simple vacuum technique for void-free mounting of fragile, porous particles without heat or pressure, (3) extensive use of background X-ray images along with characteristic X-ray images and scaler counting for semiquantitative interpretation of element distributions, compensating for topographical effects or gross variations in sample density or both, and (4) the advantageous use of peak/background ratios for quantitative measurements of element distributions at low concentrations under adverse specimen conditions.

INTRODUCTION

The unique capabilities of the scanning electron-probe microanalyzer are providing new insight into an ever-widening range of industrial problems, not only in research and process studies, but also in engineering, plant and production areas, inspection of purchased materials and components, determination of properties and behavior of industrial products in specific customer applications, characterization of materials for patent purposes, and evaluation of competitive products.

A Cambridge electron-probe microanalyzer, in operation in our laboratory for two years, has been applied to a broad range of industrial problems. Applications have included

91

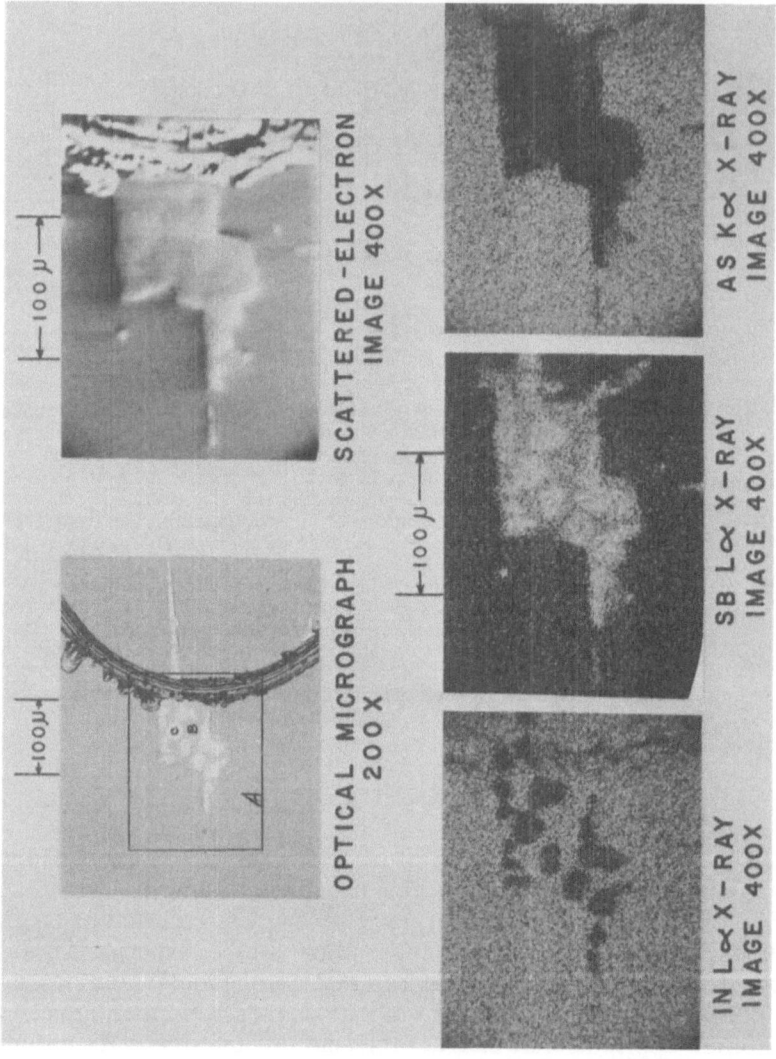

Fig. 1. Indium—antimony—arsenic alloy, phase segregations.

TABLE I

Electron-Probe Applications

Type of application	Percent of total applications
Metallurgy and corrosion	30
Catalysis	25
Semiconductor materials and devices	20
Other nonmetallic products	15
Biological	10

metal plating and corrosion problems, several catalyst systems, semiconductor materials and devices, optical films, pigment dispersion, analysis of microscopic contaminants from chemical plant systems, biological specimens, and detergent materials. In striking contrast to published electron-probe work—about 90% metallurgical or mineralogical [1]—our applications have been predominantly nonmetallurgical (Table I).

The development of our techniques and procedures has been shaped by this diversity of problems and materials. Specimens have varied widely in size, configuration, composition, texture, and physical stability. Materials having irregular surfaces, porous textures, fragile structures, and/or low electrical conductivity predominate. Most of our problems are solvable via qualitative and semiquantitative microanalytical and element-distribution data, and the more time-consuming, precisely quantitative, point-by-point methods are seldom needed.

The beam-scanning methods are most widely used, therefore, since they accumulate the greatest amount of significant data in the least time, and the X-ray images and linear concentration profiles provide vivid pictorial data presentations, which are especially effective for easy comprehension by people relatively unfamiliar with electron-probe technology. For example, the phase segregations in a three-component alloy were to be identified (Fig. 1). The pictorial results, obtained and reported in less than one hour, are virtually self-explanatory: Matrix A contains indium and arsenic, phase B contains indium and antimony, and phase C is antimony only.

SPECIMEN PREPARATION

The standard specimen holder requires that the mounting be in the shape of a cylinder $\frac{1}{4}$ in. in diameter and about $\frac{1}{8}$ in. high, with the flat top surface to be analyzed. Larger specimens, from $\frac{1}{4}$ in. up to $1\frac{3}{8}$ in. in diameter, can be mounted in the metallographic specimen holder.

Small specimens, such as pellets, granules, flakes, fragments, wires, contaminant "specks," and small semiconductor devices, are mounted inside short cylindrical rings cut from $\frac{1}{4}$-in.-diameter glass, aluminum, copper, or brass tubing. The specimen may be mounted in any manner which places the surface to be analyzed in one end-plane of the ring. The following convenient, effective methods have been developed here for mounting diverse samples.

Analysis of Existing Surfaces

For endviews or edgeviews of elongated or lamellar shapes (Fig. 2), a piece of double-coated transparent tape (Scotch Brand No. 665, adhesive on both sides) is placed on a microscope slide, and the specimen surface to be analyzed is pressed into the adhesive. A $\frac{1}{4}$-in. ring is centered over the sample and the open space filled with indium metal chips, which are soft enough to be pressed and molded around the specimen without dislocating it. The tape is then peeled off the slide and the ring, and the mounting can be solvent-cleaned for analysis.

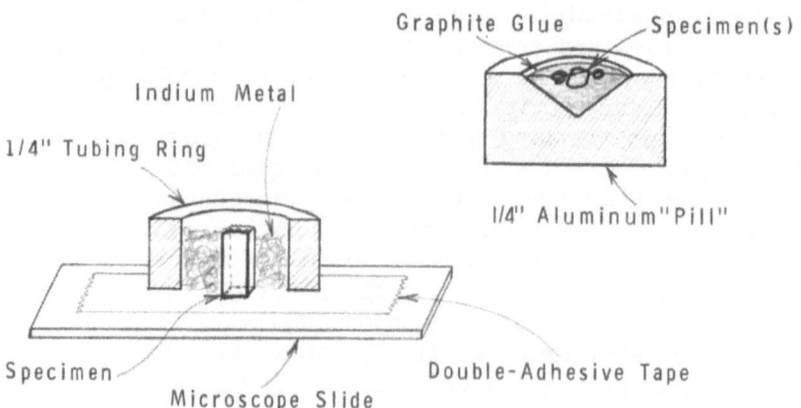

Fig. 2. Specimen mountings for analysis of existing surfaces.

For analyzing surfaces of small beads, granules, "grit," bits of metal, etc., which are to be retrieved after analysis (Fig. 2), a solid aluminum pill with a suitable depression is used. The depression is almost filled with a soluble glue (e.g., Duco cement or a solution of Formvar or Parlodion) premixed with graphite to a paste consistency. The particles are placed on the paste and pressed flush with the outer rim of the pill. The graphite-glue hardens to a stable mounting with high electrical conductivity, and the specimen can be easily dissolved free after analysis. For analysis of microscopic amounts of grit or specks, the particles may be imbedded in the flat surface of a solid aluminum pill, using a briquetting press.

A briquetting press was found useful in another type of problem, i.e., analyzing for contaminants in the bottom of a deep pit on the inside wall of a soft-metal tube. A $\frac{1}{4}$-in. disk was cut from the tube wall, with the pit at its center, and cold-pressed in a $\frac{1}{4}$-in. die, raising the pit-bottom flush with the top surface, accessible to analysis.

Analysis of Cross Sections

For small specimens to be studied in cross section, electrically-conductive epoxy resins (loaded with graphite, silver, or aluminum) are excellent imbedding media. They harden without heat or pressure and do not damage fragile specimens. The heavily-loaded epoxy silver solders provide the best thermal and electrical conductivity. When analyzing for trace elements at the edge of the imbedded specimen, the relatively high "white" X-ray emission from a silver-loaded mounting is undesirable, so graphite or aluminum-loaded epoxies are preferable. Any element to be analyzed in the specimen is avoided in the ring, the imbedding medium, and the polishing materials.

For cross-section study of small objects (such as catalyst pellets or granules), several $\frac{1}{4}$-in. rings are placed on a small metal plate and half-filled with epoxy mixture (Fig. 3). Specimens are placed in the desired orientations and epoxy added to fill the tubes, stirring somewhat to wet the specimen and work out bubbles. The epoxy binds the ring-mountings to the plate in a very convenient form for handling in grinding, polishing, cleaning, aluminizing, photomicrography, etc. Pried off the plate, the mountings are ready for analysis.

For examining sections of larger specimens, or for a wider selection of particles and structures than is provided by the $\frac{1}{4}$-in. mounting, a sheet of mylar is taped flat onto a microscope slide and covered with double-adhesive tape (Fig. 3). Particles are sprinkled onto the tape and pressed slightly into the adhesive. Individual particles of interest can be placed in desired positions and orientations. Conductive epoxy is "puddled" over the particle. The adhesive prevents particles being "swallowed up" and randomly distributed in the epoxy. Mylar and tape are removed from the hardened mounting. Grinding and polishing down about one-half the average particle diameter then yields a maximum selection of particle cross sections for analysis, using the metallographic holder.

The aluminizing step is often omitted. Satisfactory semiquantitative analyses are obtained from electrically nonconducting particles up to several millimeters in size, without aluminizing, when they are imbedded in a conducting medium. On rough, porous specimen surfaces, a thin aluminum layer often fails to form a continuous, conducting film, and adds little or nothing to specimen-current stability and reproducibility.

A mirror-smooth, metallographic-type polish may be difficult to obtain on some loaded-epoxy mountings, but this is not a serious drawback in semiquantitative work, particularly with porous, granular specimen structures.

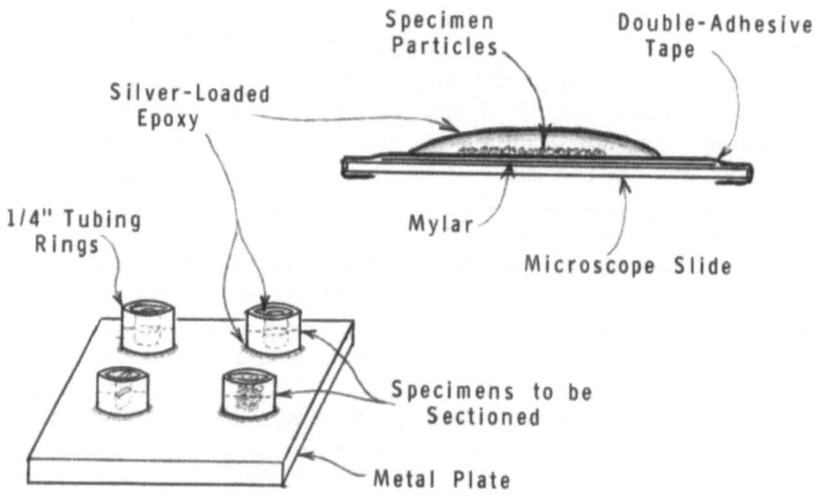

Fig. 3. Specimen mountings for analysis of cross sections.

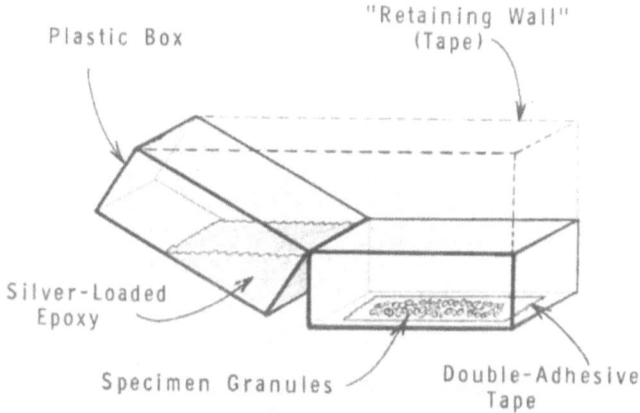

Fig. 4. Vacuum-imbedding technique for void-free mountings.

Vacuum Imbedding

With the above imbedding techniques, small bubbles and voids are not entirely eliminated. This is not usually a serious problem. However, in recent studies of detergent granules, the presence of relatively volatile organic components caused excessive voids, and the relatively low physical stability of the granules required maximum surface contact with the silver-epoxy as a heat-sink. Vacuum desiccation of the granules before using the "puddling" technique (Fig. 3) improved their epoxy-wettability and greatly reduced the number and size of voids, but the mountings were not entirely satisfactory.

A simple vacuum-imbedding technique has produced excellent results. Previously desiccated granules are placed on double-adhesive tape as for the "puddling" technique, but they are placed in the bottom of a square, transparent plastic box. The hinged box lid is set at an angle as shown in Fig. 4, and a retaining wall built up with tape. The lid is filled with epoxy-hardener mixture, and the entire assembly placed in a vacuum chamber. After a short evacuation, the volatility of the hardener causes the epoxy to foam up and flow over the sample particles. Readmission of atmospheric pressure then closes all bubbles (the hardener recondenses), and even forces the resin into voids in the granules, to support internal structures. After sectioning, the mountings are aluminized "heavily" (about 500 A thick) for added physical stability.

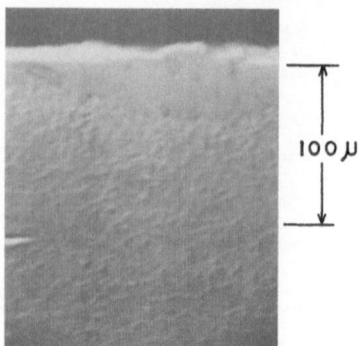

**SCATTERED – ELECTRON
IMAGE 420 X**

Fig. 5. Corrosion specimen, X-ray image analysis.

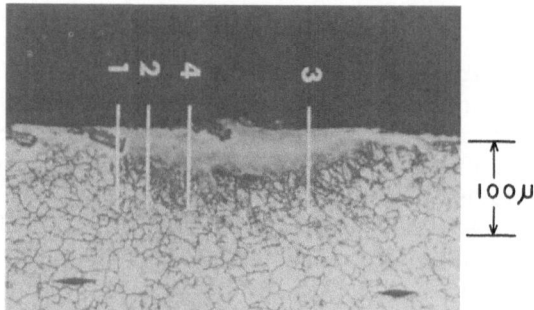

**OPTICAL MICROGRAPH
250 X**

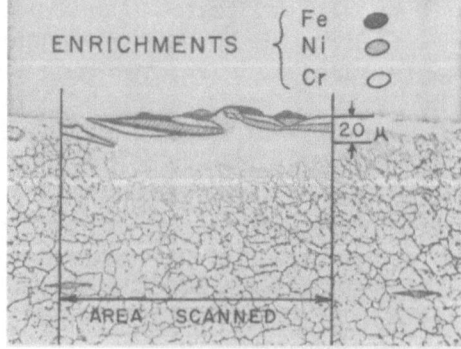

Fig. 5. (continued).

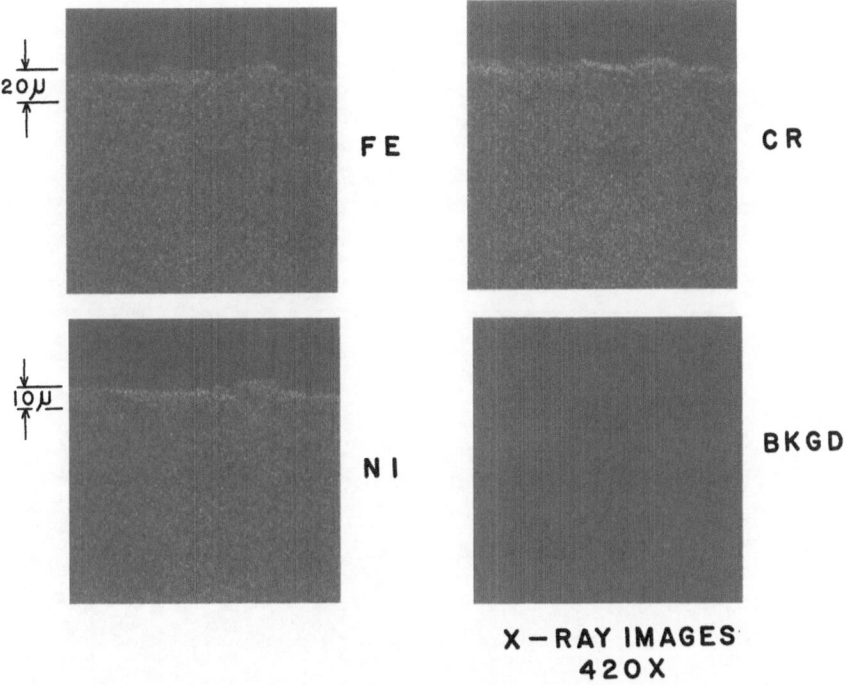

Fig. 5. (continued).

INTERPRETATION

Background X-Ray Images

While X-ray images of major elements on smooth, flat specimen surfaces can be interpreted directly in terms of concentration distributions, conditions often encountered in nonmetallurgical specimens can produce the following misleading effects:

1. When the surface is irregular, topographical shadowing effects are superimposed on the contrast due to element concentration gradients.
2. When trace concentrations are imaged, X-ray peak-to-background ratios are so low that a substantial proportion (often a majority) of the counts on the image arise from white background radiation, rather than from the characteristic X-ray line.

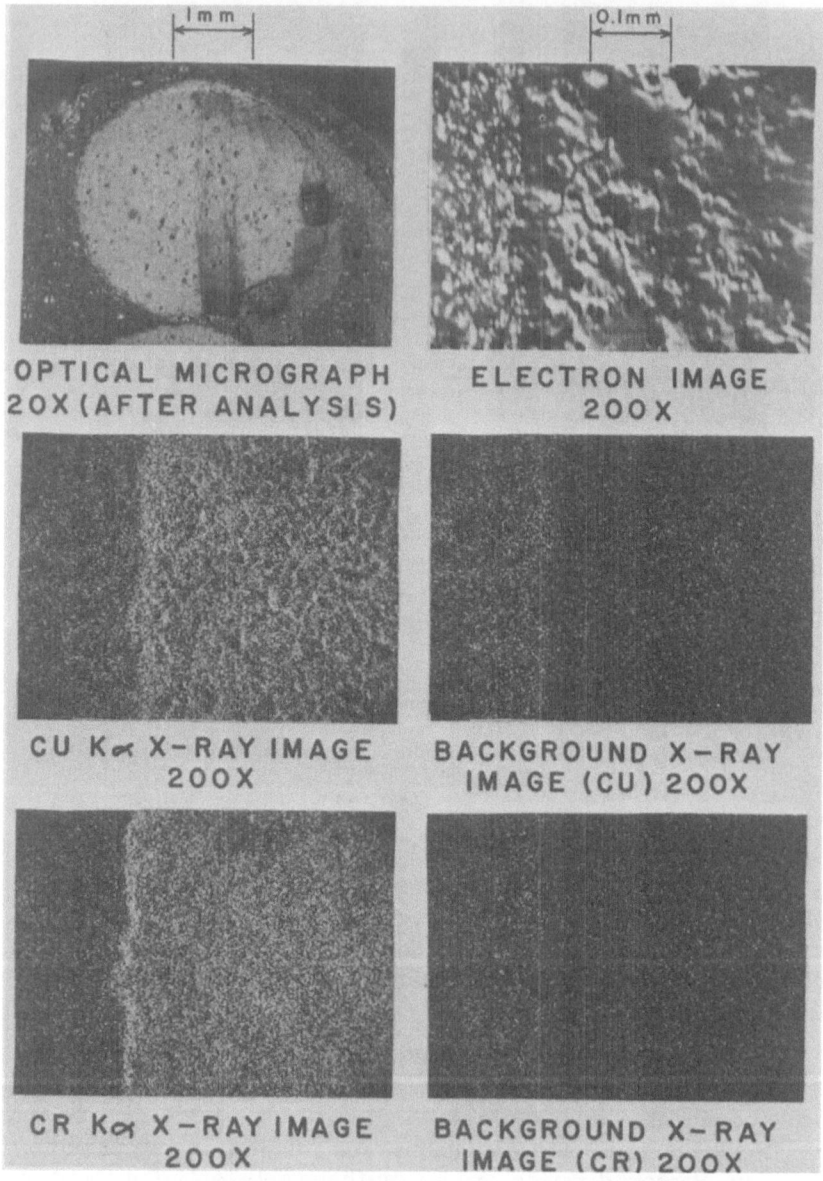

Fig. 6. Copper–chromium catalyst pellet (cross section).

3. When specimen density varies substantially over the image field, the corresponding variations in background emission can produce false indications of element enhancement or depletion.

Under such conditions, rather than resort to lengthy point-by-point methods, we have developed the following background X-ray image techniques which quickly clarify image interpretations. After photographing the element X-ray image, the spectrometer is shifted slightly off the peak, and another image is photographed of the same specimen field, but viewing only the white radiation background at a wavelength very near that of the characteristic.

For topography compensation, the background X-ray image is exposed for the same number of counts as the characteristic image, using the X-ray scaler as an exposure meter. This background image presents the contrast pattern produced by surface topography (shadowing effects) only, and is, therefore, the image which would be produced by a perfectly uniform element distribution. The X-ray image of the element itself, then, by comparison, shows the additional contrast (if any) attributable to true variations in element concentration. This type of background-image comparison is valuable even for major elemental components, if the specimen surface is very rough or irregular.

In cases where peak-to-background ratios are low, or specimen density varies appreciably over the area imaged, the background image is exposed for the same elapsed time as the characteristic image, and is thus analogous to the baseline adjacent to the peak on a strip-chart record of the X-ray emission line. It is visually subtracted out of the element image, and the remainder (if any) represents true presence and distribution of the element. This type of background compensation is necessary only for interpretation of minor elemental components.

To compensate for both topography and density variations, it follows that two background images are needed; one exposed for equal time, the other for equal counts. For trace-concentrations, as peak-to-background ratios drop near unity, however, the equal-time and equal-counts images become almost identical.

When several elements are imaged whose characteristic emission lines are in a narrow wavelength range, a single

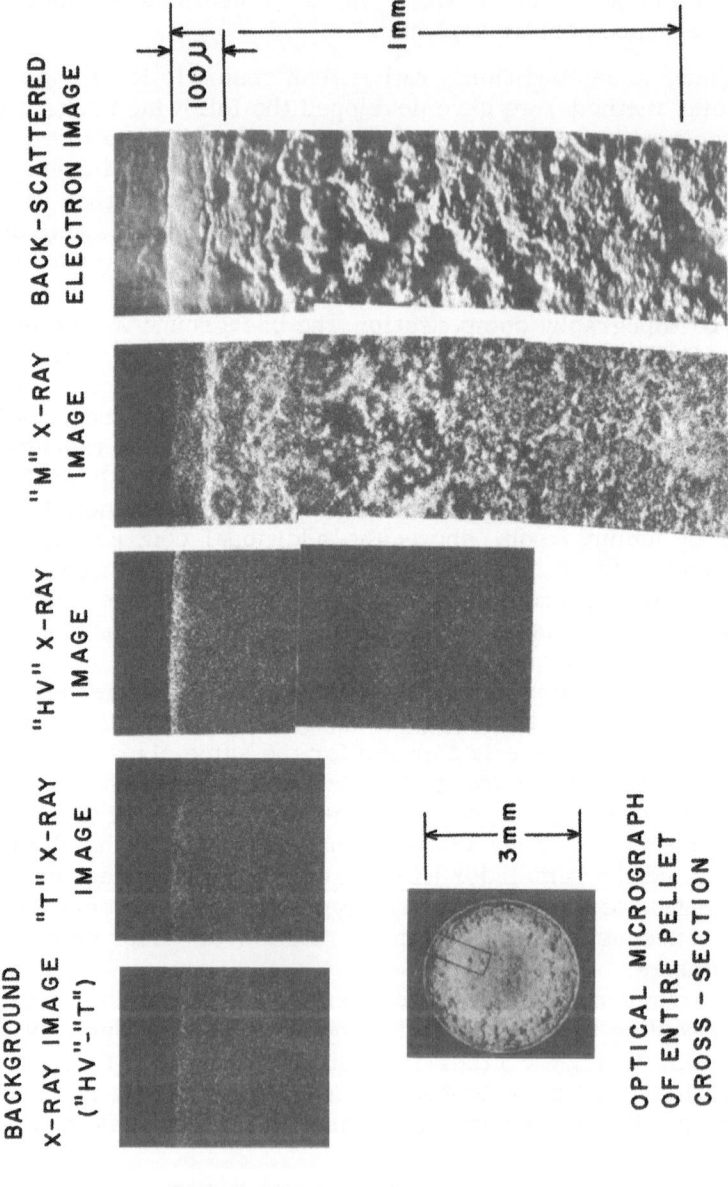

Fig. 7. M catalyst pellet (cross section).

background image serves for comparison with all of them. For widely varying characteristic wavelengths, however, separate background images are advisable (e.g., hard and soft background images), since the effects of surface contour and density variations differ appreciably for widely different wavelengths.

Figure 5 illustrates a qualitative metallographic study of a lightly etched section through a corrosion spot on a stainless steel, containing roughly equal parts of iron, nickel, and chromium. Here the specimen conditions are nearly ideal, and the X-ray images are directly interpretable. For ease of comprehension in the original study, the complicated pattern of local enrichments near the corroded metal surface was sketched onto a copy of the optical micrograph, coded according to elements.

Figure 6 shows X-ray images of a field at the edge of a cross section through a catalyst pellet, containing low percentages of copper and chromium on alumina. Apparent edge enrichments can sometimes arise from relief between specimen and mounting after grinding and polishing. In this case, however, the background X-ray images give assurance that the surface enrichments are true for both elements, and that both elements are definitely present, at reduced concentration, in the interior of the pellet. The appreciable concentrations of apparent $Cu\,K_\alpha$ and $Cr\,K_\alpha$ counts from the mounting medium (left side of images) are, however, shown to be due to enhanced white emission (at the $Cu\,K_\alpha$ and $Cr\,K_\alpha$ wavelengths) from the denser silver-loaded epoxy mounting.

Figure 7 shows a study of a similar catalyst pellet cross section, in which background-image comparisons are essential to correct interpretation. The distribution of the active metal M is shown on the M X-ray image series, with only minor topography shadowing despite the rough surface; since M is present in substantial amounts, the M characteristic line is at a fairly hard X-ray wavelength, and the alumina matrix has low density. The Hv X-ray image indicates the presence of a heavy metal Hv on the outer surface. Another element T was suspected as a possible trace contaminant on the pellet surface, and the T X-ray image shows a slight concentration of X-ray counts in the same band where the Hv occurs.

The T—Hv background image (taken midway between the characteristic X-ray lines of T and Hv, which are near the same wavelength) serves for comparison to both element images. Comparison shows that Hv is truly present at the pellet surface. The surface enhancement of T X-ray counts, however, is no greater than on the background image, attributable to the greater emission of white background radiation from the higher-density band occupied by the heavy metal Hv. The apparent detection of element T on the X-ray image is, therefore, shown to be false and misleading. Comparisons also show that the T and Hv X-ray counts shown in the interior of the pellet are attributable entirely to background, as is the depletion of counts in the lower-density band just below the surface, which contains neither M nor Hv.

It is often desirable to obtain concentration maps of relatively large areas via series of adjacent fields, matched together as in Fig. 8. The scaler is operated synchronously with the camera shutter, to record the total counts and exposure time for each image. Background-image counting rates are then useful as a reference, to make element images more quantitatively intercomparable, even though topography and specimen conductivity may change drastically from field to field, and the specimen-current reading becomes unreliable as

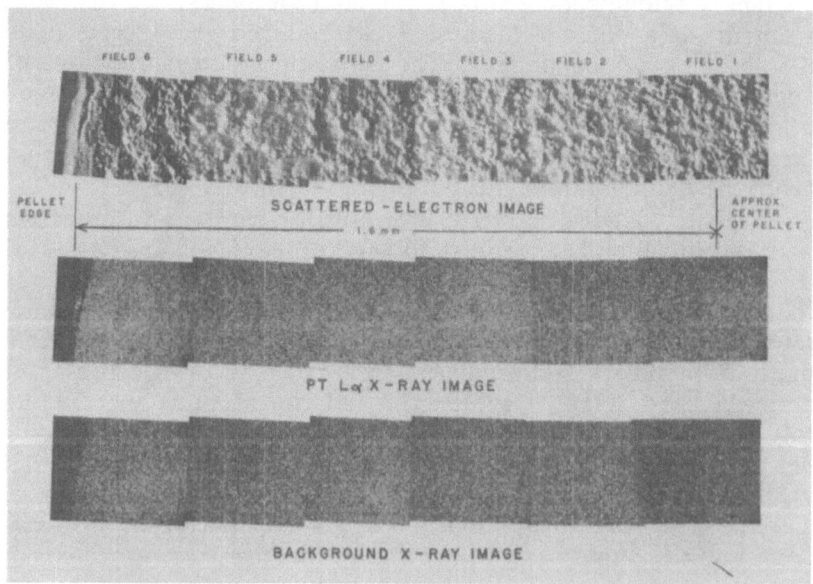

Fig. 8. Concentration map of platinum—alumina catalyst pellet (cross section).

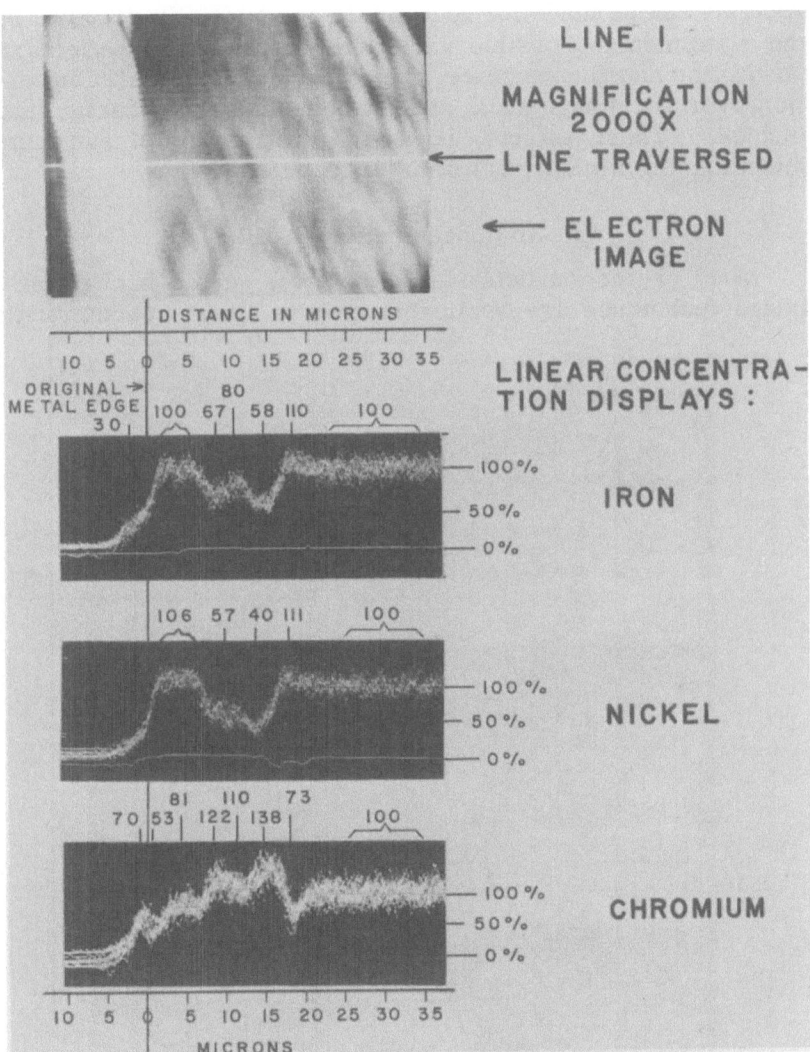

Fig. 9. Corrosion specimen, concentration profiles.

a parameter for controlling X-ray excitation. Peak and background count rates are then available also for plotting broad
variations in element concentrations, field-by-field, and corrections can be made for edge fields only partially occupied by
the specimen. Possible variations in the matrix effects of
absorption and fluorescence, due to varying chemical composition over the image area, are not compensated by background
images. These effects, however, are rarely of sufficient
magnitude to influence image interpretation.

Linear Concentration Profiles

Most of the concepts and objectives of the background-
image techniques are applicable also to linear concentration

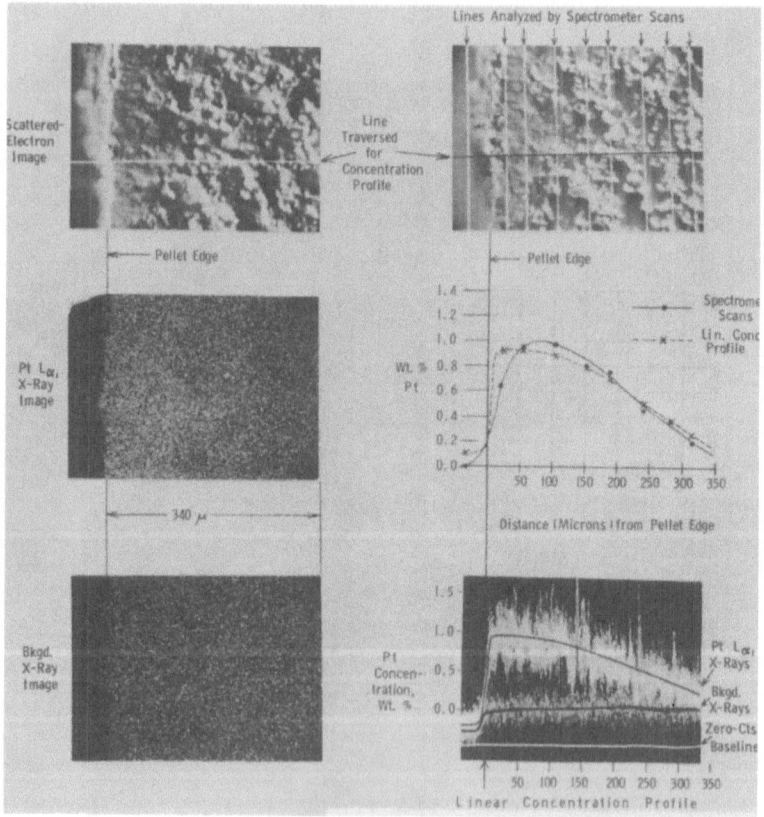

Fig. 10. Platinum distribution in platinum—alumina catalyst pellet (cross section).

profile displays and their interpretation.   Figure 9 shows concentration profiles of the three major elemental components of the ideal metallographic specimen previously discussed. One of the inclusions in the image field of Fig. 5 is here studied at higher magnification and in more quantitative detail. Since we are dealing with a relatively smooth surface and substantial concentrations, the vertical deflections in the concentration profiles, as measured from a zero-counts baseline, closely approximate the actual element concentrations. The concentrations at high and low points in the profiles are indicated in terms of percent of the matrix concentration, i.e., 100% indicates average matrix level.

For concentration profiles of very low element concentrations, however, an appreciable portion of the vertical deflection above the zero-counts baseline is due to background X-rays, as shown in Fig. 10. A background X-ray profile is therefore exposed on the same display (at reduced cathode-ray tube brightness), and the profile is calibrated and interpreted in terms of peak-to-background ratio, i.e., height of the Pt L_{α_1} trace above the background trace, rather than total height above the zero baseline. Figure 10 also illustrates how well this interpretation agrees with the results of a quantitative point-by-point (in this case, line-by-line) analysis.

Multiple concentration-profile traces are photographed to average the statistical fluctuations in counting rates.

ACKNOWLEDGMENT

The authors express their appreciation to R.B. Coffey and G. M. Haskell, who assisted in the electron-probe micro-analyses; and to R. B. Coffey, A. H. Herzog, D. E. Hill, and R. A. Rodgers, who contributed ideas for specimen-preparation techniques.

REFERENCE

1. K. F. J. Heinrich, Bibliography on electron-probe microanalysis and related subjects (second revision), E. I. du Pont de Nemours and Company, Wilmington, Delaware, October (1963).

Approximations for the Interpretations of X-Ray *K* Absorption Spectra

George R. Mitchell

Wright Junior College
Chicago, Illinois

The interpretations of X-ray absorption edges rely on approximations which are often not very completely justified. Particular attention is given an approximation used by Parratt for argon atoms in which the K-excited terms of argon are compared to the optical terms of potassium. It is pointed out that this approximation is very poor in the range of small atomic numbers, and this approximation also predicts the wrong number of X-ray term values. This approximation is investigated by applying the Hartree equations without exchange to the neon atom in the configuration $1s\,2s^2\,2p^6\,3s$ and by comparing the results with a similar calculation for sodium in the configuration $1s^2\,2s^2\,2p^6\,3s$.

INTRODUCTION

The purpose of any spectroscopy is to elucidate an energy-level structure. X-ray spectroscopy shares this purpose with any other spectroscopy. However, when one reviews the rather extensive literature on the subject, one is struck by the fact that all the interpretations are based on very approximate calculations or even intuitive guesses. This paper is concerned with some calculations which I hope give a slightly deeper insight into approximations of X-ray spectroscopy.

By way of a brief review and introduction, Fig. 1 shows the type of data one gets from absorption edge measurements. These are the *K*-absorption edges for nickel metal and potassium nickel cyanide. The interpretation of metal edge data has been made the subject of many papers. Notable among the early work are those by Jones and Mott [1], O'Bryan and Skinner [2], and Beeman and Friedman [3].

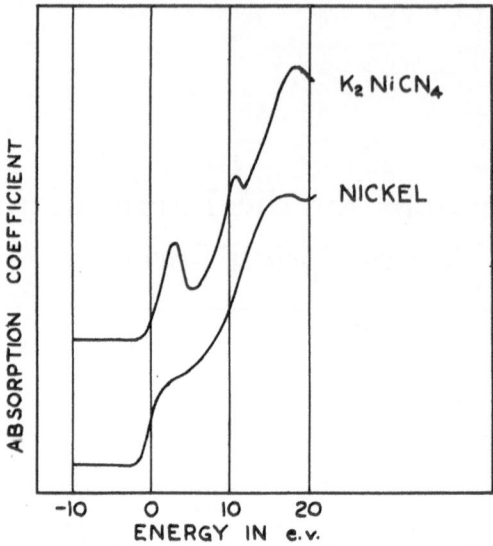

Fig. 1. K-absorption edge structure of nickel metal and nickel in potassium nickel cyanide.

To illustrate the viewpoint taken by these authors, Fig. 2 shows the interpretation of the X-ray spectra of sodium metal. According to this interpretation, one observes essentially a product of a density of states and a transition probability. The density of states is assumed to be that determined by a band approximation and does not take explicit account of a 1s vacancy.

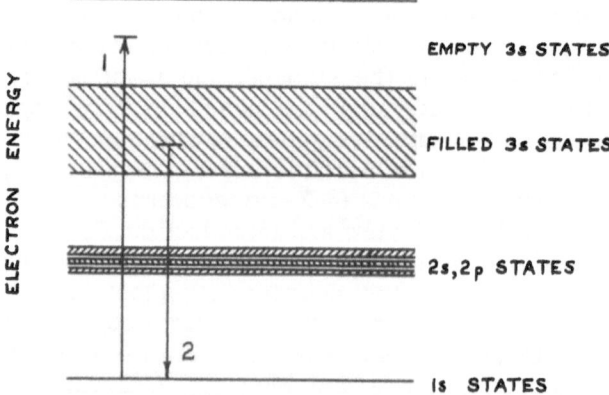

Fig. 2. Electron transitions for interpreting absorption (1) and emission (2) spectra of sodium metal.

While this explanation seems to work for sodium, recent work by Azaroff and Das [4] on the *K*-absorption spectra of copper—nickel alloys has cast doubt on the applicability of this picture to transition metals. Using a band picture, one would expect a common density of states for both the copper and nickel *K* electrons. Azaroff's data show that the edges of copper and nickel in the same alloy bear a much closer resemblance to the pure metals than to each other. An explanation based on the electronic configuration of individual atoms is proposed by these authors.

Parratt and Jossem [5] have pointed out a similar effect in the case of potassium chloride. The removal of a *K*-electron from, say, the potassium atom changes the potential so as to make the one electron energies of the solid differ greatly from those thought to apply to the solid at or near the ground state. The authors provide a qualitative discussion of this case, but, in the end, stress the need for more fundamental calculations.

THEORETICAL CALCULATIONS

I would like to elaborate on the problem of making a funda-mental calculation of the energy term values of an isolated atom having a 1s vacancy. These considerations permit a comment on an approximation which has been used in many discussions of X-ray absorption spectra. For simplicity, only

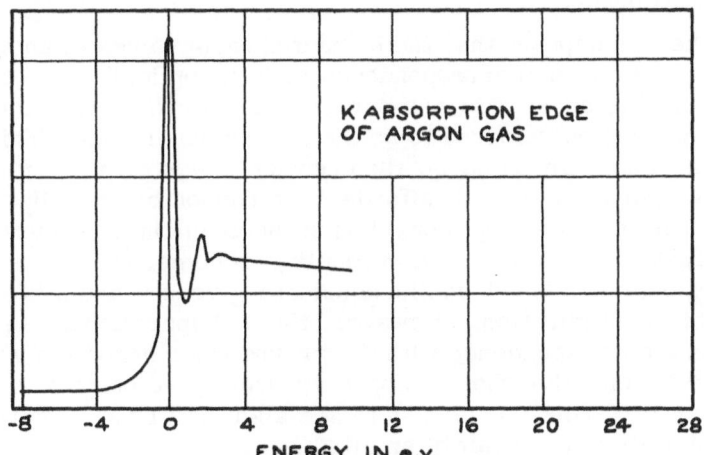

Fig. 3. *K*-absorption edge of argon gas.

inert gas atoms are considered. Again, Parratt [6] was one of
the first to work on this problem. His absorption data for argon
gas are shown in Fig. 3. His interpretation of these spectra is
in terms of single configurations. The configuration $1s\,2s^2\,3p^6$
$3s^2\,3p^6\,4p$ corresponds to the first strong maxima, and subsequent
absorptions correspond to $1s-5p$, $1s-6p$, etc. transitions. The
transitions obey the one-electron selection rule $\Delta l = \pm 1$. In
addition, Parratt needed some way of determining the energy
terms associated with these configurations. In other words, the
old question of removing a K-electron, in effect, creates an
entirely different set of energy term values from those appropriate to, say, some optical excitation of argon. The approximation Parratt used was as follows: Consider the
configuration $1s\,2s^2\,2p^6\,3s^2\,3p^6\,4p$ of the argon atom. The $4p$ electron will move in a potential of a nucleus having a charge of
$+18$ and of 17 shielding electrons.   Then, it is assumed that
this potential is not much different for a $4p$ electron in a potassium atom, which moves in a potential of a nucleus of charge
$+19$ and 18 shielding electrons.   If this is the case, then we
might expect the energy term values of K-excited argon to have a
one-to-one correspondence with the optical terms of potassium.
Historically, this remark appears to have been made first by
Skinner [7].   The idea has been extended to other atoms by a
number of authors, and, as far as it goes, there is experimental
evidence that it provides a reasonable description of the facts.
I refer particularly to the work of Beeman and Bearden [8] on
ions in solution.

The assumption that the K-excited term values of an atom
have a one-to-one correspondence with the optical terms of an
atom having an atomic number one greater than the atom in
question will be referred to as the $Z+1$ approximation. This is
only an approximation, so the question arises as to what kind
of an approximation it affords.   It cannot be strictly true.
This can be seen by considering neon in the configuration
$1s\,2s^2\,2p^6\,3s$.   Assuming an $L-S$ coupling scheme, the spins and
orbits outside closed shells couple to give a ^1S and a ^3S term
for this configuration.   However, the $Z+1$ approximation pre-
sumes a correspondence with the ground term of sodium which
is a ^2S term.   So the $Z+1$ approximation predicts the wrong
number of terms.   It is easy to see also that the $Z+1$ approximation fails in the limit of small Z.

The $Z+1$ approximation, in its simplest terms, does not

provide a means for calculating transition probabilities, nor can it provide a calculation of absolute term energies.

For all these reasons, the problems involved in applying the Hartree—Fock method, which is probably the only many-electron approximation of use, to the K-excited states of neon are examined. It turns out that even this many-electron approximation cannot be applied without the additional assumption that the K-excited term is a stationary state of the atom. The formal application of the Hartree—Fock method to neon in the configuration $1s\,2s^2\,2p^6\,3s$ is considered very briefly. The four main statements of the Hartree—Fock method as applied to atoms are:

$$H = -\sum_j \left(\frac{1}{2}\nabla_j^2 + \frac{N}{r_j} \right) + \sum_{i \neq j} \frac{1}{r_{ij}} \tag{1}$$

$$\Phi = A\,\psi_1^{(1)}\,\psi_2^{(2)}\ldots\psi_n^{(n)} \tag{2}$$

$$E' = \frac{\int \Phi^* H\,\Phi\,d\tau}{\int \Phi^*\Phi\,d\tau} = \sum_a I_a + \sum_{a \neq \beta} J_{\alpha\beta} - \sum_{a \neq \beta} K_{\alpha\beta} \tag{3}$$

$$\psi_a(r,\,\theta_j,\,\phi_j,\,s_j) = \frac{P(n_a l_a;\,r)}{r_j} \cdot Y_{l_a}^{m_a}(\theta_j,\,\phi_j) \cdot \chi(s_j) \tag{4}$$

These relations are summarized as follows: (1) The Hamiltonian neglecting spin—orbit interaction, (2) the wave function of the atom is the antisymmetrized product of one-electron orbitals, i.e., Slater determinant, (3) the energy expression, first derived by Slater, and (4) the assumption that the orbitals are of a central-field type. This is strictly true only for closed-shell atoms.

It follows that an explicit energy expression can be obtained in terms of the radial functions P. The energy expression for $\text{Ne}^*\,(^3S)\,1s\,2s^2\,2p^6\,3s$ is

$$E' = I(1s) + 2I(2s) + 6I(2p) + I(3s) + F_0(2s,\,2s) + 15F_0(2p,\,2p)$$

$$-\frac{6}{5}F_2(2p,\,2p) + 2F_0(1s,\,2s) + 6F_0(1s,\,2p) + F_0(1s,\,3s) - G_0(1s,\,2s)$$

$$- G_0(1s,\,3s) - G_1(1s,\,2p) + 12F_0(2s,\,2p) + 2F_0(2s,\,3s) - G_0(2s,\,3s)$$

$$- 2G_1(2s,\,2p) + 6F_0(2p,\,3s) - G_1(2p,\,3s) \tag{5}$$

where

$$I(nl) = -\frac{1}{2} \int P(nl; r) \left[\frac{d^2}{dr^2} + \frac{2N}{r} - \frac{l(l+1)}{r^2} \right] P(nl; r)\, dr$$

$$F_k(nl; n'l') = \int_0^\infty \int_0^\infty P^2(nl; r)\, P^2(n'l', s)\, U_k\, dr\, ds$$

$$G_k(nl, n'l') = \int_0^\infty \int_0^\infty P(nl; r)\, P(n'l'; r)\, U_k\, P(nl; s)\, P(n'l'; s)\, dr\, ds$$

$$U_k = \frac{r^k}{s^{k+1}} \qquad (\text{for } r < s)$$

$$U_k = \frac{s^k}{r^{k+1}} \qquad (\text{for } r > s)$$

The integrals $I(nl)$ are related to the same quantities in equation (3) and the F's and G's are related to the J and K integrals of equation (3), respectively. Slater has shown how this can be done by carrying out the angular and spin integrations and summations explicitly. This leaves the energy as a function only of integrals of radial functions.

The next step is to vary the radial functions independently to produce a stationary value for E'. This procedure leads to the Hartree-Fock equations:

$$\left\{ \frac{1}{2} \frac{d^2}{dr^2} + \frac{1}{r} \left[Y(r) + Y_0(1s, 1s; r) \right] - \epsilon_{1s,1s} \right\} P(1s; r)$$

$$= \chi_{1s}(r) + \epsilon_{1s2s}\, P(2s; r) + \epsilon_{1s3s}\, P(3s; r) \qquad (6)$$

$$\left\{ \frac{1}{2} \frac{d^2}{dr^2} + \frac{1}{r} \left[Y(r) + Y_0(2s, 2s; r) \right] - \epsilon_{2s2s} \right\} P(2s; r)$$

$$= \chi_{2s}(r) + \epsilon_{2s1s}\, P(1s; r) + \epsilon_{2s3s}\, P(3s; r) \qquad (7)$$

$$\left\{ \frac{1}{2} \frac{d^2}{dr^2} + \frac{1}{r} \left[Y(r) + Y_0(2p, 2p; r) \right] \right.$$

$$\left. + \frac{2}{5} Y_2(2p, 2p; r) - \epsilon_{2p2p} - \frac{1}{r^2} \right\} P(2p; r) = \chi_{2p}(r) \qquad (8)$$

$$\left\{ \frac{1}{2} \frac{d^2}{dr^2} + \frac{1}{r} \left[Y(r) + Y_0(3s, 3s; r) \right] - \epsilon_{3s3s} \right\} P(3s; r)$$

$$= \chi_{3s}(r) + \epsilon_{3s1s}\, P(1s; r) + \epsilon_{3s2s}\, P(2s; r) \qquad (9)$$

Again, we have omitted some interesting but standard discussion.

A problem which has already been mentioned arises at this point. Since there are no true stationary states other than ground states or possibly metastable states, there is no theoretical justification for the variation procedure. However, in practice, it seems possible to obtain useful wave functions for excited states in this way. Bagus [9,10] of the Argonne National Laboratory has obtained wave functions for various ionized states of neon and argon.

The final step is to find self-consistent solutions to these equations. For speed of hand calculations, these equations have been solved without the exchange terms, i.e., in the Hartree approximation. Some results of these computations are shown in Fig. 4. This figure compares the $3s$ function for neon in the $1s\,2s^2\,2p^6\,3s$ configuration with the $3s$ function for sodium calculated by Kennard and Ramberg [11]. This latter calculation was also without exchange.

The differences between the functions are slight. The indication is that the $3s$ function belonging to neon is slightly extended radially when compared to the sodium $3s$ wave function. The one-electron energies, which, by Koopman's theorem, are the ionization energies of the electrons when all other

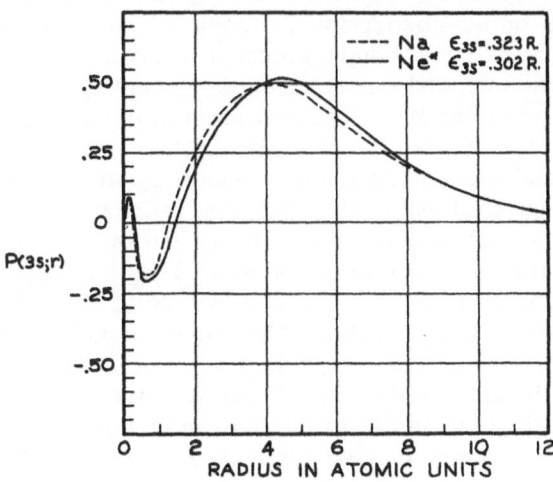

Fig. 4. $3s$ radial wave functions for sodium atom in ground state and K-excited neon in configuration $1s\,2s^2\,2p^6\,3s$.

wave functions remain unchanged, differ by only 0.021 Rydbergs or 0.29 eV. The experimental value of the ionization energy for a sodium 3s electron is 0.38 Rydbergs. The conclusion drawn is that the wave function of the 3s electron in the K-excited neon is essentially, but not exactly, that of the 3s electron in sodium in its ground state. This is substantially, but not quite, the original statement of Parratt.

One more calculation that is easily performed is the determination of the ^1S-^3S energy term difference. For this, one applies the Slater diagonal sum rule. The energy separation is about 0.1 eV with the triplet having the lowest energy. Again this confirms the speculation that the relative spin orientation of the remaining electron has a small effect.

CONCLUSION

Additional calculations, which were not possible using the $Z+1$ approximation, are now possible. The calculation of transition probabilities and of absolute energy term values may be performed. More importantly, it is hoped that further calculations on atoms will encourage the application of the Hartree-Fock method to molecules and complex ions in order to explain the observed details of these edges.

As a concluding remark, it is pointed out that even the description of an absorption process as the absorption of an X-ray quantum by a single electron transition is only an approximation. Second-order processes having an appreciable magnitude are possible. The two most important second-order processes appear to be the absorption of a photon by two electrons and the absorption of a photon by an electron plus the simultaneous excitation of a plasma oscillation quanta. The two-electron process has been observed by Schnopper [12] in his recent investigation of argon gas. The plasma process was commented upon by Sobel'man and Feinberg [13], Parratt [14], Best and Robbins [15], Leder, Mendlowitz, and Marton [16], and Mitchell [17]. However, the role played by these second-order processes is not really very clear at this time. One can be fairly certain that in the immediate neighborhood of the edge, the plasma process can be neglected, since plasma energies for solids are of the order of 15 — 20 eV.

REFERENCES

1. H. Jones and N. F. Mott, Proc. Roy. Soc. (London) A162: 49(1937).
2. H. M. O'Bryan and H. W. B. Skinner, Phys. Rev. 45: 370 (1934).
3. W. W. Beeman and H. Friedman, Phys. Rev. 56: 392 (1939).
4. L. V. Azaroff and B. N. Das, Phys. Rev. 134: A747 (1964).
5. L. G. Parratt and E. L. Jossem, Phys. Rev. 97: 916 (1955).
6. L. G. Parratt, Phys. Rev. 56: 295 (1939).
7. H. W. B. Skinner, Proc. Soc. (London) A135: 84 (1932).
8. W. W. Beeman and J. A. Bearden, Phys. Rev. 61: 455 (1942).
9. P. S. Bagus, Bull. Am. Phys. Soc. 8 (7): 535 (1963).
10. P. S. Bagus, SCF excited states and transition probabilities of some neon-like and argon-like ions, Argonne National Laboratory of Molecular Structure and Spectra, Department of Physics, University of Chicago, Chicago, Illinois.
11. E. H. Kennard and E. Ramberg, Phys. Rev. 46: 1034 (1934).
12. H. W. Schnopper, Phys. Rev. 131: 2558 (1963).
13. I. I. Sobel'man and E. L. Feinberg, J. Exptl. Theoret. Phys. (USSR) 34:·494 (1958).
14. L. G. Parratt, Rev. Mod. Phys. 31: 616 (1959).
15. P. E. Best and J. L. Robbins, Proc. Phys. Soc. (London) 77: 1046 (1961).
16. L. B. Leder, H. Mendlowitz, and L. Marton, Phys. Rev. 101: 1460 (1956).
17. G. R. Mitchell, J. Chem. Phys. 37: 216 (1962).

Infrared and Raman Spectroscopy

Construction and Performance of Highly-Efficient Micro Gas Cells for the Infrared Spectra

K. E. Stine, D. E. McCarthy, and H. J. Sloane

Beckman Instruments, Inc.
Fullerton, California

The design and use of two low-cost, highly-efficient micro gas cells are described. The first cell, having a path length of 77 mm and a volume of 2.5 cc, has a cell efficiency of 3.08 as defined by the ratio of path length to volume. The second cell, having a path length of 9 mm and a volume of 0.045 cc, has a cell efficiency of 20. This cell is designed for use with a beam condensing system.

Both cell bodies are composed of a highly reflective epoxy which produces sensitivities greater than would be predicted from physical dimensions. The increase in sensitivity indicates that cells are effectively behaving as light pipes. Sensitivities are great enough so that microgram quantities of gases produce fairly intense spectra which facilitates the use of these cells for collection and identification of gas chromatographic fractions.

INTRODUCTION

In this paper we will describe the design and use of two low-cost, highly-efficient micro gas cells for the infrared region.

Curves will be shown which illustrate the relative sensitivities of these micro gas cells as compared to the typical 10-cm gas cell. In addition, the experimental sensitivities will be compared to the sensitivities predicted from the physical dimensions of the cells.

The sensitivities were found to be great enough so that microgram quantities of gases and vapors produced fairly intense infrared spectra which facilitates the use of these cells for monitoring, collecting, and identifying gas chromatographic fractions.

Finally, a technique using these cells for monitoring,

121

collecting, and identifying components from a gas chromatograph will be described and illustrated. The IR-9 and GC-2A were used for recording spectra and making required separations.

DESIGN AND CONFIGURATION

Diverging light rays entering a fairly long gas cell will ultimately strike the internal walls, and a fraction of the incident radiation will be lost if the surface is nonreflective. In order to minimize this effect, the cells which we will describe are constructed from a polished epoxy resin that becomes highly reflective at grazing angles.

For an f/10 system, common to the Beckman line of IR instruments, a 6° offaxis angle is the largest deviation from the central ray in the sampling area. This corresponds to an 84° incident angle on the internal walls of the gas cells. Reflectivity tests on the polished resin have shown that 75% of the light at 84° is reflected. Thus, transmission for the various light rays should range from 74% for the most divergent rays to 100% for the parallel rays. An average of 90% would not be improbable.

A high polish is placed on the internal walls of the cells at the time they are cast by using molds possessing mirrored surfaces which form the internal portion of the cells.

Therefore, highly-polished, internal surfaces are incorporated in the minimum-volume gas cell shown in Fig. 1. The cell body is 77 mm long, with a beam area of (2.5 × 13) mm, producing a cell volume of 2.5 cm^3.

At the top of the cell are two ports which have been kept as small as possible to keep dead volume to a minimum. Into the two ports are incorporated $\frac{1}{8}$-in. Swagelok fittings for attaching either flexible or metal lines. When the lines are removed, small septa can be placed in the fittings and the caps tightened to form a seal, permitting syringe injections to be made. These two types of connections allow the cell to be used statically or in a flow-through manner.

One end of this cell is designed to fit directly into the spectrophotometer cell holder, as shown in Fig. 2. Adjustment of the cell in the light path for maximum energy is made by means of two Allen screws. No beam condensing system is required.

Fig. 1. The minimum-volume gas cell. Cell volume is 2.5 cm^3.

The second cell, which we call the micro-volume gas cell, is shown in Fig. 3. The dimensions of this cell are considerably smaller than those of the first cell described. The beam dimensions are approximately 1 mm × 5 mm, and the length is 9 mm, resulting in a volume of 45 μl (0.045 cm^3). The micro-volume gas cell is used in conjunction with a standard Beckman three-lens beam condenser.

A single port is provided for filling this cell, and fillings must be done through a needle, or in the case of a flow-through system, two needles may be used. A very small septum is placed into the port, and a drilled, threaded screw above the septum provides the sealing pressure. The dimensions of the screw, septum, and port are such that when the screw head is butted against the cell body, the septum is held under the proper amount of compression. Too much compression is actually capable of forcing the septum into the beam area of the cell. In addition, syringe needles tend to plug when pushed through highly-compressed septum material. Of course, if too little compression is used, leakage will occur.

The cell is located in a holder by several locating pins and held by means of a permanent magnet cast into the cell body. The cell and holder fit into the sampling area of the beam condenser. Figure 4 shows this cell mounted in a cell holder

Fig. 2. The minimum-volume gas cell and the spectrophotometer cell holder.

Fig. 3. The micro-volume gas cell. Cell volume is 0.045 cm^3.

Fig. 4. The micro-volume gas cell shown mounted in a cell holder and beam condenser in the sampling compartment of an IR-9.

and beam condenser in the sampling compartment of an IR-9. Vertical adjustment of the cell in the light path is made using adjusting screws also serving as the legs of the cell holder. Lateral adjustment is made by a larger, knurled screw which, in turn, moves a track on which the cell holder and cell slide. The cell holder employs the three-point kinematic mount system allowing ready reproducibility of positioning.

In order to produce a cell capable of withstanding sub-ambient temperatures, the windows of both cells are cemented in place with an epoxy whose coefficient of expansion closely matches that of some window and cell-body materials. We have not as yet subjected these cells to extreme temperatures, but plan to do so shortly to establish their temperature limits. However, we would estimate that temperatures at least as low as 0°C could be tolerated, to permit cooled trapping directly in the cell. Leak tests with punctured septa in place have

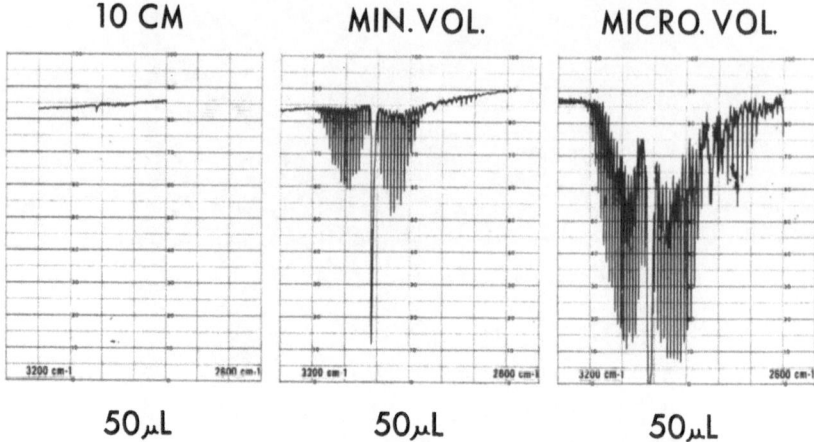

Fig. 5. Relative sensitivities to methane of the 10-cm, minimum-volume, and micro-volume gas cells in the 2600–3200–cm^{-1} range. 50 μl (0.050 cm^3) methane was injected into each cell.

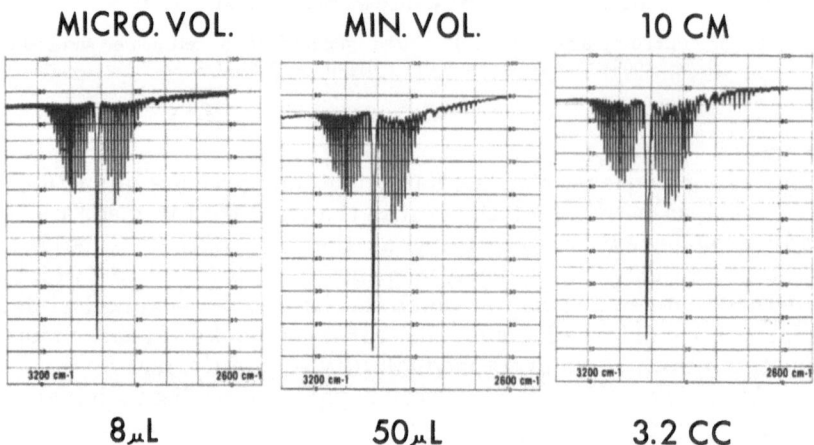

Fig. 6. Relative sensitivities to methane of the 10-cm, minimum-volume, and micro-volume gas cells in the 2600–3200–cm^{-1} range. 8 μl (0.008 cm^3) methane was injected into the micro-volume cell; 50 μl (0.050 cm^3) into the minimum-volume cell; and 3.2 cm^3 into the 10-cm cell.

indicated that the cells will hold a pressure slightly above one atmosphere before leakage occurs through the septum or around an inserted needle. As with most infrared cells, preliminary tests have shown that adsorption of polar materials by the cell walls seems to occur.

TRANSMISSION

The minimum-volume gas cell transmits 25–40% depending on the instrument used and the window material. The micro gas cell with KBr or AgCl windows transmits 45–60% relative to the beam condenser. As for any microwork in which energy losses are involved, the benefit produced by the use of a versatile, high-energy instrument is obvious.

SENSITIVITY AND CELL EFFICIENCIES

In order to determine the relative sensitivities of the two cells, methane (CH_4) was injected into each cell and into a typical 10-cm gas cell having a volume of 200 cm^3. A 50-μl (0.050-cm^3) injection into each cell produced the three spectra shown in Fig. 5. As can be seen, the C—H stretching bands, which are barely observable in the 10-cm gas cell, become more intense in the minimum-volume cell, and are extremely intense in the micro gas cell.

Another method of illustrating the relative sensitivities is shown in Fig. 6 where injections of various amounts of CH_4 were placed into each cell to produce spectra of the same intensity. Although equal intensities are produced, only $\frac{1}{400}$ as much sample was required in the micro gas cell as was used in the 10-cm gas cell.

Included in Table I are the cell efficiencies (C.E.) as defined by path length divided by volume, the ratios of the cell efficiencies compared to the 10-cm cell, and the cell efficiency ratios found by comparing the methane intensities shown in previous slides.

From the data tabulated in this slide, we can readily see that the sensitivity of the minimum-volume gas cell is about 8% greater than that predicated from its physical dimensions, while the micro gas cell is about 3% below its predicted sensitivity.

TABLE I

Comparison of Sensitivity and Cell Efficiency of Three
Types of Gas Cells for Infrared Spectra

	10-cm cell	Minimum-volume cell	Micro-volume cell
Theoretical cell efficiency (path length/volume)	0.05	3.08	20.0
Cell efficiency ratio as compared to 10-cm cell	1.0	61.6	400
Cell efficiency ratio found	1.0	67.0	386
CH_4 detection limits (μg)	130	2	0.4

Whether or not the increased sensitivity of the minimum-volume cell is due to internal reflections or some other effect, such as pressure broadening, would be impossible to state positively at this time. However, as previously mentioned, this cell was designed to enhance the reflection phenomena.

The detection limits for methane, which we shall define here as that weight of gas required to produce an absorbance of 0.05, is calculated to be $130\,\mu$g, $2\,\mu$g, and $0.4\,\mu$g for the 10-cm, minimum-volume, and micro gas cells, respectively. These figures would appear to be well within the weight ranges normally encountered in gas chromatographic effluents.

APPLICATIONS TO GAS CHROMATOGRAPHY

In order to test the feasibility of using these cells to monitor, collect, and identify gas chromatographic fractions, a synthetic gas mixture was prepared containing 10% V/V of methane, ethane, ethylene, and acetylene in helium.

An infrared spectrum of this mixture was recorded, as shown in Fig. 7, using a 10-cm gas cell at 100 mm pressure. The chromatogram of the mixture is also shown in Fig. 7.

Using the minimum-volume gas cell as a flow-through cell, it then should be feasible to monitor the gas chromatographic effluent fractions as they emerge by setting the infrared

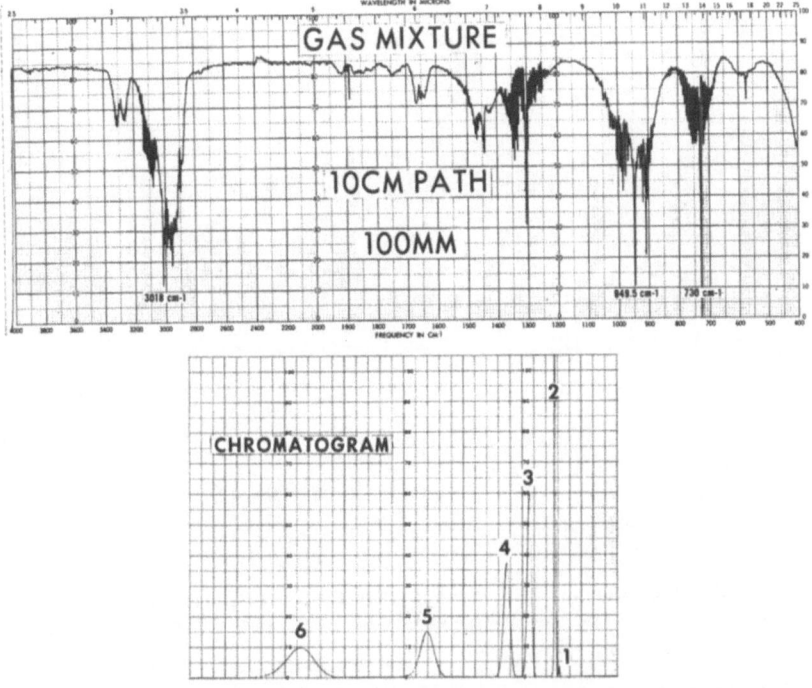

Fig. 7. The infrared spectrum and chromatogram of a mixture of methane, ethane, ethylene, propylene, and acetylene in helium. (10-cm gas cell at 100-mm pressure.)

instrument to a suitable wavelength. After selecting the wavelength to be monitored, the wavelength drive was disconnected so that the wavelength would remain fixed, while the chart was driven at some constant rate.

The minimum-volume gas cell was placed in the instrument using septa in the two filling ports. With $\frac{1}{8}$-in. tubing, a needle was attached to the tail of a silica gel column mounted in the GC-2A. The needle was then inserted into one port of the gas cell. A second needle was inserted into the remaining port to serve as an exit fot the gas chromatographic effluent.

With the monochromator set at 3018 cm^{-1}, which is a region of absorption of at least one component, a 5-cm^3 sample of the gas mixture was injected into the gas chromatograph while simultaneously starting the spectrophotometer chart drive.

The upper left-hand portion of Fig. 8 shows the absorption at 3018 cm^{-1} in the C—H stretching region, recorded by the

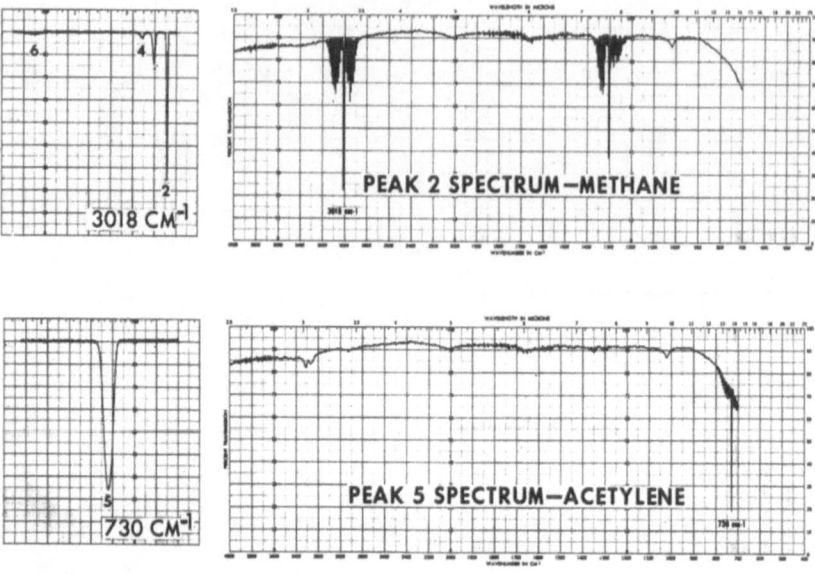

Fig. 8. (a) Absorption at 3018 cm⁻¹ in the C—H stretching region. (b) Peak 2 spectrum—methane. (c) Absorption at 730 cm⁻¹ (d) Peak 5 spectrum—acetylene.

IR-9 as a function of time. Peaks 2, 3, 4, and 6 all absorb at this wavelength indicating that all contain CH groups.

Next, a second sample was injected into the chromatograph and, when the point of maximum absorption of Peak 2 was reached, both needles were simultaneously pulled from the gas cell, trapping the most concentrated portion of the chromatographic peak. The trapped material was then scanned, and the material was identified as methane (the upper right-hand curve).

In a similar manner, the fifth chromatograph peak was trapped by monitoring the effluent at 730 cm⁻¹ (Fig. 8), and was identified as acetylene.

Figure 9 shows the resulting spectrum from trapping Peak 4, monitored at 949.5 cm⁻¹. In this manner all the chromatographic peaks can be monitored, trapped, and identified, providing, of course, that the spectra produced are sufficiently intense.

Although this gas mixture was a rather idealized case, the results should serve to illustrate the usefulness of these cells and the technique described for identifying chromatographic peaks. The maximum infrared absorbance obtainable will be

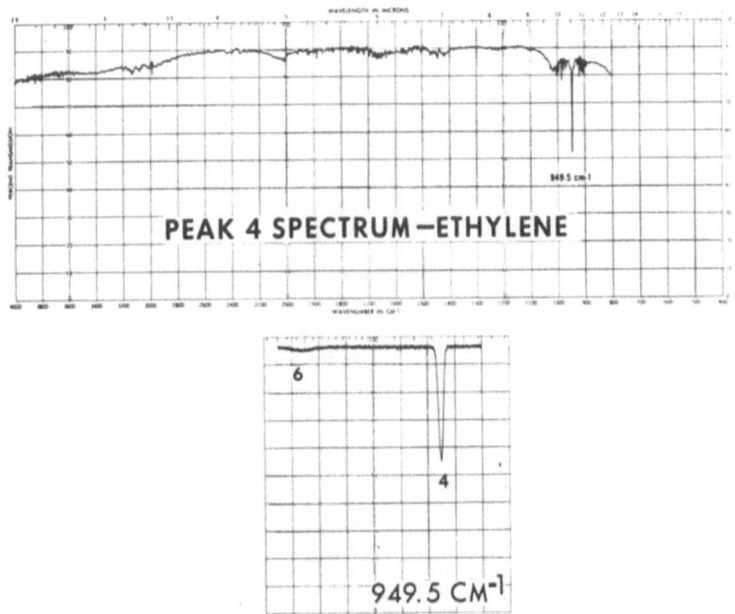

Fig. 9. (a) Peak 4 spectrum—ethylene. (b) Absorption at 949.5 cm⁻¹.

dependent on a number of factors in addition to the cell efficiency, such as sample size, concentration, temperature, column efficiency, and carrier gas flow rate.

CONCLUSIONS

The sensitivities of the described cells have been shown to correlate well with values predicted from their physical dimensions and are great enough to allow use of the cells for trapping and identifying microgram amounts of gases and vapors. The feasibility of using these cells with a gas chromatograph has been established.

Operation of the cells in a flow-through manner also suggests other possibilities, such as using an infrared spectrophotometer as a more specific GC detector by monitoring gas chromatographic effluents at wavelengths characteristic of a particular functional group.

This study, although somewhat limited, should once again point out the value of combined IR-GC techniques.

Electronic Phase Null Photometric System of the Series 2000 Infrared Spectrophotometer

Charles W. Warren and Albert W. Chapple

Instruments and Communications, Inc.
Wilton, Connecticut

A new electronic phase null system has been developed which provides precise measurement of sample-to-reference ratio continuously from fractional to 100% transmittance. The Series 2000 system is based on measurement of the phase angle of algebraically summed sample and reference signals having a 90° phase difference. The system yields high photometric accuracy and reproducibility, even with samples having low infrared transmission. No optical attenuator is used.

The new Series 2000 infrared spectrophotometer utilizes a photometric system employing a unique electronic phase null method (patent applied for) of measuring transmittance or absorbance values. First, the instrument itself will be described; second, the electronics system philosophy will be outlined; third, the phase null measuring technique will be explained, and last, some significant advantages of the Series 2000 electronic phase null system will be indicated.

The spectrophotometer is a double-beam, filter-grating monochromator instrument with linear wavenumber presentation over the fundamental region from 4000 to 625 cm^{-1}. Constant resolution scanning with slit override is provided. Additional performance flexibility can be achieved through the use of optional instrument accessories, such as the source intensity monitor, and the ordinate scale expansion—zero suppression accessory. Except for the preamplifier, solid state electronics is used throughout.

Figure 1 shows a front view of the instrument which is mounted on a floor stand on casters. At the top right is the

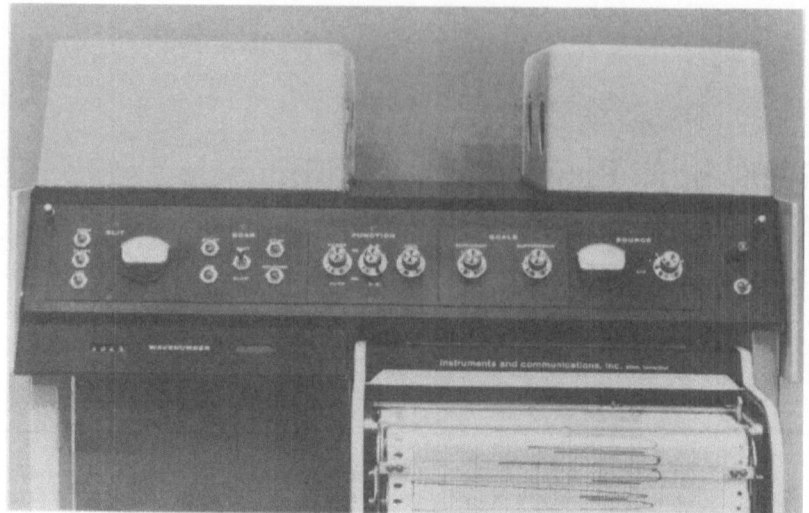

Fig. 1. Series 2000 infrared spectrophotometer.

source section, with the large sample area in the center.
Sample and reference beams are recombined in the section at
the top left. Underneath are the monochromator, detector, and
scanning mechanism. Below to the right is the variable-speed
strip chart recorder. The control panel swings down to provide
access to the solid state circuitry which consists largely of
plug-in modules. The primary control functions include fast
and slow scan (7 and 28 min), slit override, double beam—single

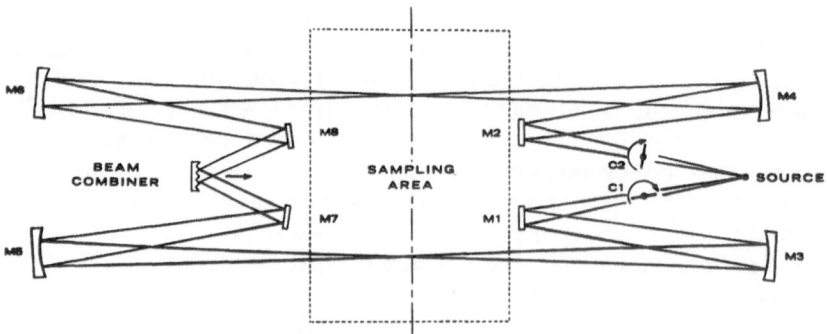

Fig. 2. Optical diagram (top view) of Series 2000 infrared spectrophotometer.

beam switch, filter control, zero control, 100% transmittance adjustment, wavenumber counters, and, on the recorder, chart-speed selector.

Figure 2 illustrates the top portion of the Series 2000 optical system. The path taken by either the sample or the reference beam is equivalent. In both beams, radiation from a high-temperature helical globar source is chopped by two thin bladed choppers mounted 90° apart on the same shaft. Chopping the radiation before passing through sample allows transmittance measurements to be made at elevated or reduced sample temperatures without error or necessity for radiation correction.

The optical design provides ready access to a central image point for both reference and sample beams, allowing optimum positioning of all types of sample cells and related accessories or special equipment. Multiple internal reflection accessories are easily accommodated without undue energy loss by lateral movement of the entire source-chopper section.

Reference and sample beams are recombined in this section through a beam combiner. The radiation is introduced to the entrance slit of the monochromator by means of a 45° mirror shown in Fig. 3.

Figure 3 illustrates the monochromator and detector layout. Radiation passes through one of four selected interference filters and then through the entrance slit. The monochromator is a Czerny—Turner type utilizing two precision replica gratings back to back. After leaving the exit slit, mono-chromatic radiation is directed into an ellipsoid which then presents energy to the evacuated thermocouple detector.

The Series 2000 infrared spectrophotometer differs funda-mentally from other infrared instruments in that it uses a new

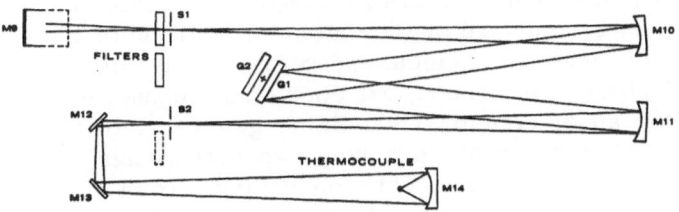

Fig. 3. Optical diagram — monochromator.

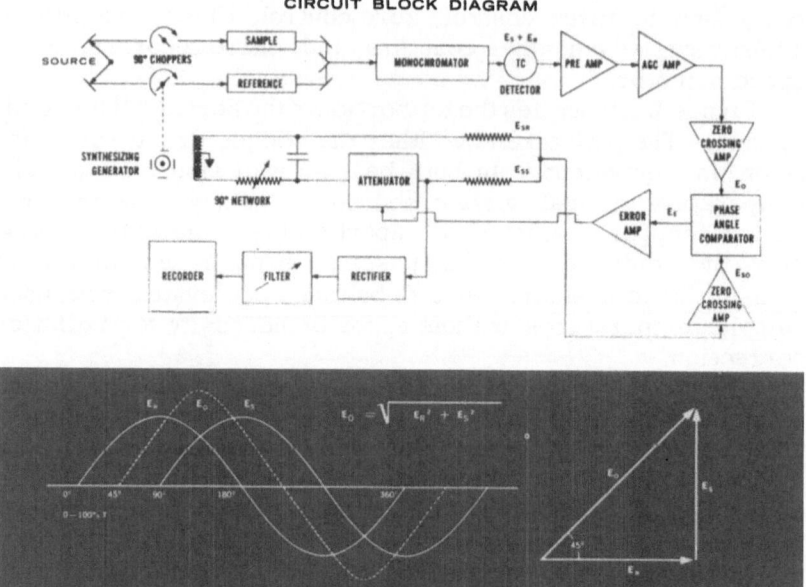

Fig. 4. Series 2000 circuit block diagram, with diagrams of related mathematical
relationships.

electronic phase null system for measuring sample-to-reference ratio, rather than either an optical null system or a phase
measurement system.   In effect, the phase null system
measures time information rather than amplitude information.
The phase null system has the capability of responding with
high photometric accuracy and reproducibility even with murky
samples.   No optical attenuator or mechanical commutator is
used.   Most important, the system is dependent solely on
measurement of phase angle and not signal amplitude.

Figure 4 shows the electronics block diagram.   Two basic
channels of information are provided in the Series 2000 system.
One channel is based on the algebraic addition by the thermocouple of sinusoidal sample and reference signals E_s and E_r 90°
out of phase.   A second and equivalent channel is created by
coupling an electrical synthesizing generator to the chopper
shaft.   This generator provides two outputs again phased 90°
apart, which are summed by an adding circuit.   The addition
of the two signals in either the optical or synthetic channel can
be represented vectorially as the addition of sample and

reference energy, as shown in the triangle below. The sinusoidal sample and reference signals produce a resultant sinusoid E_o whose phase angle θ equals the arc tan of E_s divided by E_r, and whose amplitude is equal to $(E_s^2 + E_r^2)^{1/2}$. The combined signal emanating from the radiation detector is amplified by a Nuvistor tube preamplifier and again by an AGC type amplifier. Then, it is fed to a high-gain, zero-crossing amplifier, whose principal function is to amplify the signal tremendously in order to define precisely the point in time at which it has crossed the zero axis. This point in time will vary from 0 to 45° as the transmittance varies from 0 to 100%. The synthetic channel contains both a synthetic reference signal E_{sr} and a synthesized sample signal E_{ss}. The synthesized sample signal is added through an attenuator to the synthesized reference signal producing a resultant sinusoidal signal E_{so}. As before, vector addition of these two signals produces a phase angle. A second zero-crossing amplifier is employed to determine this phase angle precisely.

We now have two bits of phase angle, or time information, which we wish to compare. In the Series 2000 system, a null will be produced when the phase angle of the signal E_o from the optical channel equals the phase angle of signal E_{so} from the synthetic channel. When a sample absorbs energy, the phase angle of E_o will change. If the phase angles of E_o and E_{so} are unequal, we may attenuate the synthesized sample channel in order to bring the system back to null. The attenuator is adjusted in proportion to a DC error signal output from a phase null detector system. This signal is directly proportional to the difference in magnitude of the two phase angles. Therefore, we have a closed loop system, which automatically determines the true sample-to-reference ratio, which is recorded.

The phase null detector utilizes a method of repetitively sampling a ramp voltage, whose level changes as the point in starting time of the ramp changes. Information from the optical channel is used to trigger a ramp voltage created by an RC network with a transistor switch arranged to short a capacitor to ground. This causes the ramp voltage to rise from zero exponentially and reproducibly. The ramp will always be started at a point in time which is a true measure of the phase angle from the optical channel. The starting time of the ramp will move through 0 to 45° as the sample transmittance varies from 0 to 100%. We now allow the synthetic channel signal to trigger the sampling of the ramp voltage at very high speed (about 50—100 μsec). A transistor switch is arranged to connect a storage capacitor to the ramp capacitor for this very short time. If the ramp is at, say, 10 V at the time of sampling, 10 V will be transferred to the storage capacitor.

Now, if the voltage on our storage capacitor happens to be higher than that on the ramp capacitor, the current will flow in the opposite direction. We now have a method of detecting the ramp voltage which contains extremely little of the basic signal frequency ripple. Any change in the starting time of the ramp will represent a change in this sampled voltage. This sampling point then becomes our null. Any deviation either forward or backward in time of the phase angle from the optical channel must be followed by the attenuator in the synthetic channel which proportions the synthetic sample signal to the reference in order to maintain null. The time at which the ramp is sampled is determined by the relative phase angle of the optical reference signal and the synthesized reference signal.

This is the basic method of phase null measurement in the Series 2000. It is a demodulating system operating at a low frequency which provides a phase null solely dependent upon the positions in time of the sample and reference signals. The method of sampling the ramp capacitor voltage, or null point, essentially eliminates ripple. The total change in the ramp voltage leaves an ample signal available for measurements near 0% transmission. A further advantage in the use of the ramp sampling system is that a very narrow equivalent output pass band may be built in by setting upper and lower voltage limits which span the chosen voltage sampling point. If through some disturbance the ramp sampling is too early or too late, we effectively limit the noise in the measurement.

In summary, the electronic phase null measurement system developed for the Series 2000 infrared spectrophotometer has a number of significant advantages.

1. The problems of optical null systems, including non-linearity and dead zone with high sample absorbance, are eliminated.
2. Since the system is dependent solely on measurement of phase angle information and not signal amplitude, there is no practical problem with nonlinearity or saturation in the amplifier system.

3. The problems encountered in earlier phase measurement systems are avoided by the use of a true null system of phase measurement.
4. The demodulating technique not only provides a precise means of null detection, but also provides this information with very little ripple and an effective band pass.
5. Mechanical simplicity is derived by avoiding the use of an optical attenuator servo, moving optical parts, or electromechanical contacts.
6. The use of solid state electronics permits most functional circuits to be plug-ins, allowing simple servicing by substitution.

The Identification of Fibers and Fabrics by Internal Reflection Spectroscopy

Paul A. Wilks, Jr., and Mayhew R. Iszard

Wilks Scientific Corp.
South Norwalk, Connecticut

Synthetic and natural fibers have been analyzed by infrared spectroscopy in the past only after special sample preparation, such as mulling and pelletizing in KBr or by casting a film. Internal reflection spectroscopy offers a new method of examining and identifying fibers and fabrics directly with little, if any, special treatment. Textile fiber and fabric, internal-reflection, sample holders are described, and infrared spectra of a number of representative samples are shown.

INTRODUCTION

With the increasing variety of textile fibers (both synthetic and natural) in use today, there exists a growing need for a rapid means of identifying them in the form of individual filaments, as well as in finished yarns and fabrics. Infrared spectroscopy offers the most positive method of identifying such materials chemically, but because of the physical form of fibers and fabrics, it is difficult to obtain representative spectra by conventional transmission procedures. Among the methods that are currently being used are the casting of a film from a melt or solution, and the grinding of sample fibers into a powder and pressing with KBr into a transparent pellet. The first method can be used only with synthetic fibers. Both approaches require extensive sample preparation before a useful spectrum can be produced.

The newly developed technique of multiple internal reflection spectroscopy [1,2] greatly simplifies the sampling problem and permits the direct recording of infrared spectra of fiber and fabric samples with no sample preparation other than cutting a piece of fabric to size or winding a length of fiber around the sample holder. Both natural and synthetic fibers can be handled in this fashion. Qualitative and quantitative data on most mixed yarns and fabrics can also be obtained by the internal reflection technique. It is the purpose of this report to add a supplemental infrared method based on internal reflection to those presented in the recent publication of the American Association of Textile Chemists and Colorists on fiber identification [3].

THEORY

Internal reflection spectroscopy is made possible by the fact that, when a beam of light is internally reflected from the surface of a transmitting medium, a portion of the energy in the beam apparently passes outside the surface (Fig. 1a), and then is returned into it during the process of reflection. When a material of lower index of refraction than the transmitting medium is brought in contact with the surface, energy will be absorbed at those wavelengths where the material absorbs.

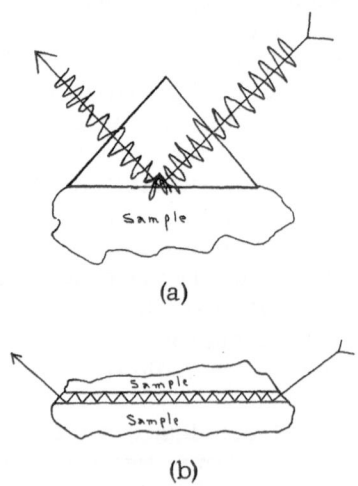

(a)

(b)

Fig. 1. (a) Diagram of attenuated total reflectance, and (b) diagram of frustrated multiple internal reflectance.

Thus, an infrared spectrum of a material may be produced by internal reflection that is nearly identical to a conventional transmission spectrum.

The internal reflection effect is relatively weak. The amount of penetration into a sample is influenced by several factors — the angle of incidence, the ratio of the reflective index of the transmitting medium to that of the sample, and the wavelength of the radiation, to name a few — but it is generally of the same order of magnitude as the wavelength being reflected. Furthermore, the degree of contact between the sample and the reflecting face of the prism greatly influences the energy exchange between the radiation beam and the sample.

The initial attempts at applying single internal reflection equipment to the study of fibers were not too successful because the area of contact between the reflecting surface and a bundle of fibers is necessarily limited. However, when the prism is truncated, the beam makes a number of reflections between the two parallel surfaces (Fig. 1b) and the chances of the individual rays striking fibers are greatly enhanced. As the following spectra show, even the hardest synthetic fibers provide enough contact to produce strong absorption bands.

PROCEDURE

The procedure for obtaining the spectra is as follows.

1. Preparing the sample:

    a. For fabrics—cut two pieces approximately 45 by 20 mm in size (Fig. 2).

    b. For long single filaments or yarns—wind a band about 20 mm wide around both halves of the sample holder (Fig. 3).

    c. For uncombed fibers—form a smooth pad and fasten to each plate with pressure-sensitive tape (Fig. 4).

2. Place the internal reflector plate in position over the sample on the main portion of the sample holder, then place the cover plate in position, again with the sample between it and the plate, and tighten the retaining screws. The degree of pressure has a large effect on the amount of absorption in the spectrum. This fact can be used to good advantage in achieving optimum band intensities.

Fig. 2. A fabric sample for internal reflection spectroscopy analysis.

Fig. 3. A yarn sample for internal reflection spectroscopy analysis.

Fig. 4. An uncombed fiber sample for internal reflection spectroscopy analysis.

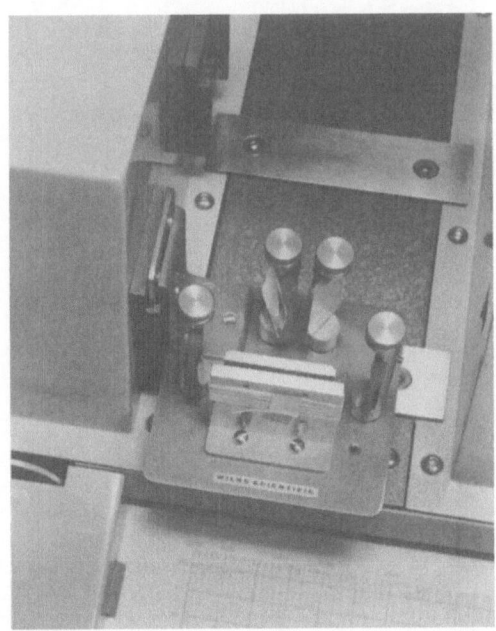

Fig. 5. Sample holder in sampling position in the internal reflection spectrometer.

3. Place the sample holder in the sampling position in the internal reflection spectrophotometer (Fig. 5) or in the internal reflection attachment in a conventional spectrometer (Fig. 6).

The spectra were run in a Wilks Scientific Model 8A internal reflection spectrophotometer (based on a Perkin-Elmer Model 137 prism spectrophotometer). In most cases, a 2-mm-thick KRS-5 plate was used as the internal reflector plate. The entrance face angle was 65°, giving an internal reflection angle of 25°.

RESULTS

Figure 7 shows the spectra of several vegetable fibers. Note that they are all basically cellulosic. The cotton and flax spectra are so similar that it does not appear practical to differentiate between them by means of infrared. There is a

Fig. 6. Sample holder in sampling position in the internal reflection attachment of a conventional spectrometer.

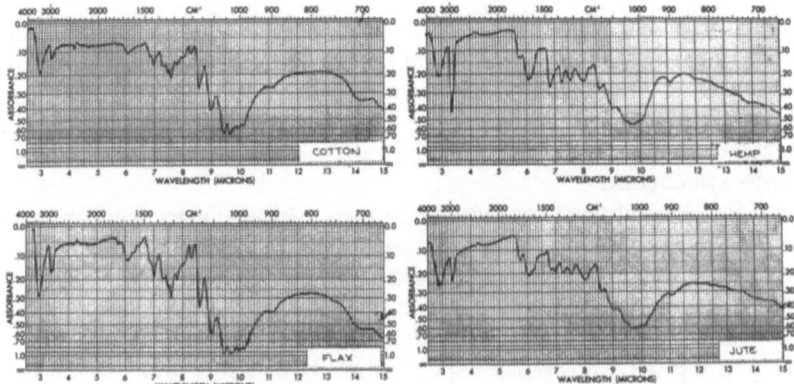

Fig. 7. Internal reflectance spectra of cotton, flax, hemp, and jute.

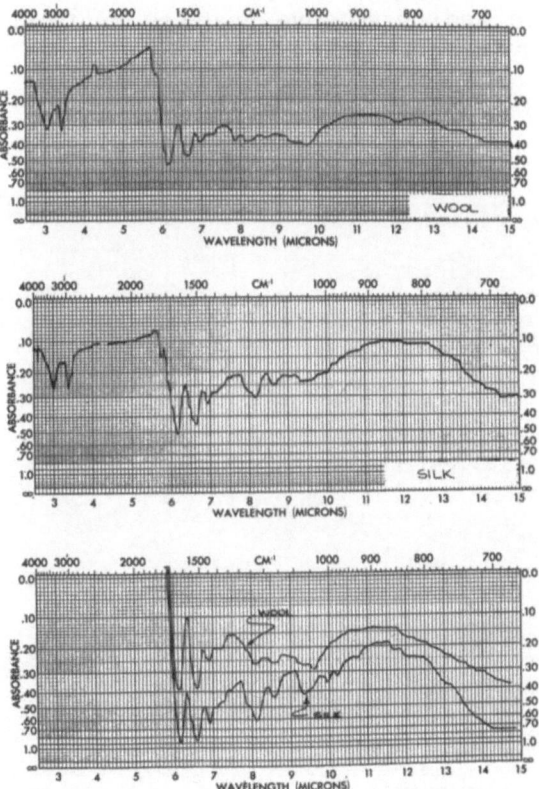

Fig. 8. Internal reflectance spectra of wool and silk.

significant difference between the cotton and hemp curves.

Figure 8 shows the spectra of the two principal animal fibers, silk and wool. They appear quite similar because their structure is basically protein. However, there are differences in the 7—11 μ region, which are more apparent in the expanded scale spectrum.

Man has shown considerably more imagination than nature in developing fiber; hence, the spectra of the various synthetics are much more varied, making identification by infrared analysis much simpler. The curves of the synthetics cover a number of typical fibers in use today (Fig. 9).

Figure 10 shows the spectra of two other synthetics and a blend (mixture of one or more different fibers). There is, of course, an endless variety of such blends, and with experience the spectroscopist will be able to make both qualitative and quantitative analyses on them by means of internal reflection spectroscopy.

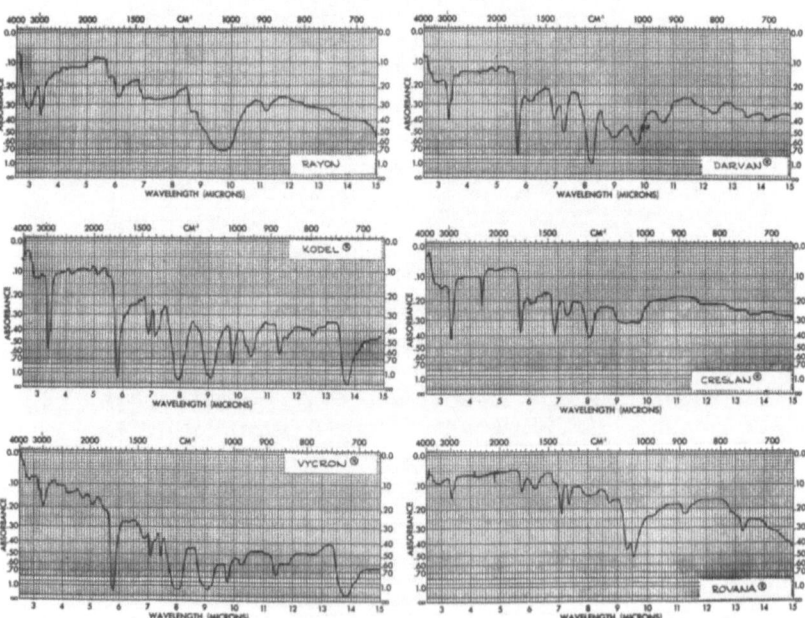

Fig. 9. Internal reflectance spectra of some typical synthetics.

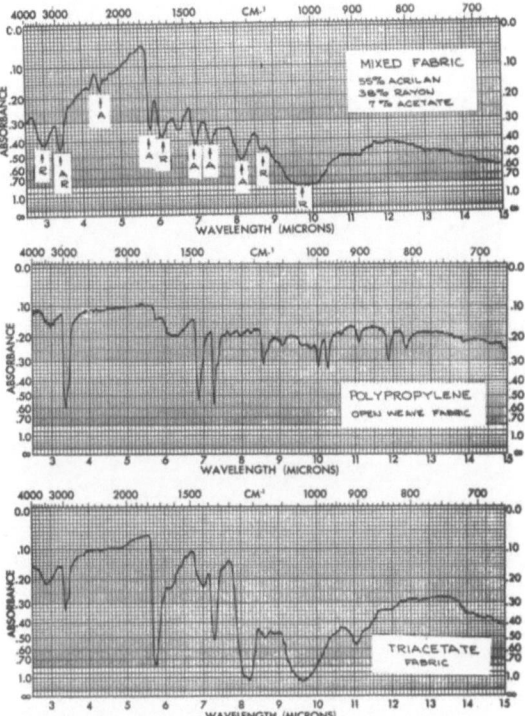

Fig. 10. Internal reflectance spectra of two common synthetics and the spectrum of a blend of acrilan, rayon, and acetate.

These spectra indicate that the internal reflection technique is capable of producing infrared spectra of fibers and fabrics that are nearly identical to transmission spectra. Unlike transmission procedures, no special sample preparation is necessary in the internal reflection method other than cutting samples to size.

REFERENCES

1. Fahrenfort, Spectrochim. Acta 17: 698–709 (1961).
2. Harrick, Ann. N.Y. Acad. Sci. 101(3): 928–959 (1963).
3. Fibers in textiles: identification. Proc. Am. Assoc. Textile Chemists Colorists, in: Am. Dyestuff Reptr., pp. 28—47, Oct. 28 (1963).

Multicomponent Infrared Analysis
by the Absorbance Ratio Method

Arthur S. Wexler

Dewey and Almy Chemical Division
W. R. Grace and Company
Cambridge, Massachusetts

Improved precision and higher accuracy (by elimination of cell path as a parameter) are achievable in quantitative infrared analysis of multicomponent mixtures by the absorbance ratio method. This involves solution of simultaneous equations of the type

$$A_i/A_h = \frac{1}{a_{hk}c_k} \sum_j a_{ij} c_j$$

for concentration ratios c_1/c_k, c_2/c_k, ... The coefficients in these equations are absorptivity ratios which are dimensionless constants. The analytical variable to measure is the absorbance ratio A_i/A_h of each component to a reference component k. The absorbance ratio is less susceptible to random sample and instrument variations than is the absorbance. A major advantage of the absorbance ratio method is the ability to do quantitative analysis in the very thin cells or films of material, since path length does not have to be specified.

INTRODUCTION

In multicomponent infrared analysis, a commonly-used technique is the solution of a set of simultaneous equations containing absorptivities as coefficients and concentrations as unknowns to solve for in terms of measured absorbances. The accuracy of the technique depends on the accuracy of the coefficients and the freedom from overlap of analytical wavelengths. Precision depends on the precision to which absorbances at each analytical wavelength can be measured. Precision is a function of several instrumental variables including spectral slit width, gain, and scan speed. Precision also is limited

151

by sample-handling factors including concentration, cell path, and temperature.

It has been found that a substantial increase in precision is obtainable in many cases by use of the absorbance ratio method for setting up the required simultaneous equations. Instead of solving for concentrations in sets of equations of the type

$$A_i = \sum_j A_{ij} = \sum_j a_{ij}bc_j \tag{1}$$

one solves for concentration ratios in sets of equations of the type

$$A_i/A_h = 1/A_h \sum_j A_{ij} = 1/a_{hk}bc_k \sum_j a_{ij}bc_j \tag{2}$$

where A_i is the absorbance at ith frequency, A_h is the absorbance at frequency h (due to component k), and a_{hk} is the absorptivity of reference component k at frequency h.

For this approach to be successful, an analytical wavelength for a selected reference or standard component free from overlap should be available such that

$$A_h = a_{hk}bc_k \tag{3}$$

In solving simultaneous equations of the type in (2), one obtains as solutions concentration ratios $c_1/c_k, c_2/c_k, c_3/c_k, \ldots$ c_n/c_k, where c_k is the concentration of the selected reference component or standard, and may be either an unknown or a known depending on whether it is a component or an added internal standard.

The optical path term b is cancelled out in the ratio set of equations (2). Each absorptivity term a_{ij} is replaced by an absorptivity ratio a_{ij}/a_{hk} where a_{ij} is the absorptivity due to the jth component, at frequency i, and a_{hk} is the absorptivity due to reference component k at frequency h. This is a frequency at which reference component k alone effectively absorbs (over the background) so that

$$a_{hk} \gg a_{h1}, a_{h2}, a_{h3}, \ldots \tag{4}$$

Gain in precision is achieved because, in many cases, changes in instrumental and sampling variables affect each of the analytical absorbances to the same degree with the result that, while the absorbance at each analytical wavelength undergoes a significant increase or decrease, the absorbance ratio of each component to the reference component undergoes a much smaller change.

Gain in accuracy as well as precision is obtainable by determining the absorptivities as ratios, thereby dispensing with the thickness or cell path term. The absorptivity ratios are determined by least squares or graphical treatment of data involving measurement of ratios of absorbances of each known component to the reference component. Binary known mixtures of each component with the reference component at several concentrations are prepared, and the required absorbances are measured. The absorbance ratio is plotted against concentration ratio to determine whether the ratio is a linear function of concentration. The slope of the linear curve is the absorptivity ratio required.

Accurate results are achieved provided that the absorbance dependency on concentration is linear and that the off-diagonal terms (absorptivity ratios) are small relative to the diagonal terms. Small departures from linearity can be handled by approximation procedures, provided good calibration data is available enabling careful plots of absorbance ratio to concentration ratio to be made.

Before discussion of results, it will be helpful to compare the mathematical expressions involving concentration, absorptivity, and absorbance for binary, ternary, and multicomponent mixtures of higher orders. The binary mixture will be discussed in detail because of the importance of two-component mixtures in setting up a multicomponent analysis by the absorbance ratio method.

ABSORBANCE RATIO METHOD — BINARY PAIRS

Case 1. Mutual Band Overlap

This example is cited for purposes of discussion and entry into the required mathematics.

If Beer's law is followed, we may write for a two-component mixture

$$A_i = \sum_j a_{ij} c_j$$

$$A_1 = a_{11} c_1 + a_{12} c_2 \tag{5}$$

$$A_2 = a_{21} c_1 + a_{22} c_2 \tag{6}$$

Division of equation (5) by equation (6) yields

$$\frac{A_1}{A_2} = \frac{a_{11} c_1 + a_{12} c_2}{a_{21} c_1 + a_{22} c_2} \tag{7}$$

Several ways to solve this equation are available. If we set $x = c_1/c_2$, then

$$\frac{A_1}{A_2} = \frac{a_{11} x + a_{12}}{a_{21} x + a_{22}} = R \tag{8}$$

where $R = A_1/A_2$ and

$$x = \frac{a_{12} - R a_{22}}{R a_{21} - a_{11}} \tag{9}$$

If we set $x_1 = c_1/(c_1 + c_2)$ and $x_2 = c_2/(c_1 + c_2)$, then $x_1 + x_2 = 1$. Substitution in equation (1) yields

$$\frac{A_1}{A_2} = \frac{a_{11} x_1 + a_{12} x_2}{a_{12} x_1 + a_{22} x_2} \tag{10}$$

$$R = \frac{a_{11} x_1 + a_{12} (1 - x_1)}{a_{21} x_1 + a_{22} (1 - x_1)} \tag{11}$$

Solution for x_1 yields

$$x_1 = \frac{a_{12} - R a_{22}}{R(a_{21} - a_{22}) - (a_{11} - a_{12})} \tag{12}$$

This is the same solution as obtained by Ish-Shalom et al. [1]. The absorbance ratio equation (9) can be set up by measurement of individual absorptivities of pure components. Thereafter, the thickness term can be eliminated, since it is only necessary to measure absorbance ratios to solve for composition.

Case 2. No Band Overlap—$a_{11} \gg a_{12}$ and $a_{22} \gg a_{21}$

This sample case reduces mathematically to

$$\frac{A_1}{A_2} = \frac{a_{11}}{a_{22}}\, x \qquad\qquad (13)$$

which is the equation of a straight line of slope a_{11}/a_{22} and intercept 0.

Case 3. Single Overlap—$a_{11} \gg a_{12}$

This case (Fig. 1) is the most frequently occurring type with the exception of Case 1. Since the term involving a_{12} may be neglected, we can write

$$\frac{A_1}{A_2} = \frac{a_{11}c_1}{a_{21}c_1 + a_{22}c_2} \qquad\qquad (14)$$

Inverting and cancelling out like terms, we obtain

$$\frac{A_2}{A_1} = \frac{a_{22}}{a_{11}}\, x^{-1} + \frac{a_{21}}{a_{11}} \qquad\qquad (15)$$

$$x^{-1} = \frac{1}{x} = \frac{c_2}{c_1}$$

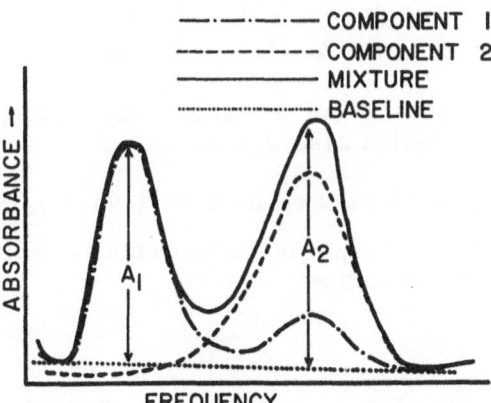

Fig. 1. Two-component mixture with single overlap. A_1 is the absorbance at frequency 1, and A_2 is the absorbance at frequency 2, each measured from a suitable baseline as shown.

This is the equation of a straight line of slope a_{22}/a_{11} and intercept a_{21}/a_{11}. The most reliable way to obtain these values is to make up several (preferably six) concentration ratios over a range of about fivefold, measure absorbance ratios directly by the baseline method, and then treat the data by a least-squares analysis.

The absorbance ratio method can be used to determine the concentration ratio for a binary pair by solving for x, x_1, or x^{-1} in equations (9), (12), (13), and (15). For the last case, we set

$$x^{-1} = \frac{a_{11}}{a_{22}} \frac{A_2}{A_1} - \frac{a_{21}}{a_{22}} \tag{16}$$

Case 4. Isobestic Point— $a_{11} \gg a_{12}$ and $a_{21} = a_{22}$

We can write

$$\frac{A_1}{A_2} = \frac{a_{11}c_1}{k(c_1 + c_2)} \tag{17}$$

$$k \equiv a_{21} \equiv a_{22}$$

$$\frac{A_1}{A_2} = k' \frac{c_1}{c_1 + c_2} = k'' P_1 \tag{18}$$

where P_1 is the percent of c_1. Alternatively,

$$\frac{A_2}{A_1} = \frac{a_{21}}{a_{11}} + \frac{a_{22}c_2}{a_{11}c_1} = k\left(1 + \frac{c_2}{c_1}\right) \tag{19}$$

This is an equation of a straight line of slope a_{22}/a_{11} and intercept a_{21}/a_{11}. In this case, $a_{22}/a_{11} = a_{21}/a_{11}$.

Case 5. Three Unknowns with No Overlap

For determining c_1, c_2, c_3 of components 1, 2, 3, we can use absorbance ratios as follows:

$$\frac{A_1}{A_2} = \frac{a_{11}c_1}{a_{22}c_2} \qquad\qquad \frac{c_1}{c_2} = \frac{a_{22}A_1}{a_{11}A_2}$$

$$\frac{A_1}{A_3} = \frac{a_{11}c_1}{a_{33}c_3} \qquad\qquad \frac{c_1}{c_3} = \frac{a_{33}A_1}{a_{11}A_3} \tag{20}$$

Case 6. Three Unknowns, General Case, with One Band Free from Overlap

$$A_1 = a_{11}c_1 + a_{12}c_2 + a_{13}c_3$$

$$A_2 = a_{21}c_1 + a_{22}c_2 + a_{23}c_3 \tag{21}$$

$$A_3 = a_{31}c_1 + a_{32}c_2 + a_{33}c_3$$

Suppose $a_{11} \gg a_{12}$ and $a_{11} \gg a_{13}$, so that we can write

$$A_1 \equiv a_{11}c_1 \tag{22}$$

We can divide A_2 and A_3 by A_1, obtaining

$$\frac{A_2}{A_1} = \frac{a_{21}}{a_{11}} + \frac{a_{22}c_2}{a_{11}c_1} + \frac{a_{23}c_3}{a_{11}c_1} \tag{23}$$

$$\frac{A_3}{A_1} = \frac{a_{31}}{a_{11}} + \frac{a_{32}c_2}{a_{11}c_1} + \frac{a_{33}c_3}{a_{11}c_1} \tag{24}$$

This is a pair of simultaneous equations with two unknowns and is readily solved yielding concentration ratios c_2/c_1 and c_3/c_1.

This can readily be extended to n equations with n unknowns which can be solved by standard techniques.

Case 7. Three Unknowns Solved by Adding an Internal Reference Standard

If a suitable reference band is not present in the mixture, it can readily be obtained by use of an internal standard. We then have for three components plus an internal standard

$$A_1 = a_{11}c_1 + a_{12}c_2 + a_{13}c_3 + a_{1s}c_s \tag{25a}$$

$$A_2 = a_{21}c_1 + a_{22}c_2 + a_{23}c_3 + a_{2s}c_s \tag{25b}$$

$$A_3 = a_{31}c_1 + a_{32}c_2 + a_{33}c_3 + a_{3s}c_s \tag{25c}$$

$$A_4 = a_{4s}c_s \tag{25d}$$

where c_s is the concentration of the internal standard. Division of equations (a), (b), and (c) by (d) yields

$$\frac{A_1}{A_4} = \frac{a_{11}c_1}{a_{48}c_8} + \frac{a_{12}c_2}{a_{48}c_8} + \frac{a_{13}c_3}{a_{48}c_8} + \frac{a_{18}}{a_{48}}$$

$$\frac{A_2}{A_4} = \frac{a_{21}c_1}{a_{48}c_8} + \frac{a_{22}c_2}{a_{48}c_8} + \frac{a_{23}c_3}{a_{48}c_8} + \frac{a_{28}}{a_{48}} \tag{26}$$

$$\frac{A_3}{A_4} = \frac{a_{31}c_1}{a_{48}c_8} + \frac{a_{32}c_2}{a_{48}c_8} + \frac{a_{33}c_3}{a_{48}c_8} + \frac{a_{38}}{a_{48}}$$

A system of six components with one band free from over-lap is represented mathematically in the standard form by equations (27a)–(27g) and in the absorbance ratio form by equations (28a)–(28e).

$$A_1 = a_{11}bc_1 + a_{12}bc_2 + a_{13}bc_3 + a_{14}bc_4 + a_{15}bc_5 + a_{16}bc_6 \tag{27a}$$

$$A_2 = a_{21}bc_1 + a_{22}bc_2 + a_{23}bc_3 + a_{24}bc_4 + a_{25}bc_5 + a_{26}bc_6 \tag{27b}$$

$$A_3 = a_{31}bc_1 + a_{32}bc_2 + a_{33}bc_3 + a_{34}bc_4 + a_{35}bc_5 + a_{36}bc_6 \tag{27c}$$

$$A_4 = a_{41}bc_1 + a_{42}bc_2 + a_{43}bc_3 + a_{44}bc_4 + a_{45}bc_5 + a_{46}bc_6 \tag{27d}$$

$$A_5 = a_{51}bc_1 + a_{52}bc_2 + a_{53}bc_3 + a_{54}bc_4 + a_{55}bc_5 + a_{56}bc_6 \tag{27e}$$

$$A_6 = a_{61}bc_1 + a_{62}bc_2 + a_{63}bc_3 + a_{64}bc_4 + a_{65}bc_5 + a_{66}bc_6 \tag{27f}$$

Suppose

$$A_1 \equiv a_{11}bc_1 (a_{11} \gg a_{12}, a_{13}, a_{14}, a_{15}, a_{16}) \tag{27g}$$

Division of equations (b), (c), (d), (e), and (f) by equation (g) yields

$$\frac{A_2}{A_1} = \frac{a_{21}}{a_{11}} + \frac{a_{22}c_2}{a_{11}c_1} + \frac{a_{23}c_3}{a_{11}c_1} + \frac{a_{24}c_4}{a_{11}c_1} + \frac{a_{25}c_5}{a_{11}c_1} + \frac{a_{26}c_6}{a_{11}c_1} \tag{28a}$$

$$\frac{A_3}{A_1} = \frac{a_{31}}{a_{11}} + \frac{a_{32}c_2}{a_{11}c_1} + \frac{a_{33}c_3}{a_{11}c_1} + \frac{a_{34}c_4}{a_{11}c_1} + \frac{a_{35}c_5}{a_{11}c_1} + \frac{a_{36}c_6}{a_{11}c_1} \tag{28b}$$

$$\frac{A_4}{A_1} = \frac{a_{41}}{a_{11}} + \frac{a_{42}c_2}{a_{11}c_1} + \frac{a_{43}c_3}{a_{11}c_1} + \frac{a_{44}c_4}{a_{11}c_1} + \frac{a_{45}c_5}{a_{11}c_1} + \frac{a_{46}c_6}{a_{11}c_1} \tag{28c}$$

$$\frac{A_5}{A_1} = \frac{a_{51}}{a_{11}} + \frac{a_{52}c_2}{a_{11}c_1} + \frac{a_{53}c_3}{a_{11}c_1} + \frac{a_{54}c_4}{a_{11}c_1} + \frac{a_{55}c_5}{a_{11}c_1} + \frac{a_{56}c_6}{a_{11}c_1} \tag{28d}$$

$$\frac{A_6}{A_1} = \frac{a_{61}}{a_{11}} + \frac{a_{62}C_2}{a_{11}C_1} + \frac{a_{63}C_3}{a_{11}C_1} + \frac{a_{64}C_4}{a_{11}C_1} + \frac{a_{65}C_5}{a_{11}C_1} + \frac{a_{66}C_6}{a_{11}C_1} \qquad (28e)$$

Two approaches are available for determining the absorptivity ratios in equation (28). The usual procedure is to calculate the individual absorptivities by determining the absorbance of each separate component at known concentrations in a solvent at the required frequencies, employing cells of carefully-measured path lengths. These individually-determined absorptivities may be inserted as coefficients in equations (27) and (28).

A better procedure is to measure absorbance ratios of each component to the standard or reference component prepared as binary mixtures of known composition. For example, a mixture of component 1 (the reference component) and component 2 is handled by means of the following equations:

$$\frac{A_2}{A_1} = \frac{a_{21}}{a_{11}} + \frac{a_{22}C_2}{a_{11}C_1} \qquad (29a)$$

$$\frac{A_3}{A_1} = \frac{a_{31}}{a_{11}} + \frac{a_{32}C_2}{a_{11}C_1} \qquad (29b)$$

$$\frac{A_4}{A_1} = \frac{a_{41}}{a_{11}} + \frac{a_{42}C_2}{a_{11}C_1} \qquad (29c)$$

$$\frac{A_5}{A_1} = \frac{a_{51}}{a_{11}} + \frac{a_{52}C_2}{a_{11}C_1} \qquad (29d)$$

$$\frac{A_6}{A_1} = \frac{a_{61}}{a_{11}} + \frac{a_{62}C_2}{a_{11}C_1} \qquad (29e)$$

The absorbance ratios at each of six frequencies for several mixtures, prepared by weighing each component on the analytical balance to an accuracy of at least 1 part per thousand, are determined and are plotted graphically against concentration ratio.

If the correlation is linear, the slope of the curve yields the ratio a_{i2}/a_{11} and the intercept yields the ratio a_{i1}/a_{11}. These ratios are inserted as coefficients into equation (28).

Once the required coefficients have been determined, several test mixtures of carefully-weighed compositions are analyzed and the measured absorbance ratios are compared with the calculated values obtained by plugging the known concentration ratios into equation (28). Alternately, if a suitable computer such as an IBM 1620 is available, one may solve the system of simultaneous equations after programming the absorptivity ratio matrix to yield spectrophotometrically determined concentration ratios for comparison with the ratios prepared by weighing.

Small discrepancies can be spotted and errors detected and eliminated by the analysis of test mixtures in this way. Two sources of error or discrepancy are readily detected by this check-off procedure. One is error in the off-diagonal terms, especially in regions of rapidly changing absorbances on the steep side of absorption bands. The off-diagonal term can be determined more accurately by adjusting discrepancies observed in suitable test mixtures of all components in such cases.

A second source of error is departure from nonlinearity of the relation between absorbance and concentration. Improved accuracy can be obtained in such cases by use of the better fitting coefficients obtained from graphical plots of binary mixtures of each component with the reference component.

INHERENT PRECISION OF THE ABSORBANCE RATIO METHOD

The precision of the absorbance ratio method is a function of concentration ratio as shown in Fig. 2 and Table I.

The percent error equivalent to an absorbance error of 0.005 is plotted against component concentration for two-component systems with absorptivity ratios of 1, 2, and 4 in Fig. 2. The method of calculation for the case with a ratio of unity is shown in Table I. An absorbance error of plus 0.005 in the smaller of the two bands under comparison has been assumed. An error of 0.005 in the other band will be compensating in the plus direction and trivial in the negative direction, and therefore is neglected in this discussion. In each case, the absorbance of the stronger band has been assumed to be unity.

For the case of a ratio of 1.000 in both absorptivity and concentration, errors of 0.125, 0.187, and 0.372% can be

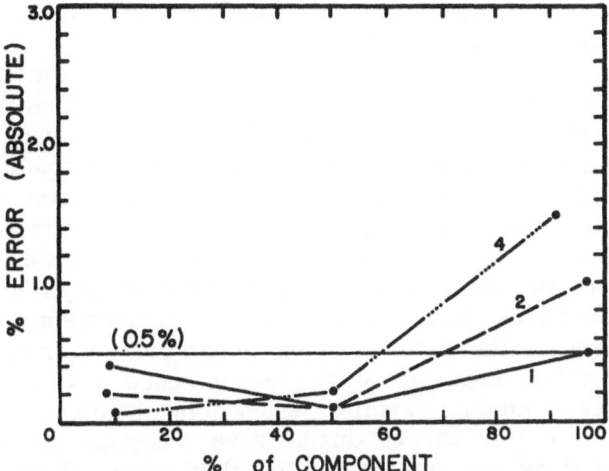

Fig. 2. Composition error equivalent to 0.005 absorbance error for absorbance ratios of 1, 2, and 4.

TABLE I

Precision of the Absorbance Ratio Method as a Function of Concentration Ratio

(The absorptivity ratio has been taken as unity in this example.)

R	$R + 1$	%	$R*$	$R + 1$	% †	% Error, absolute	% Relative error
0.02	1.02	98.03	0.025	1.025	97.56	0.47	0.48
0.05	1.05	95.23	0.055	1.055	94.8	0.45	0.47
0.10	1.10	90.91	0.105	1.105	90.49	0.42	0.47
0.25	1.25	80.00	0.255	1.255	79.68	0.32	0.40
0.50	1.50	66.66	0.505	1.505	66.44	0.22	0.33
0.75	1.75	57.14	0.755	1.755	56.98	0.16	0.28
1.00	2.00	50.00	1.005	2.005	49.88	0.12	0.24
2.00	3.00	33.33	1.980	2.980	33.56	0.23	0.69
3.00	4.00	25.00	3.960	3.960	25.25	0.25	1.00
5.00	6.00	16.65	5.890	5.890	16.97	0.34	2.00
10.00	11.0	9.09	9.520	10.520	9.51	0.42	4.40

*Ratio of stronger band at an absorbance of unity to weaker band with a plus 0.005 absorbance error in the weaker band.

† $\% = \dfrac{100}{R + 1}$.

expected for absorbances of 1.000, 0.666, and 0.333, respectively, with an error of 0.005 in one peak.

The significance of the error distributions shown in Table I and Fig. 2 is that components in a mixture can be estimated to better than 0.5% absolute, even with absorbance ratios greater than ten between the reference peak and the component peak. It is necessary to scan at several thicknesses in direct spectrophotometry to achieve satisfactory precision when band intensities of components vary so greatly. For example, a mixture of two components of about equal absorptivity in a 20 to 1 ratio of concentration will require measurement at two cell thicknesses in the ratio of about 20 to 1 for optimum results. By the ratio method, a single thickness is satisfactory.

The ratio method permits the use of much thinner cells than is advisable for direct quantitative analysis because cell thickness does not enter into the calculations. Thus, cells of 0.01 mm or even thinner can be employed. The ratio method affords wide latitude as to solvent selection. In the first place, solvent can be eliminated by using short cell paths. Secondly, it is possible to use solvents which would ordinarily be precluded because of excessive background absorption in longer cell paths ordinarily employed in direct spectrophotometry.

EXPERIMENTAL PROCEDURE

Methylcyclohexane, o- m- and p-xylene, and cumene were obtained from Eastman Kodak (white label grade); the toluene and benzene used were Fisher Certified Reagents; the cyclohexane used was MC & B spectroquality reagent. All these were used as received.

The absorption cells were calibrated by the fringe method.

All measurements were made and programmed on a double-beam Beckman IR-9 grating spectrophotometer at the following settings:

Absorbance. 0-1
Single beam—double beam ratio . . . about 1.2 at 1000 cm^{-1}
Slit . . . . .at 1.2 mm at 1000 cm^{-1}
Gain. 0.70
Period . . . . either 1 or 2
Scanning speed . . . . .either 20 or 40 cm^{-1}/min
Frequency scale. . . . . .25 cm^{-1}/in.

Base lines were drawn from suitable points (Fig. 1). Peak heights were measured with plastic rulers to the nearest 0.2 mm. Usually a peak could be estimated to the nearest 0.5 mm (0.002 absorbance). The transmission scale was judged by comparison with single-beam readings to be linear within 0.2% from 100 — 10% transmission.

The methylcyclohexane–toluene system was examined in a 0.150-mm cell without any diluting solvent. The six-component system was examined in an 0.0454-mm cell with cyclohexane as diluting solvent.

Absorbance ratio data for the six-component system was punched on IBM cards and processed with an IBM 1620 computer to obtain concentration ratios by solution of sets of five simultaneous equations.

RESULTS

The absorbance ratio data for the methylcyclohexane–toluene system is presented in Table II for the frequencies 1177 and 1264 cm^{-1}.

TABLE II

Analysis of Methylcyclohexane — Toluene Mixtures by Absorbance Ratio Method

	Absorbance ratio (1177/1264)	Percentage of toluene		
		Input	Found*	Difference
1	0.0276	6.75	9.74	+ 3.0
2	0.065	12.07	14.35	+ 2.3
3	0.099	17.35	18.15	+ 0.8
4	0.196	27.60	27.80	+ 0.2
5	0.673	53.3	53.35	+ 0.05
6	1.268	67.5	67.40	- ·0.1
7	1.495	70.8	71.1	+ 0.3
8	2.330	79.2	79.1	− 0.1
9	2.585	80.7	80.75	+ 0.05
10	4.590	88.0	88.1	÷ 0.1
11	5.505	90.0	89.9	− 0.1
12	13.660	95.4	95.6	+ 0.2

*% T = 100 WR / 1 + WR where WR is the weight ratio of toluene to methylcyclohexane. WR = (A_{1177}/ A_{1264} + 0.04) 1.60.

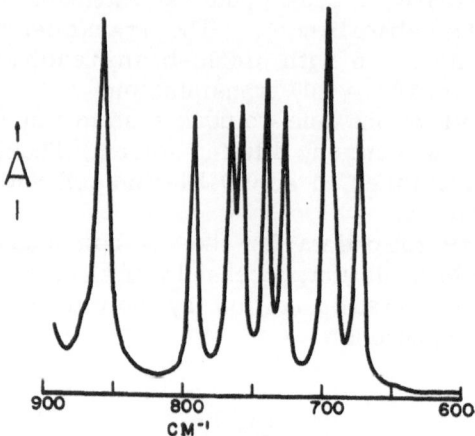

Fig. 3. Spectrogram of six-component mixture in cyclohexane.

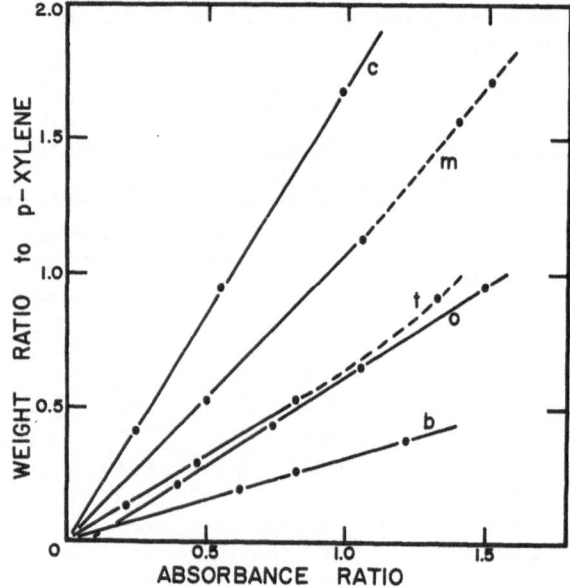

Fig. 4. Absorbance ratio versus weight ratio with p-xylene as reference component.
c is cumene, m is m-xylene, t is toluene, o is o-xylene, and b is benzene.

The standard deviation of scan of the m-xylene—p-xylene system (1.15 to 1.00 weight ratio) at 767.5 and 794.5 cm^{-1} was observed to be 0.55 in absorbance and 0.05 in absorbance ratio units. Changing the pen period by one step in the direction of increased response resulted in a 2.8% increase in absorbance at each frequency. The absorbance ratio increase was less than 0.3%, changing from 1.0517 to 1.0545. This corresponds to the concentration increment of 0.1% absolute in the absorbance ratio method and a 2.8% increase in the conventional absorbance method.

Scan reproducibility data for the six-component system (Fig. 3) of p-xylene, m-xylene, cumene, o-xylene, toluene, and benzene are shown in Table III. Absorptivity ratio data for the six-component system with p-xylene as the reference component are presented in Table IV and Fig. 4.

Analytical results for nine trials of the six-component system are presented in Table V and Fig. 5.

The relative standard deviation of scan for each of five pairs of the six-component system was 0.75% in absorbance units and 0.44% in absorbance ratio. The latter figure cor-

TABLE III

Reproducibility of Nine Infrared Scans of a Six-Component Mixture

Component	Frequency, cm^{-1}	Absorbance values			Absorbance ratio*	
		Average	Standard deviation	Relative standard deviation (%)	Average	Relative standard deviation (%)
p-xylene	794.5	0.686	0.0039	0.62		
m-xylene	767.5	0.280	0.0011	0.41	0.406	0.34
cumene	759	0.685	0.0049	0.71	0.997	0.53
o-xylene	741	0.739	0.0040	0.55	1.076	0.32
toluene	728	0.720	0.0064	0.90	1.050	0.30
benzene	675	0.790	0.0102	1.30	1.153	0.68
average				0.75		0.44

*The absorbance ratio is the ratio of the absorbance of each component divided by the absorbance of p-xylene. This sample was run in an 0.0454-mm cell with approximately an equal weight of cyclohexane as diluting solvent.

TABLE IV

Absorptivity Ratio Matrix for Six-Component Mixture

| | Frequencies employed | | | | |
| Component ratios | 767.5/794.5 | 759/794.5 | 741/794.5 | 728/792.5 | 675/794.5 |
	Absorptivity ratios*				
m-xylene/p-xylene	0.913†	0.040	0.000	0.000	−0.013
cumene/p-xylene	0.140	0.581	0.000	0.000	−0.017
o-xylene/p-xylene	0.004	0.028	1.500	0.058	−0.012
toluene/p-xylene	0.000	0.010	0.035	1.530	−0.008
benzene/p-xylene	0.001	0.012	0.000	0.008	3.300

*Absorptivity ratios as a first approximation are equivalent to the slopes of the curves in Fig. 4.
† The value 0.860 was used for examples 6a and 9a in Table V.

responds to a constituent error of less than 0.1%, with the exception of benzene which has a corresponding error of 0.2%.

DISCUSSION

The relation between absorbance ratio and weight ratio for the toluene—methylcyclohexane system is represented closely by the linear equation

$$T/M = 1.60 \ \frac{A_{1177}}{A_{1264}} + 0.04 \tag{30}$$

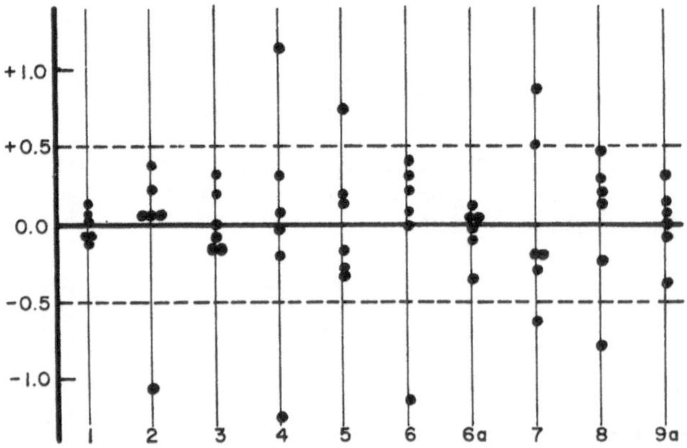

Fig. 5. Percent deviation of each component in a six-component mixture for nine trials.

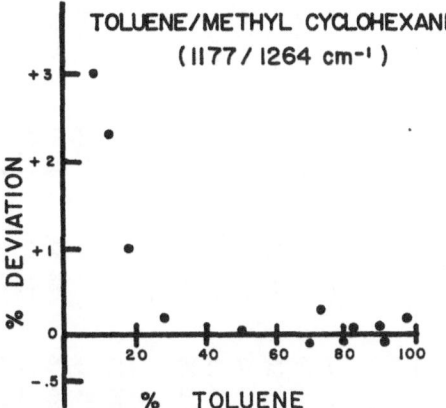

Fig. 6. Percent deviation of toluene as a function of concentration.

where T/M is the weight ratio of toluene to methylcyclohexane.

This equation fits well above 20% toluene concentration (Fig. 6). Below this concentration, a better fit is obtained with the following linear equation:

$$T/M = 1.86 \left(\frac{A_{1177}}{A_{1266}} + 0.009 \right) \tag{31}$$

The reasons for the failure of equation (30) at low-toluene concentration are the shift in background absorption due to methylcyclohexane, and the shift in baseline points from 1205 and 1170 cm^{-1} to 1187 and 1173 cm^{-1}. The background due to methylcyclohexane is actually negative by 0.0478 absorbance units in a 0.150-mm cell at 1177 cm^{-1}. Equation (31) corrects for this effect. The deviations using equation (31) are - 0.48, + 0.03, + 0.35, 0.00, and +2.60% for toluene concentrations of 6.75, 12.07, 17.35, 27.00, and 55.9%, respectively.

The data in Table V indicate that it is possible to obtain concentrations of four of the components of the six-component system to better than 0.3% absolute, using the absorptivity matrix of Table IV. More erratic results were obtained with two of the components, cumene and m-xylene. Run number 8, which is a virtual repeat of run number 4, yielded each of these two components to better than 1% absolute. Run number 6 was recalculated as 6a by changing one of the coefficients. This change was dictated by the nonlinearity observed with the

TABLE V

Infrared Analysis of Six-Component Mixture by the Absorbance Ratio Method

Compound	1	2	3	4	5	6	6a	7	8	9a
				Deviations from inputs in percent absolute						
p-xylene	−0.02	+0.25	−0.01	−0.21	−0.29	+0.33	+0.13	−0.18	+0.31	+0.15
m-xylene	−0.07	+0.38	−0.12	+1.15	+0.76	−1.14	+0.05	+0.87	+0.47	+0.33
cumene	+0.05	−1.11	+0.32	−1.25	+0.20	+0.42	−0.35	+0.52	−0.77	−0.39
o-xylene	+0.03	+0.17	+0.20	+0.33	−0.32	+0.22	+0.06	−0.62	+0.21	−0.09
toluene	+0.11	+0.17	−0.10	−0.04	−0.17	+0.08	−0.01	−0.29	−0.23	−0.01
benzene	−0.09	+0.17	−0.13	+0.08	+0.16	+0.00	−0.09	−0.17	+0.15	+0.02
				Inputs in percent						
p-xylene	18.79	19.12	23.83	17.07	21.68	16.82	16.82	21.33	16.35	16.55
m-xylene	18.70	19.88	21.52	15.84	7.33	26.49	26.49	7.40	15.33	26.64
cumene	30.09	28.59	14.94	38.51	34.73	28.58	28.58	33.83	36.37	26.91
o-xylene	13.07	13.06	15.87	11.37	14.44	11.20	11.20	15.40	11.35	12.18
toluene	13.14	13.09	15.73	11.28	14.33	11.19	11.19	14.88	14.09	11.73
benzene	6.21	6.29	8.25	5.89	7.48	5.81	5.81	7.02	6.16	5.97

m- and p-xylene pair above a weight ratio of 1 to 1. Run 9a, which is virtually a repeat of run 6, was calculated on the same basis with similar results.

Some difficulty was encountered in accurately determining the coefficient 0.140 in this table due to the absorption of cumene at 767.5 cm^{-1} overlapping with absorption due to m-toluene. The actual figure is an estimate rather than a precise value.

The absorptivity ratio for the binary pair m- and p-xylene is somewhat uncertain as reflected by the curvature in Fig. 4 for this pair. The average relative standard deviation of nine trials for m-xylene was computed as + 4.0% compared with values of + 0.03, - 0.7, + 0.3, - 0.5, and + 0.2% for p-xylene, cumene, o-xylene, toluene, and benzene, respectively.

More accurate results could be obtained for m-xylene by using a more detailed calibration curve relating absorbance ratio to weight ratio for the binary mixture of m- and p-xylene. An approximation procedure would then be employed in which an average absorptivity ratio would be used to get an initial result. From the initial result, a more reliable absorptivity ratio would be selected from the curve for the binary mixture. Recalculation would then lead to an improved figure. If desired, a third set of calculations can be made to reduce the discrepancy to a very small figure.

No further work was attempted for this paper, as the system was selected for demonstration purposes rather than for determination of the ultimate accuracy of the absorbance ratio method.

So far, no really negative results have been obtained by the absorbance ratio method. The poorest result is the one reported for m-xylene with an average relative standard deviation of 4% in a six-component mixture.

CONCLUSION

The absorbance ratio method is capable of analyzing multi-component mixtures with a precision of a few tenths of a percent. Since path length is not a factor in the calculations, a source of inaccuracy is eliminated by the ratio method. The absorbance ratio is subjected to less random variation in scan conditions and sampling, when, as is frequently the case, the disturbances equally affect the band pairs used in ratio computations.

The absorbance ratio procedure is ideal for multicomponent liquids which may interact with or be insoluble in suitable infrared solvents, because solvents may be omitted by employing very short cell paths. The absorbance ratio is, of course, especially adaptable to materials of poorly-defined thickness, such as films and layers of insoluble polymers. The absorbance ratio method is rapid because a single thickness is sufficient for most analyses.

ACKNOWLEDGMENTS

The suggestions and assistance of A. M. Schneider, A. L. Stockett, and R. F. Russo in provision of computer facilities are gratefully acknowledged.

REFERENCE

1. M. Ish-Shalom, J.D. Fitzpatrick, and M. Orchin, J. Chem. Educ. 34, 496–499 (October, 1957).

Sensitivity of Calculated Wave Numbers of a Normal Coordinate Treatment to Assumed Molecular Geometry*

Robert R. Hart [†]

Illinois Institute of Technology
Chicago, Illinois

Normal coordinate treatments rarely include a consideration of the effect on the treatment of experimental uncertainties in bond lengths and interbond angles. At the same time, the agreement between calculated and observed wave numbers is taken as the major criterion of the validities of the potential-energy constants obtained, of the form of internuclear potential-energy function assumed, and of such band assignments as may have been based on the treatment. To help understand this effect, numerical results were obtained for the changes of calculated wave numbers resulting from small alterations in bond lengths and interbond angles. The molecules used were thirteen types of substituted methanes, for which potential-energy constants had previously been obtained in this laboratory. The use of such molecules also helps in understanding how seriously the common assumption of tetrahedral structure for substituted methanes affects the many normal coordinate treatments of them that have appeared in the literature. The significance of the unexpectedly large changes obtained for normal coordinate treatments of more complex and structurally less well-determined molecules is discussed.

INTRODUCTION

In carrying out normal coordinate treatments of substituted methanes, the actual, nontetrahedral structure of the molecule is often approximated [1,2] by a tetrahedral structure. Another approximation, the semitetrahedral, obtains the elements of the G matrix (the kinetic-energy matrix, expressed relative to

*One of the investigations carried out in partial fulfillment of requirements for the doctoral degree.
†National Science Foundation predoctoral fellow. Present address: Bell Telephone Laboratories, Room 1A-366, Dept. 1124, Murray Hill, New Jersey.

symmetry coordinates) from the elements of the g matrix (the
kinetic-energy matrix, expressed relative to internal coordinates) calculated on the basis of nontetrahedral structure and
from the symmetry coordinates appropriate to tetrahedral
structure; thus, the symmetry coordinate taken as redundant
is not identically equal to zero. The expressions so obtained [3]
for the G matrix elements and their derivations are simpler
than in the nontetrahedral case.

Overend and Scherer [4] have investigated the accuracy of
the tetrahedral approximation, by obtaining the Urey—Bradley
force constants of the monohalogen-substituted methanes and
deuteromethanes, using both tetrahedral and nontetrahedral
structure. The present report contains results bearing on the
same and similar questions—the sensitivities of calculated wave
numbers to the tetrahedral and semitetrahedral approximations,
and to experimental uncertainties in the internuclear distances.
The molecules treated were of C_{3v} symmetry and included
monohalogen-substituted   methanes,   trihalogen-substituted
methanes, methanes triply substituted with one halogen and
singly substituted with another, and their deuterated analogs.

RESULTS

Using the general valence force-field, potential-energy
constants previously obtained in this laboratory assuming
tetrahedral structure [5-10], and using interbond angles and
internuclear distances obtained from the microwave spectra,
the wave numbers of thirteen types of molecule of the general
form CXY_3 (CBr_3H, CBr_3D; CCl_3D, CCl_3H; CF_3H, CF_3D, CF_3Cl;
CH_3I, CD_3I; CF_3Br, CF_3I; CD_3Br, CH_3Br) were recalculated
using the tetrahedral approximation, the semitetrahedral approximation, and the actual, nontetrahedral structure of the
molecule. Some of the computer methods and programs used
have been described elsewhere [11,12]. The average and maximum departures of the microwave Y-C-Y angle of these molecules from the tetrahedral angle of 109° 28' were, respectively,
1° 28' and 2° 9'. The expressions used for the nontetrahedral
F and G matrix elements were independently rederived. The
G_{13}^a, G_{23}^a, G_{33}^a, and G_{22}^e elements so obtained were compared with
the expressions given by Ramaswamy, Sathianandan, and
Cleveland [13]. The remaining elements of the G matrix were
compared with the expressions of Ziomek and Piotrowski [3].

The elements of the F matrix (the potential-energy matrix, expressed relative to symmetry coordinates) were compared with the expressions given by Bethke and Wilson [14]. The re-derived expressions and those given in the literature agree, provided the following is noted: The use by Ramaswamy et al. and by Bethke et al. of an a_1-type symmetry coordinate not orthogonal to the redundant coordinate, results [15] in misplaced factors of the quantity denoted (in their respective notations) by γ and b in their expressions for a_1-type matrix elements; in the work of Ramaswamy et al., the definitions of the angles α and β are interchanged; in the work of Ziomek et al., the subscripts X and Y have been interchanged, and a V appears instead of an X in the expression for G_{11}^a. The expressions for the semitetrahedral and nontetrahedral F and G matrix elements were also checked by verifying that they reduced to the tetrahedral expressions. Similarly, the computer programs based on them yielded, in the special case of tetrahedral angles, the same numerical results as did programs based on the simpler tetrahedral expressions. The programs used in the actual, nontetrahedral molecular structure were also checked by using them to recalculate numerical values of F and G, which had been previously and independently obtained by hand for a nontetrahedral molecule.

The average, average percent, median, median percent, maximum, and maximum percent differences between the corresponding wave numbers recalculated on the basis of tetrahedral and nontetrahedral structures are shown in the first row of Table I. The occurrence of large differences in one or two wave numbers in each molecule, and of smaller differences in the bulk of the wave numbers, is reflected in the differences between the averages and medians, and in the fact that the maximum difference is five times the average.

It is not my intention to enter into the difficult question of the significance of potential-energy constants and normal coordinate treatments [16], but rather to provide some data bearing on that question. However, the following comments seem reasonable. Firstly, in none of the molecules considered did the use of the tetrahedral approximation result in the failure of the normal coordinate treatment as a check on the assignments. This may be partly fortuitous since, in several cases, the separation between observed wave numbers was less than, or only slightly greater than, the maximum difference given in Table I. Secondly, the differences are well outside

TABLE I

Differences Between Corresponding Wave Numbers Recalculated Using Different Methods

Comparison of methods	Average		Median		Maximum	
	cm^{-1}	%	cm^{-1}	%	cm^{-1}	%
Tetrahedral $\div$ Nontetrahedral	8.9	1.1	4.4	0.72	44*	4.3†
Semitetrahedral — Nontetrahedral	2.6	0.33	2.0	0.21	10‡	1.4‡
Tetrahedral — Tetrahedral¶	4.1	0.38	0.83	0.077	13	0.94

*In CH_3I.
†In CBr_3H.
‡In CF_3I.
¶Differences between corresponding wave numbers recalculated using tetrahedral structure, and using tetrahedral structure assuming all C—H and C—D bond lengths to be 1 pm (0.01 A) greater than before. The results include data from only the ten molecules containing one or more C—H or C—D bonds. The a_1-type wave numbers did not change when the C—X distance was altered; these unaltered wave numbers were likewise not included in these results.

experimental error, and are comparable to the effects due to the anharmonicities. Thirdly, since the force constants are, roughly speaking, proportional to the square of the wave numbers, one might expect that the percent differences between force constants obtained in two normal coordinate treatments—one using the tetrahedral and the other using the nontetrahedral structure—would be about twice the percentages given in Table I. (The precise mathematical sense in which this holds may be derived from the equality between the determinant of FG and the product of its eigenvalues.) Some confirmation for this may be obtained by comparing the 1.1% average difference of the table with the results of Overend and Scherer [4]. They obtained an average difference of 2.4% in the case of their physically most-meaningful constants (C—X and C—Y stretching), and differences as great as several hundred percent in their other constants. Without asserting that these differences are larger than the other ambiguities [16] to which the potential-energy function is subject, it would nevertheless seem preferable in a careful normal coordinate treatment; to avoid the additional uncertainty arising from the tetrahedral approximation, especially as this uncertainty, unlike the others, can usually be eliminated in a straightforward and unambiguous manner.

Differences between corresponding a_r–type wave numbers recalculated by the semitetrahedral and nontetrahedral methods are given in the second row of Table I. The e–type F and G matrices and, hence, the e–type wave numbers are identical in the two methods. From a comparison of the first and second rows, it appears that the semitetrahedral approximation is considerably better than the tetrahedral.

The differences in the wave numbers recalculated using tetrahedral and nontetrahedral structures correspond to changes in linear dimensions which average about 3 pm (0.03 A) in the Y—Y internuclear distance. The uncertainties in the equilibrium C—H internuclear separations in the methyl halides due to zero-point oscillations, are about 1 pm (0.01 A) [17,18]. Since the computer programs prepared for the preceding work could easily be adapted to this purpose, it was decided to see how sensitive the calculated wave numbers were to changes in the C—H bond length. The results are given in the third row of Table I.

The table does not present separate analyses of the data from monohalogen-substituted methanes, trihalogen-substituted methanes, and methanes triply substituted with one halogen and singly substituted with another; such analyses revealed no significant differences among these three kinds of molecule.

DISCUSSION

The foregoing comments apply specifically only to the many normal coordinate treatments of halogen-substituted methanes that have been previously reported, both from this laboratory [5–10, 19] and from others. However, as molecules of intermediate complexity for which considerable data are available, these have been used as a testing-ground for the transferability of potential-energy constants between molecules containing similar chemical bonds, and for force fields, most recently the Urey—Bradley field [4]. Thus, results bearing on them may have a wider interest.

In addition, it appears to me that the unexpectedly large changes in calculated wave numbers that result from small changes in assumed geometry allow one to raise a larger question. Normal coordinate treatments never include a consideration of the effects of structural uncertainties; at the same time, the agreement between calculated and observed wave numbers is taken as the principal criterion of the validities of the normal coordinate treatment, of the form of internuclear potential-energy function assumed (e.g., Urey—Bradley), and of such band assignments as may have been based on the treatment.   Yet, the structural uncertainties of molecules subjected to normal coordinate analyses are often many times larger than the changes dealt with above. This is especially the case with molecules on the frontiers of our understanding; structural data transferred from other molecules or based on educated guesses must be used. A similar comment holds for the initial set of force constants used in the normal coordinate treatment.   If, in addition, the molecule contains many atoms, the spectral lines that the normal coordinate treatment is helping to assign are many and closely spaced, and the effects of changes in assumed structure might be correspondingly complex and unexpected.

A typical normal coordinate treatment of such a molecule is the treatment by Nakamoto, Fujita, Condrate, and Morimoto [20] of the 1,1-dithiocarbamato complex,

$$H_2NC \langle \begin{smallmatrix} S \\ S \end{smallmatrix} \rangle Pt$$

Here, the angles are known to a likely error of perhaps 5°; the internuclear distances have, perhaps, likely errors ranging from 1 pm (0.01 A) for the C—H and C—N internuclear distances, to 5 pm (0.05 A) for the C—S distance, and 10 pm (0.10 A) for the S—Pt distance. Using these likely errors together with the results presented in Table I, and recalling that the first line corresponds to average angular changes of 1° 28', and that the third line corresponds to linear changes of 1 pm (0.01 A), one may estimate an average uncertainty of 35 cm^{-1} or 4% in the calculated wave numbers due simply to structural uncertainties, with a maximum effect several times as large. This may be compared with an average departure of calculated from observed wave numbers of 14 cm^{-1} or 1.4%, and a maximum departure of 68 cm^{-1} or 6.4%, obtained by these authors for nineteen in-plane vibrations of this complex and its totally deuterated analog in the range 288—1628 cm^{-1}.

A few comments seem appropriate.   Firstly, the significance of a reported agreement between calculated and observed wave numbers, that is well below the effects of structural uncertainties alone, is not clear. Indeed, in the absence of an estimate of those effects, such an agreement seems somewhat deceptive. After all, if the wave numbers were recalculated using the reported potential-energy constants, together with a structure differing by the likely structural errors from that used by the above authors, this agreement would be several times worse than that reported.

Secondly, if the percent effects on the force constants are taken, as was successfully done above, as twice the percent effects on the calculated wave numbers, one would assign an average uncertainty of 8% in the force constants to the structural uncertainties alone. Such an estimate of uncertainty could well appear in conjunction with the above authors' (Nakamoto et al.) table of reported force constants. Instead, the only indication of their reliability is the number of significant figures given— three for the stretching force constants.

Thirdly, in using this normal coordinate treatment as an aid in making assignments, it should be kept in mind that the maximum effect of the structural uncertainties on the calculated wave numbers could easily be as large as 100 cm^{-1}. This is not an unimportant effect if assignment of ten bands within a range of 1350 cm^{-1} is to be made.

Of course, the results in Table I were obtained for entirely different molecules, and their use for this complex is questionable. They are used because Nakamoto et al., in accord with prevailing practice, give no estimate of the effect of structural uncertainties on their treatment, and because, as far as I know, no theoretical or numerical investigation of these effects exists for any other molecules.

CONCLUSION

I do not wish to deprecate the good work of Nakamoto et al., which I am using simply as an example of current practice; but I do wish to suggest that the effects of structural uncertainties on calculated wave numbers and on the force constants obtained from normal coordinate treatments may be sufficiently large as to merit some attention. In current practice, these effects receive no attention whatsoever.

ACKNOWLEDGMENTS

I wish to thank Dr. Monomohan Mazumder for his rederivation of the expressions for the nontetrahedral G matrix elements, and to acknowledge the advice and help of Mr. Edward A. Piotrowski. This work was part of a research program which has been aided by a grant from the National Science Foundation. The author is grateful for this assistance.

REFERENCES

1. W. F. Edgell and L. Parts, J. Am. Chem. Soc. 78: 2358 (1956).
2. T. Shimanouchi and I. Suzuki, J. Mol. Spectroscopy 6: 277 (1961).
3. J. S. Ziomek and E. A. Piotrowski, J. Chem. Phys. 34: 1087 (1961).
4. J. Overend and J. R. Scherer, J. Chem. Phys. 33: 446 (1960).
5. A. G. Meister, S. E. Rosser, and F. F. Cleveland, J. Chem. Phys. 18: 346 (1950).
6. J. P. Zeitlow, F. F. Cleveland, and A. G. Meister, J. Chem. Phys. 18: 1076 (1950).
7. C. E. Decker, A. G. Meister, and F. F. Cleveland, J. Chem. Phys. 19: 784 (1951).
8. P. F. Fenlon, F. F. Cleveland, and A. G. Meister, J. Chem. Phys. 19: 1561 (1951).
9. P. R. McGee, F. F. Cleveland, A. G. Meister, C. E. Decker, and S. I. Miller, J. Chem. Phys. 21: 242 (1953).
10. H. B. Weissman, R. B. Bernstein, S. E. Rosser, A. G. Meister, and F. F. Cleveland, J. Chem. Phys. 23: 544 (1955).
11. R. R. Hart, Comparison of computer methods of solving the eigenvalue problem of molecular spectroscopy, J. Franklin Inst. 279: 1 (1965).
12. R. R. Hart, R. W. Estin, and E. A. Piotrowski, General normal coordinate treatment program, filed with and obtainable from Quantum Chemistry Program Exchange (QCPE), Indiana University, Bloomington, Indiana; see also QCPE Newsletter, No. 8 (January 1965), and R. Hart, Spectroscopia Mol. 13: 2 (1964).
13. K. Ramaswamy, K. Sathianandan, and F. F. Cleveland, J. Mol. Spectroscopy 9: 107 (1962).
14. G. W. Bethke and M. K. Wilson, J. Chem. Phys. 26: 1118 (1957).
15. S. Sundaram and F. F. Cleveland, J. Chem. Phys. 32: 166 (1960).
16. S. Brodersen, Some of the principal problems in the determination of the potential function, in: Molecular Spectroscopy, Butterworth's, London (1962), pp. 27–32.
17. S. L. Miller, L. C. Aamodt, G. Dousmanis, C. H. Townes, and J. Kraitchman, J. Chem. Phys. 20: 1112 (1952).
18. C. C. Costain, J. Chem. Phys. 29: 864 (1958).
19. K. Sathianandan, K. Ramaswamy, and F. F. Cleveland, J. Mol. Spectroscopy 8: 470 (1962).
20. K. Nakamoto, J. Fujita, R. A. Condrate, and Y. Morimoto, J. Chem. Phys. 39: 423 (1963).

Vibrational Spectra of $C_{10}Cl_{12}$ and $C_{10}Cl_{10}O$

S. Sundaram

Illinois Institute of Technology
Chicago, Illinois

The infrared spectra of the compounds $C_{10}Cl_{12}$ and $C_{10}Cl_{10}O$ have been studied in solutions of carbon tetrachloride and carbon disulfide. Also, KBr pellets of the two molecules were used in the infrared study. Films of the samples and the dispersions in paraffin oil were also investigated for comparison. The observed wave numbers are compared with the data of previous studies, and some conclusions regarding the molecular structure have been given.

INTRODUCTION

For several years since the chlorocarbon compound $C_{10}Cl_{12}$ was obtained by Prins [1] in the reaction between 1,1,2;3,3,4,5,5-octachloropentene and aluminum chloride, this compound was believed to have an empirical composition $C_{10}Cl_{12}$. This situation remained the same even after the experiments of Newcomer and McBee [2] and of Gilbert and Groleta [3] who reacted hexachlorocyclopentadiene with aluminum chloride to yield this compound. The hydrolysis of hexachlorocyclopentadiene yielded the ketonic compound $C_{10}Cl_{10}O$. Earlier investigations by McBee et al. [4] on the properties of the chlorocarbon $C_{10}Cl_{12}$ have yielded certain useful results. Zijp and Gerding [5] have made dipole moment and spectral studies of $C_{10}Cl_{12}$. The structure and identity of $C_{10}Cl_{12}$ have been the subject of controversy for some years. Originally it was believed to be the so-called Diels—Alder dimer (perchloro-3a,4,7,7a-tetrahydro-4,7-methanoindene) of hexachlorocyclopentadiene, similar to the unsaturated dimer of the unchlorinated cyclopentadiene. McBee et al. [6] later prepared this dimer

179

and identified it as a compound different from $C_{10}Cl_{12}$. Furthermore, the melting point (485°C) and other data did not support
the belief that this chlorocompound could be the above-
mentioned dimer. The present investigation was undertaken
to elucidate the controversial structure of this highly interesting
and complex molecule. It was also the object of this investigation to study the influence of different solvents on the sample
and different states of the sample. Hence, the experiments
that were performed fall into different categories, depending
on the techniques used. Furthermore, it was felt that a study
of the ketonic compound $C_{10}C_{12}O$ along with the chlorocarbon
$C_{10}Cl_{12}$ would provide more information on the structure and
aid in the analysis of the observed spectra.

EXPERIMENTAL

The infrared spectra of the chlorocompound $C_{10}Cl_{12}$ and the
ketonic compound $C_{10}Cl_{10}O$ were studied in detail under four
different categories of experiments.

In the first series of experiments, saturated solutions
of the compounds were prepared using carbon tetrachloride
and carbon disulfide as solvents. The spectra of these solutions
were recorded using Perkin-Elmer Model 21 Spectrophotometer
with sodium chloride optics in the region 600 to 4000 cm^{-1}
and calcium fluoride optics in the region 2000 to 4000 cm^{-1}.
Similar experiments also were done using a Beckman IR 9
instrument from 400 to 4000 cm^{-1}.

In the second category of experimental techniques, potassium bromide disks of the compounds were prepared, and the
spectra were obtained using the Perkin-Elmer Model 21
Spectrophotometer and the Beckman instrument.

In the third series of experiments, small amounts of the
solutions of the two compounds were placed on plates, and the
solvents were allowed to evaporate. The infrared spectra of
the fine films thus obtained were studied using the Perkin-
Elmer Model 21 Spectrophotometer.

The series was completed by running the spectra using the
technique of mulling the sample in paraffin oil nujol.

Calibration of the instruments was carried out using
polystyrene film and indene sample.

TABLE I

Wave Numbers for $C_{10}Cl_{12}$ from Infrared Data

CCl_4 solution	CS_2 solution	Pellet	Data of Zijp and Gerding [5]
523s	523s	-	-
655vs	655vs	654s	-
664vvw	665sh	-	-
768w	770w	767w	-
821s	822s	819s	820s
887s	888m	887s	886m
966s	967m	963s	962s
1058vvs	1059vvs	1055vs	1056vs
1075w	1077w	-	-
1130vs	1132vs	1130vs	1131vs
1149vvs	1150vvs	1147vs	1146vs

RESULTS

Table I shows the wave numbers obtained for $C_{10}Cl_{12}$ from the infrared spectra of the compound in solutions of carbon tetrachloride and carbon disulfide. For the sake of comparison, the corresponding infrared data for the potassium bromide disks and the infrared data of Zijp and Gerding [5] are also shown in the same table. There are eleven wave numbers observed in each of the solutions and eight wave numbers in the case of the potassium bromide disks. Only two very weak bands were observed in the case of the spectrum with the film, and hence they are not shown here. In the case with the nujol mull, the background was such that no clear location of the positions of the bands was possible.

It is to be noticed that, except for one of the bands, i.e., the one at 665 cm^{-1}, all are clear, sharp bands. Most of the bands have appreciable intensities and the half-widths of all of them are less than 20 cm^{-1}. The only earlier experiment providing some infrared data on this chlorocarbon is by Zijp and Gerding [5]. They have recorded the spectrum in the region 700 to 2100 cm^{-1} using the samples in CCl_4 and CS_2 solutions, as well as using the solid phase of the sample in the form of potassium bromide pellets. They report six remarkably sharp bands whose characteristics are in agreement with the observations in the present investigation. No spectral record is shown in their paper, and therefore direct comparisons could not be made. However, the positions and estimated intensities of the six bands reported by these authors are in very good agreement with the corresponding ones in this study. This is evident from the data shown in Table I.

Another interesting feature, which was noticed in the nature of the spectra of this compound in solutions and in the solid phase in postassium bromide pellets, is that there are practically no differences between phases in either the positions or the intensities of the bands. The differences in wave numbers for the solid phase and the molecules in solution are, in almost all cases, very small. The average of the differences is comparable to experimental uncertainties in fixing the positions of the bands.

It may also be seen that there are no absorption bands observed for this molecule either around 1600 cm^{-1} or beyond. The absence of any absorption above 1200 cm^{-1} is very significant, as will be pointed out in the following discussion. Furthermore, it is also significant to find that, for a molecule with as many as twenty-two atoms, the number of observed vibrational wave numbers is relatively small.

It is interesting to compare the results with those obtained by Zijp and Gerding [5] in their studies of the Raman spectrum of the molecule $C_{10}Cl_{12}$ in solution with carbon tetrachloride. Table II gives the wave numbers and the relative intensities for the Raman spectrum. There are seventeen observed wave numbers, and most of them are very weak. With the exception of five, all have very low intensities and, hence, only the wave number displacements are shown. These small intensities are responsible for the absence of any polarization data, and perhaps, as pointed out by Zijp and Gerding [5], the study of the Raman spectrum is not complete. (We propose to study this under more favorable conditions for enhanced intensity. At the present time, these conditions have not been determined, partly due to lack of any experimental details given by the previous investigators.) Obviously, since the spectrum of the solution itself is so weak, the determination of the Raman spectrum of the solid phase presents more difficulties. It is to be noted that there are no strong lines about 1600 cm^{-1} or beyond 1250 cm^{-1} in the Raman spectrum.

TABLE II

Observed Wave Numbers for $C_{10}Cl_{12}$
(Raman and Infrared Spectra)

Raman	Infrared	Raman	Infrared
162 (4)	-	720	770w
192	-	-	822s
252	-	892	888m
280	-	945 (4)	967m
362 (8)	-	1016 (?)	1059vvs
388 (6)	-	1089 (2)	1077w
520	523s	1126	1132vs
604	655vs	1159	1150vvs
618 (?)	665sh	1232 (?)	-

Table II also shows the comparison between the wave numbers obtained from Raman and infrared spectra. The differences are not too great for some of the wave numbers observed in the two spectra.

The infrared spectral data for the ketonic compound $C_{10}Cl_{10}O$ have been summarized in Table III. There are sixteen bands observed in the region 400 to 4000 cm^{-1} in solutions and in the solid phase in the form of a potassium bromide pellet. The data for the evaporated film and the solid in the mull are also presented. Again, as in the case of the chloro-compound, the infrared bands are intense and sharp. The spectra of the various phases are almost identical. Also, some of the bands are not very significantly different from those in the infrared spectrum of $C_{10}Cl_{12}$. The total number of wave numbers observed for this molecule is greater than that for $C_{10}Cl_{12}$. This is expected because of the lesser symmetry the compound $C_{10}Cl_{10}O$ possesses compared to the compound $C_{10}Cl_{12}$.

DISCUSSION

The following discussion is applicable to both the molecules $C_{10}Cl_{12}$ and $C_{10}Cl_{10}O$. If one takes into account each of the several experimental observations mentioned earlier, significant information about the structure and the nature of the forces in the molecules can be obtained.

TABLE III

Infrared Data for $C_{10}Cl_{10}O$

Solution	Pellet	Film	Mull
509s	510s	-	-
638s	640s	638s	638s
659s	663vs	660s	660s
701m	705m	700m	700m
783m	782m	-	785w
819s	820s	-	-
849s	850s	848m	848s
932m	932s	930m	930s
956m	958m	956vw	956w
987m	994m	990w	988w
1056s	1060vs	1056s	1056s
1076m	1076w	-	-
1122s	1116sh	1131vw	1122w
1145s	1153s	1150s	1150s
1173m	1168s	1164s	1164s
1205vw	1205w	1220m	1220m

Both the molecules have been found to have practically identical wave numbers for the solid phase and in solution. This is a significant result, as it indicates that the molecules in solution almost preserve their identity as in the solid. Consequently, one may conclude that the interactions between the molecules must be weak. In this connection, one may recall the appreciable differences in the wave numbers between water and ice, as have been observed by several investigators.

The most noteworthy feature—the absence of absorption bands near 1600 cm^{-1}—strongly suggests that there is an absence of double bonds (C=C) in both molecules. In the spectra of the compounds involving a C=C bond, the Raman lines corresponding to the C=C stretching vibration are of considerable intensity.

Further evidence for the absence of the C=C bond is presented in the ultraviolet spectrum of $C_{10}Cl_{12}$ in n-hexane, as observed by Zijp and Gerding [5]. While they have recorded a strong absorption maximum at 2195 A, no absorption was found at or above 2300 A, as is characteristic of all fully-chlorinated, cyclic monoenes and dienes.

For the twenty-two-atom molecule $C_{10}Cl_{12}$, one would

expect sixty vibrational wave numbers—$(3 \times 22) - 6 = 60$; and for $C_{10}Cl_{10}O$, fifty-seven vibrational wave numbers. The experiments reveal that the observed wave numbers are fewer than expected. This fact suggests that these molecules possess a high degree of symmetry and the consequent degeneracies. In fact, some preliminary studies on diffraction of X-rays by $C_{10}Cl_{12}$ have indicated the possibility of a highly-symmetrical cubic structure for the solid.

It was pointed out earlier that the wave numbers for the molecule $C_{10}Cl_{12}$ in the infrared and the Raman spectra do not differ very much. It seems probable in this case that the interactions between equivalent groups may be small. One may use the symmetric and antisymmetric (with respect to the center of symmetry) vibrations of the identical CCl_2 groups to illustrate this point. If the difference between the Raman and infrared wave numbers for these modes is small, low coupling between the two groups is indicated. (Perhaps, one has to compare this with the experimental observations on dihaloethylenes.)

In conclusion, the experimental observations favor a highly-symmetric structure with only single bonds in the carbon frame for the two molecules. A detailed discussion of the assignments and a force constant calculation will be published elsewhere.

ACKNOWLEDGMENT

The author is thankful to Mr. Edward A. Piotrowski for his very valuable assistance in the entire experimental part of this investigation.

REFERENCES

1. H. J. Prins, Rec. Trav. Chim. 65: 445 (1946).
2. J. S. Newcomer and E. T. McBee, J. Am. Chem. Soc. 71: 952 (1949).
3. E. E. Gilbert and S. L. Groleta, U. S. Patent 2 616 928 (1952).
4. E. T. McBee, C. W. Roberts, J. D. Idol, Jr., and R. H. Earle, Jr., J. Am. Chem. Soc. 78: 1511 (1956).
5. D. H. Zijp and H. Gerding, Rec. Trav. Chim. 77: 682 (1958).
6. E. T. McBee, J. D. Idol, Jr., and C. W. Roberts, J. Am. Chem. Soc. 77: 4375 (1955).

Spectroscopic Study of the Molecular Complex HMX:DMF*

Alex Castelli and Delbert J. Cragle

Spectroscopy Group, Propellants Laboratory
Picatinny Arsenal
Dover, New Jersey

The possibility of hydrogen bonding involving the aldehydic hydrogen of DMF in the crystalline complex molecule HMX:DMF has been investigated. Spectroscopic methods were employed. DMF-d_1 with deuterium as the aldehydic hydrogen was employed. The behavior of the aldehydic hydrogen-stretching frequency of DMF indicates that such hydrogen (deuterium) bonding does not exist in the complex molecule or in liquid DMF.

A comparison of the infrared spectra of DMF (g) and HMX:DMF (s) indicates that the shift in the carbonyl-stretching frequency is too large to be accounted for by the usual amide resonance mechanism. The behavior of the carbonyl band may be explained in terms of a bond between the oxygen atom of DMF and some positive site on HMX.

INTRODUCTION

The precipitate formed by the evaporation of a solution of dimethylformamide (DMF) which is saturated with sym-cyclotetramethylene tetranitramine (HMX) consists of a complex of HMX and DMF in a one-to-one molar ratio [1,2]. Two mechanisms have been mentioned to account for this molecular pairing. The first is that the complex is a solvate in a one-to-one molar ratio [1]. The second is that the complex is a clatharate of DMF molecules trapped in the HMX crystal vacancies. The latter mechanism has been discounted [2].

Recently, the results of thermochemical measurements on the nature of this complex were reported [3]. A value of

*This work was presented at the Spectroscopy Seminar of the University of California on October 28, 1963.

187

approximately 5.2 kcal/mole was observed for the enthalpy of the reaction:

HMX:DMF(solid) $\xrightarrow{\text{DMF (liquid)}}$ HMX(solvated) + DMF(liquid)

Furthermore, it was cited that such a value is in the range of enthalpy magnitudes that are typical of hydrogen bonding. Also, it was observed that the structures of DMF and HMX allow the possibility that a bond involving the aldehydic hydrogen atom of DMF may be responsible for such a linkage in this complex.

The object of this work is to establish whether or not the complex molecule HMX:DMF is held together by bonding involving the aldehydic hydrogen atom of DMF and to search for evidence of any other bonds between these two molecules.

EXPERIMENTAL

The samples of DMF (reagent grade) and DMF-d_1 (b.p. 153—155°C) were obtained from Merck (U.S) and Merck, Sharp & Dohme (Canada), respectively. HMX-I was the form of HMX used throughout this study. The sample of HMX-I was prepared at Picatinny Arsenal. Its spectrum showed it to be the same as that of the HMX-I used in the work of Bedard et al. [2].

The vibrational spectra were measured with a Perkin-Elmer Model 21 infrared spectrophotometer equipped with rock-salt optics. Spectra of both varieties of gaseous DMF at STP were obtained with a 10-cm gas cell. Spectra of the liquids were obtained as laminar films 0.05 mm thick. Spectra of the solids were obtained as 0.5% KBr pellets, 0.7 mm thick. All spectra were obtained at room temperature. The absorbance band in the 3500 cm^{-1} region in the spectra of liquid DMF and liquid DMF-d_1 is due to water. The spectra were rerun with dry DMF and DMF-d_1. The presence of water was not found to alter any conclusions reported here.

RESULTS AND DISCUSSION

Spectroscopic Evidence of Absence of Aldehydic Hydrogen Bonding in the HMX:DMF Complex Molecule

In vibrational spectroscopy one of the most effective indications that a molecule is hydrogen-bonding involves

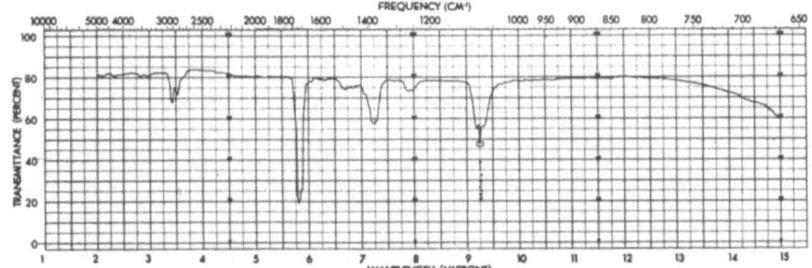

Fig. 1. Infrared spectrum of gaseous DMF.

measuring the frequency with which the hydrogen atom vibrates along the bond joining it to the rest of the molecule [4]. This frequency shall be designated ω_s (C–H) where C–H and s refer to the carbon—hydrogen stretching frequency of the aldehydic group in DMF. If DMF and HMX molecules are linked together in the complex by hydrogen bonding, the ω_s (C–H) of DMF in the vibrational spectrum of the complex molecule should be less than the ω_s (C–H) of gaseous DMF molecules at STP.

In order to employ vibrational spectra in this problem, the first step is to assign the ω_s (C–H) in the spectrum of gaseous DMF and to see if and how its value in the complex molecule is altered. However, the assignment of ω_s (C–H) in the spectra of either DMF or solid HMX:DMF is too uncertain to be of any value in solving the problem because of the other C—H bands present in the spectrum of the complex which might be masking the aldehydic C—H band. The infrared spectra of gaseous DMF and of the solid complex molecule HMX:DMF are shown in Figs. 1 and 2. Absorbances in the region between 2700 cm^{-1} and 3000 cm^{-1} of both spectra are due to C—H bands.

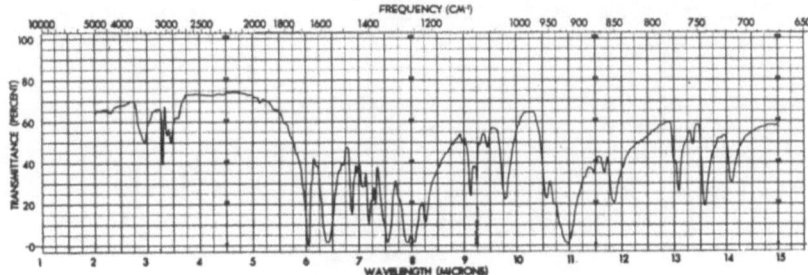

Fig. 2. Infrared spectrum of 1:1 M complex of HMX and DMF.

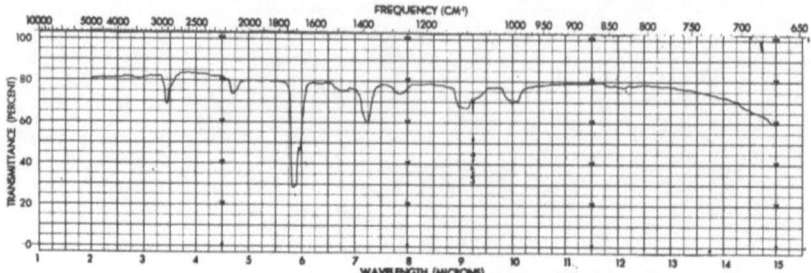

Fig. 3. Infrared spectrum of gaseous DMF-d₁.

A solution to the problem caused by the many C—H bands was found by using DMF-d₁ in which the deuterium atom replaces the aldehydic hydrogen atom of DMF causing the ω_s (C–D) to split out from the ω_s (C–H) region. The degree to which ω_s (C–D) splits out from ω_s (C–H) was predicted by means of a theory relating changes in vibrational frequency with changes in the masses of the vibrating atoms [5,6]. With utilization of this theory and temporary assumption that vibrational coupling effects are small, the value of ω_s (C–D) should be approximately $2900/\sqrt{2}$ or 2050 cm⁻¹. Figure 3, which shows the vibrational spectrum of DMF-d₁, reveals the presence of a new band at 2110 cm⁻¹. Concurrently, the band which was at 2850 cm⁻¹ in the spectrum of gaseous DMF can now be assigned to ω_s (C–H), since it is absent in the spectrum of DMF-d₁. As such, it appears very likely that the band at 2110 cm⁻¹, which is the only substantial change in the vibrational stretching region, is ω_s (C–D) of the deuterium–labeled aldehydic group. The ω_s (C–D) of the aldehydic group, which appears at 2110 cm⁻¹, is

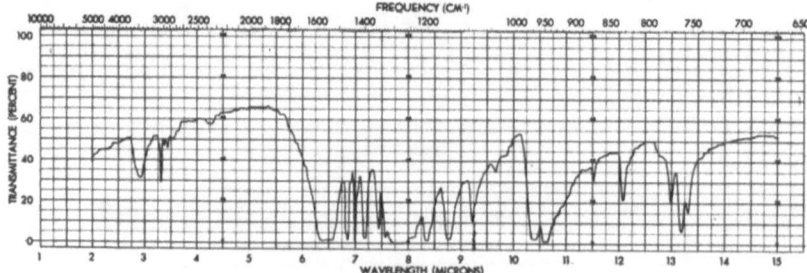

Fig. 4. Infrared spectrum of HMX-I.

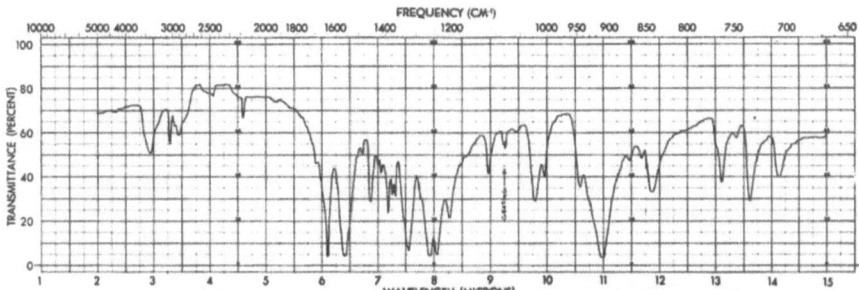

Fig. 5. Infrared spectrum of 1:1 M complex of HMX and DMF-d$_1$.

now seen to be quite free of any overlap (Fig. 3), so that possible spectral shifts due to such hydrogen (deuterium) bonding should be observable.

Figure 4 shows the vibrational spectrum of HMX-I in the region of 5000–670 cm^{-1}. In this spectrum, the 1910–2120 cm^{-1} range, where ω_s (C–D) is expected to appear in the spectrum of the complex, is relatively free from any absorbance. Also, the spectra of other known forms of HMX crystals [2] show that this range is relatively free from any strong absorbance bands.* As such, any new band appearing between 1910 and 2120 cm^{-1} in the HMX:DMF-d$_1$ spectrum may be confidently assigned to ω_s (C–D) from DMF-d$_1$.

Figure 5 shows the spectrum of HMX:DMF-d$_1$. Clearly, in going from gaseous to complexed form, ω_s (C–D) of DMF has not shifted to lower frequency. It must be concluded that the case for hydrogen bonding in the complex by means of the aldehydic hydrogen from DMF has failed an important test.

Comparison of Fig. 3 with Fig. 5 indicates that quite possibly a change takes place in the intensity distribution of ω_s (C–D) between the gaseous and complexed states. Nevertheless, the change in the band center of ω_s (C–D) is not sufficient to alter our conclusion.

Resonance Phenomena and Bonding in HMX:DMF

It is well-known that the spectra of amide molecules exhibit a band in the vicinity of 1700 cm^{-1}, most of whose character

*The small differences between the spectrum of HMX-I that is shown here and the spectrum of HMX-I obtained by Bedard et al. [2] are primarily due to the fact that, in this work, the HMX-I was suspended in KBr, and, in the work of Bedard et al., the HMX-I was suspended in an alkane medium.

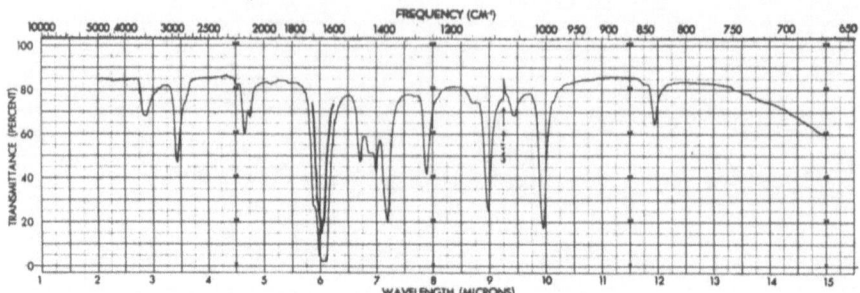

Fig. 6. Infrared spectrum of liquid DMF-d$_1$.

originates in the vibrational motion of the carbon and oxygen atoms along their valence bond. Since this group is primarily of double-bond character, its frequency will be designated ω (C=O).

In the gaseous state, ω (C=O) appears at 1695 cm^{-1} in DMF-d$_1$ and at 1715 cm^{-1} in DMF, while in the solid complex molecules ω (C=O) appears at 1634 cm^{-1} in HMX:DMF-d$_1$ and 1653 cm^{-1} in HMX:DMF. These two shifts may be accounted for in terms of the well-known amide resonance structures [4]:

$$\underset{I}{\overset{\displaystyle >N-C\overset{O}{\underset{H}{\diagdown}}}{}} \longleftrightarrow \underset{II}{\overset{\displaystyle >^{\oplus}N=C\overset{O^{\ominus}}{\underset{H}{\diagdown}}}{}}$$

It is found that in amides ω (C=O) shifts to lower frequencies in passing along the series: gas→dilute nonpolar solution→ liquid and solid (the same ω (C=O) for liquids as well as solids)

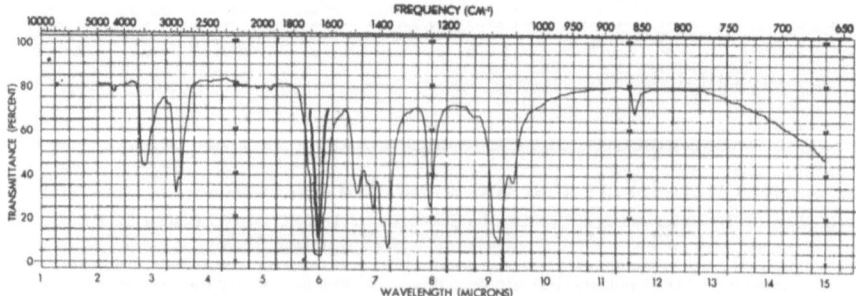

Fig. 7. Infrared spectrum of liquid DMF.

TABLE I

Frequencies, Shifts, and Ratios of ω (C=O) Bands

Compound	Gas	Shift	Liquid	Shift	Complex	Total shift	$\left\lvert \dfrac{\Delta\omega\,(C=O)}{\omega\,(C=O)} \right\rvert$[*]
			cm^{-1}				
DMF	1715	−43	1672	−19	1653	−62	0.011
DMF-d_1	1695	−42	1653	−19	1634	−61	0.012

[*]Refers to the -19 cm^{-1} shift.

[8]. These spectral shifts may be shown, through the Gordy equation, to be consistent with the effects of an increased contribution from the polarized form II in the condensed state (see Appendix). The fraction of the shift of ω (C=O) that may be attributed to these polarization effects may be taken as about 43 cm^{-1} from the difference between values for ω (C=O) in the gaseous and liquid states (see Figs. 6 and 7 for spectra of liquid DMF-d_1 and DMF) as shown in Table I. This leaves 19 cm^{-1} or approximately one-third of the total shift unaccounted for. The shift, however, is in the direction of increased bonding at the oxygen atom of the amide. Furthermore, the ratios $\lvert \Delta\omega\,(C=O)/\omega\,(C=O) \rvert$, which are 0.011 and 0.012 for DMF and DMF-d_1, respectively, are in the range of electrostatic bonds for oxygen atoms [7]. Moreover, the nitro-nitrogen atom in HMX, with its positive charge resulting from the contributions from III, IV, and possibly V in the resonance structure,

is a likely site for such electrostatic interaction.

The possibility that the -19 cm^{-1} shift is due to some DMF—DMF mechanism must also be considered. In N-methyl-acetamide, a -18 cm^{-1} shift of ω (C=O) is observed in crossing a transition point in the range 15 to -60°C in the solid state [8]. The likelihood that HMX molecules are interposed between DMF molecules tends to discount this consideration. If on the other hand, the complex consists of layers of DMF molecules between HMX layers, or of some similar arrangement, then the possibility remains that a DMF—DMF mechanism is responsible for the -19 cm^{-1} shift.

In addition to bonding between amide—oxygen and nitro—nitrogen atoms, bonding can occur between positive nitrogen atoms and negative oxygen atoms in DMF and HMX, respectively. Unfortunately, the assignment of one or two bands in the vibrational spectrum of DMF or DMF-d$_1$ to a motion which might clearly reflect this bonding, such as an out-of-plane movement of the positively-charged nitrogen atom, is at best difficult and may be impossible, due to the high degree of mixing of vibrations which is known to occur in these molecules [8].

In principle, it is also possible to detect this bonding in terms of some spectral feature of HMX. Vibrations involving N—NO$_2$ groups, for which a satisfactory assignment seems to be available [9], would appear to provide the opportunity for doing this. However, this assignment is of little help in dealing with this problem, since substantial changes that are attributable merely to the various crystalline forms of HMX [2] occur in the region for which the assignment is given. Thus, the assignment of small spectral shifts in N—NO$_2$ bands due to complexation of DMF with HMX is too uncertain to permit interpretations concerning N—NO$_2$ bonding.

APPENDIX

Resonance Inferences Directly From Spectral Shifts

The Gordy equation [10] provides a quick method for inferring the occurrence of transitions to states which involve a change in resonance contribution. This possibility arises when such transitions are accompanied by a shift in the frequency of a normal vibration whose coordinates indicate it to be predominately of a stretching character. By converting the Gordy equation to

$$\omega = D \left[a N (\chi_A \chi_B / d^2)^{3/4} + b \right]^{1/2}$$

where $D = \frac{1}{2} \pi c \mu^{1/2}$, c is the velocity of light, μ is the reduced mass of atoms A and B, χ_A is the electronegativity of atom A, χ_B is the electronegativity of atom B, N is the order of the bond between A and B, d is the length of the bond between A and B, ω is the stretching frequency of bond A—B, and a and b are constants, the observed lowering in the value of ω (C=O) in going from DMF (g) to DMF (l) implies an increased contribution from a structure such as II whenever such resonance structures may be written.

ACKNOWLEDGMENTS

Gratitude is expressed to C. Lenchitz, who acquainted us with this problem, and to M. Halik and W. Fredricks, who assisted in the work.

REFERENCES

1. C.D. Bockman, Jr., Armour Research Foundation, private communication, cited in M. Bedard et. al., Can. J. Chem. 12: 2278 (1962).
2. M. Bedard, H. Huber, J.L. Myers, and G.F. Wright, Can. J. Chem. 12: 2278 (1962).
3. C. Lenchitz and R. Velicky, Eighteenth Calorimetry Conference, Bartlesville, Oklahoma, Oct. 17 (1963).
4. L. Pauling, The Nature of the Chemical Bond, Cornell Univ. Press, Ithaca (1960).
5. E. Teller, cited in W.R. Angus et al., J. Chem. Soc. (London) 971 (1936).
6. O. Redlich, Z. physik. Chem. B28: 371 (1935).
7. G. Pimentel and A.L. McClellan, The Hydrogen Bond, Freeman and Co., San Francisco (1960).
8. T. Miyazawa, T. Shimanouchi, and S. Mizushima, J. Chem. Phys. 24: 408 (1956).
9. L.J. Bellamy, The Infrared Spectra of Complex Molecules, Methuen and Co., London (1956).
10. W. Gordy, J. Chem. Phys. 14: 305 (1946).

Intramolecular NH···Halogen Hydrogen-Bond Strengths in Five- and Six-Membered Chelate Rings

P. J. Krueger and D. W. Smith

University of Alberta
Calgary, Alberta, Canada

The fundamental symmetric and asymmetric NH_2, NHD, and ND_2 stretching vibrations are compared in o- and p-haloanilines, -benzylamines, and -benzamides in dilute solution. Ortho-substituted benzylamines exhibit rotational isomerism, with free NH_2 groups and intramolecularly-bonded NH_2 groups. The cis—trans isomerism of the NHD group in mono-deuterated-o-haloanilines and o-benzamides indicates that the intramolecular NH···X hydrogen-bond strength increases in the former in the order X = F < Cl < Br < I, but that the reverse order holds for the latter (as well as for o-halobenzylamines). These results are discussed in terms of the halogen size, the halogen electronegativity, and the geometry of the chelate system. In mono-deutero-o-haloanilines, ND···X bonds appear to be weaker than NH···X bonds.

INTRODUCTION

Recent work in this laboratory has dealt with the effect of nonbonded interactions between neighboring groups in molecules as revealed in their infrared spectra. A systematic study of the fundamental symmetric and asymmetric NH_2 stretching vibrations in o-, m-, and p-substituted anilines in dilute CCl_4 solution led to an interpretation of many "anomalous" frequencies, intensities, and band widths for the o-isomers in terms of the contributions to the transition moment in each mode and the hybridization of the nitrogen atom [1]. Weak intramolecular interactions of the hydrogen-bonding type are of particular interest, since even these can modify the properties of a molecule substantially. As Jaffe has pointed out [2], the fact that the proton–donating group and the proton-accepting

197

group cannot separate permits the observation of weaker intramolecular than intermolecular hydrogen bonds.

Intramolecular hydrogen bonds with covalently-bound halogen atoms are well known, but the order in which the halogen atoms fall for increasing AH···X hydrogen-bond strength in various systems is still subject to some controversy. Although earlier measurements based on the trans-gauche frequency displacement of the OH stretching band in 2-haloethanols led to the conclusion that the OH···X hydrogen-bond strength increased in the order F<Cl<Br<I parallel to the frequency displacement [3], the frequency shift on hydrogen-bond formation is not necessarily a direct reflection of hydrogen-bond strength [4], particularly for weak, intramolecular hydrogen bonds. Extensive thermodynamic studies [5] have confirmed that in the 2-haloethanols the classical factor dominates, i.e., the intramolecular OH···X hydrogen-bond strength increases with increasing halogen electronegativity.

Many publications have dealt with the intramolecular hydrogen bond in o-halophenols since Wulf, Liddel, and Hendricks [6] detected two OH stretching bands for these compounds, which Pauling [7] interpreted as cis—trans isomerism with the cis isomer stabilized by a hydrogen bond. Conclusions based on frequency displacements again lead to erroneous results. On the basis of integrated absorption intensities of the cis and trans bands in the o-halophenols and in 2,4,6-trisubstituted phenols with two different halogens in the 2 and 6 positions, Baker and Kaeding [8] have concluded that the strength of the hydrogen bond increases in the order I<F<Br<Cl. This order was attributed to both the varying size of the halogen atoms and to an orbital—orbital repulsive interaction, which was held to increase in the order Cl<Br<I. The interaction would be between the OH bonding orbital and the donated, lone-pair orbital of the halogens, and would be due to the small amount of directional character in these lone-pair orbitals, and to interaction with lower-lying, completed, electronic shells. Recently, Sandorfy et al. [9] have obtained an almost identical halogen order (F<I<Br<Cl) on the basis of competitive intermolecular hydrogen bonding to ether in dilute CCl_4 solution. These authors conclude that the intramolecular OH···F hydrogen bond is present, but very weak. In contrast, the NMR spectra of Allan and Reeves [10] show no such inter-

action, which was rationalized on the basis of the small size of the fluorine atom held at a fixed distance from the OH group.

It should be noted that the size of the halogen atom and the fixed OH···X distance and geometry probably play an important part in intramolecular hydrogen bonds, whereas they are absent in intermolecular hydrogen bonds (except in the solid state). This paper describes the results of spectroscopic studies designed to investigate the factors involved in intramolecular NH···X hydrogen bonds in o-haloanilines, -benzylamines, and -benzamides. The first series involves a five-membered chelate ring and an amino group with "s-character" of about 0.530 (0.500 corresponds to sp^3 hybridization, and 0.577 to sp^2 hybridization) [1]; the second series involves an essentially tetrahedral NH_2 group, insulated from the intramolecular electronic effects of the aromatic ring and the halogen atom, and leads to a six-membered chelate ring; whereas the third series involves a much more acidic amino group (essentially sp^2) in a six-membered chelate ring.

The technique selected was based on the observable cis—trans isomerism of the NHD group in compounds with mono-deutero, primary-amino groups. Moritz [11, 12] has demonstrated that this is a sensitive method for the detection of nonequivalent N-H bonds in compounds containing the NH_2 group. With acid amides and aromatic amines, conjugation of the nitrogen atom with the carbonyl group and the aromatic ring, respectively, leads to some degree of double-bond character in the C-N bond with partial restriction of rotation about it. For an NHD group attached to an aromatic ring, cis—trans isomerism acquires significance if an ortho-substituent is present, but not with meta- or para-substituents. Thus, Moritz has shown that the NH stretching frequency of the NHD group in partially-deuterated aniline falls at $3330\,cm^{-1}$, almost in the center between the asymmetric and symmetric NH_2 stretching frequencies (3478 and $3395\,cm^{-1}$, respectively). Mono-deutero-p-nitroaniline shows corresponding bands at 3508, 3458, and $3413\,cm^{-1}$, whereas in partially-deuterated o-nitroaniline, the asymmetric and symmetric NH_2 bands at 3518 and $3397\,cm^{-1}$, respectively, are separated by two new bands due to the NHD group at $3477\,cm^{-1}$ (NH "free") and 3446 (NH "bonded" to NO_2). The separation ($\Delta = 31\,cm^{-1}$) between the two NH bands for the two isomers reflects the strength of the intramolecular hydrogen bond. Partially-deuterated primary amines with nonequivalent N-H bonds also show the expected symmetric and asymmetric ND_2 bands, as well as the two ND bands arising from the NHD group in the two isomers of the mono-deutero amine. Bagratishvili, Tsitsishvili, and Bezhashvili [13] have independently reported similar work on o-nitroaniline. The technique has also been used to demonstrate bilateral hydrogen bonding in picramide [14], as well as in other applications [15].

EXPERIMENTAL PROCEDURE

Spectroscopic measurements were made with a Beckman IR-7 infrared spectrophotometer with a NaCl foreprism and a 75-line/mm grating blazed at 12μ, and operating in the fourth order. Slit widths were calculated to be about 1 cm^{-1} in both the NH and ND stretching regions. The calibration of the instrument has been described previously [16], and the accuracy of frequency measurements is believed to be limited largely by the widths of the absorption bands themselves. Measurements were made at ambient temperatures, using 2.0-cm, matched, fused-quartz, absorption cells and "spectro" quality CCl_4 and C_2Cl_4 as solvents. Amine concentrations of the order of 0.01−0.05 M were used. Benzamide concentrations were of the order of 0.001 M because of limited solubility. Weak bands suggesting a slight amount of intermolecular association (even at below 5×10^{-4} M) were observed in the amide spectra, but these do not interfere with the conclusions.

The low-frequency side of ν_s in amides is overlapped by the first overtone of the "carbonyl" frequency, and the low-frequency side of ν_{as} is overlapped by a relatively small sharp band (or bands) believed to be due to cyclic association of these polar molecules. Intensities were determined by numerical integration of the high-frequency half of these bands, and doubling the results. Half-band widths $\Delta\nu_{1/2}^a$ for these compounds are likewise twice the width of the nonoverlapped side of these bands.

All the benzamides were made from the corresponding benzoic acids via the acid chlorides. The anilines and benzylamines were commercial products purified by fractional distillation or recrystallization.

Partial deuteration was achieved by repetitive equilibration with 99% methanol-d (Merck, Sharp, and Dohme). Final traces of methanol were removed under high vacuum. No deuterium analyses were performed.

RESULTS AND DISCUSSION

All the frequency measurements in CCl_4 solution for partially-deuterated o-haloanilines and for partially-deuterated o- and p-halobenzamides are summarized in Table I and Figs. 1 and 2. The equilibria in question are

There is no evidence to suggest that any benzamide molecules have the carbonyl group cis to the X atom, which would be expected to be a very unstable conformation.

For the partially-deuterated benzylamines, the following equilibria apply:

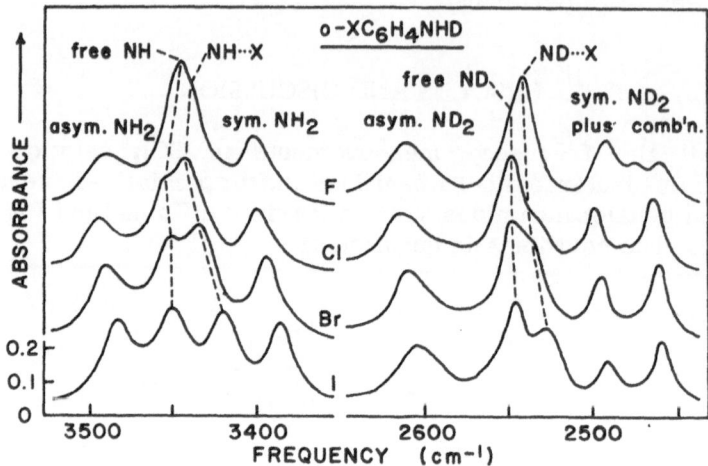

Fig. 1. Partially–deuterated o–haloaniline spectra in dilute CCl_4 solution. Concentrations and level of deuteration vary.

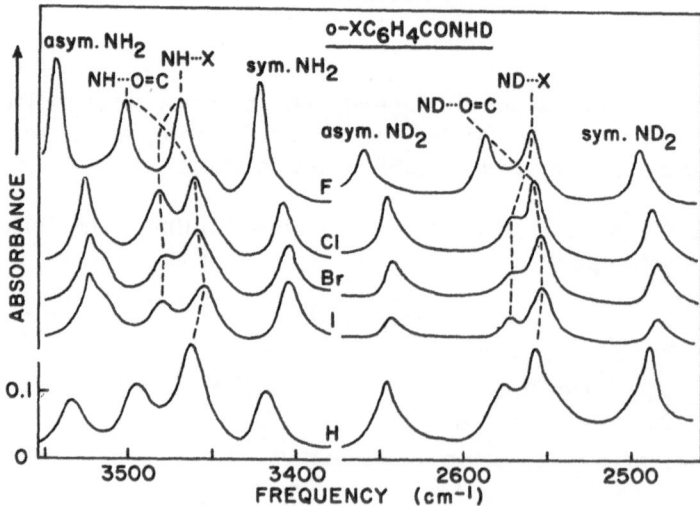

Fig. 2. Partially–deuterated o–halobenzamide spectra relative to partially–deuterated benzamide spectrum, all in dilute CCl_4 solution. Concentrations and level of deuteration vary.

$$[4]$$

Unlike the o-haloanilines where all the NH_2 groups are under the influence of the adjacent halogen atom due to the conjugation of the nitrogen atom with the ring and the energetically-favored pseudoplanarity of the NH_2 group with the ring, in the o-halobenzylamines rotation about the C-C bond allows the NH_2 group to be cis or trans with respect to the X atom. Undeuterated samples therefore show two pairs of NH stretching bands, interpreted as symmetrical and asymmetrical NH_2 bands of the trans and cis isomers [equilibrium (4)]. These measurements, made in tetrachloroethylene solution, are summarized in Table II and Fig. 3. Comparative data on the corresponding p-halobenzylamines are included.

A partially-deuterated o-halobenzylamine would be expected to show the original two pairs of symmetrical and asymmetrical NH_2 bands due to the two residual undeuterated isomers, plus a "free" NH band of the NHD group approximately centered between the high-frequency symmetrical and asymmetrical (NHD group trans re X), plus two further bands straddling the midpoint between the low-frequency symmetrical and asymmetrical components (NHD group cis re X, with further rotational isomerism about the C-N bond leading to "free" NH and NH···X bands). A replica of this seven-band system might be expected in the ND stretching region. Since these bands are rather wide and overlap badly, resolution difficulties precluded accurate measurement of band separations. However, the expected bands are present as can be seen in Fig. 4.

The cis/trans ratio in the o-halobenzylamines, estimated from the relative peak absorbances of the doublet asymmetric NH_2 bands (assuming equal extinction coefficients) is 3.8, 2.0, and 1.3 for X = F, Cl, and Br, respectively. A certain amount of NH···X distance and geometry adjustment is possible in these systems, since rotation about the C-C and C-N bonds is possible, and since a six-membered chelate ring results in a very strong NH···F hydrogen bond despite the small size

TABLE I

Fundamental Stretching Frequencies of NH_2, NHD, and ND_2 Groups in Dilute CCl_4 Solution (in cm^{-1})

Compounds	ν_{as} (NH$_2$)	ν (NHD)			ν_s (NH$_2$)	ν_{as} (ND$_2$)	ν (NDH)			ν_s (ND$_2$)†
		"free"	NH···X	Δ*			"free"	ND···X	Δ'*	
Deuterated anilines:										
o–F	3493.5	3445.5	3439.5	~6 (0)	3404	2612.5	2550.5	2542.5	~8 (0)	2492.5 / 2475.5 wk. / 2460.5
o–Cl	3494	3453.5	3442.5	10,5(11)	3400.5	2616	2549.5	2540.5	9 (0)	2499 / 2480.5 wk / 2464.5
o–Br	3490.5	3450.5	3435	15,5(15)	3395.5	2611.5	2548.5	2536	12,5(12)	2495 / 2461.5
o–I	3485	3450.5	3420.5	30 (29)	3387.5	2606.5	2546	2526.5	19,5(22)	2491 / 2459

Deuterated
benzamides:

Benzamide	3538	3498.5	3467.5	31	3421	2651	2582	2562	20	2493.5
p–F	3538.5	3498.5	3467	31.5	3422	2653	2585	2562	23	2495.5
p–Cl	3537.5	3500	3466.5	33.5	3421	2652.5	2583.5	2562	21.5	2494.5
p–Br	3538	3499	3465	34	3421.5	2649.5	2584.5	2560.5	24	2495
p–I	3538	3497	3466	31	3422	2652.5	2580	2560.5	19.5	2493.5
o–F	3548.5	3506‡	3473	33	3426	2663.5	2591¶	2563.5	27.5	2499
o–Cl	3531	3465‡	3485.5	20.5	3412	2650.5	2562¶	2577	15	2491.5
o–Br	3527	3463‡	3484	21	3408.5	2647.5	2558.5¶	2576.5	18	2488.5
o–I	3527.5	3459.5‡	3483	23.5	3408	2648	2557.5¶	2577	19.5	2487.5

*Values in parentheses from Hambly and O'Grady [15].

†In the deuterated aniline spectra, two bands (and sometimes a third weak band) occur in this region. This has been ascribed to Fermi resonance of ν_s (ND$_2$) with a combination band of the aromatic system [15]. Califano and Moccia [17] suggest the assignment (νC–N+δND$_2$). Only a single band appears in deuterated o– and p– benzamides.

‡Indicates NH···O=C.

¶Indicates ND···O=C.

TABLE II

Fundamental NH$_2$ Stretching Frequencies in
Halobenzylamines in Dilute C$_2$Cl$_4$ Solution (in cm^{-1})

Substituent	ν_{as}		ν_s	
	trans	cis	trans	cis
o-F	3423.5	3398.5	3348	3333
o-Cl	3403	3394	3345	3330
o-Br	3404.5	3391.5	3354.5	3349.5
p-F	3399.5		3331.5	
p-Cl	3401		3331.5	
p-Br	3400.5		3336	

of the fluorine atom. For these compounds, the extinction coefficient of ν_{as} in the cis conformer clearly must exceed that of $\nu_{as,}$ in the trans conformer because of this interaction. This intensification is most marked where the strongest hydrogen bonds are formed (Fig. 3). Unless the steric interaction between the methylene hydrogen atoms and the halogen atom in the trans conformation varies significantly with changes in the size of X in this series, these results show that the NH$\cdots$X hydrogen-bond strength increases with increasing halogen electronegativity.

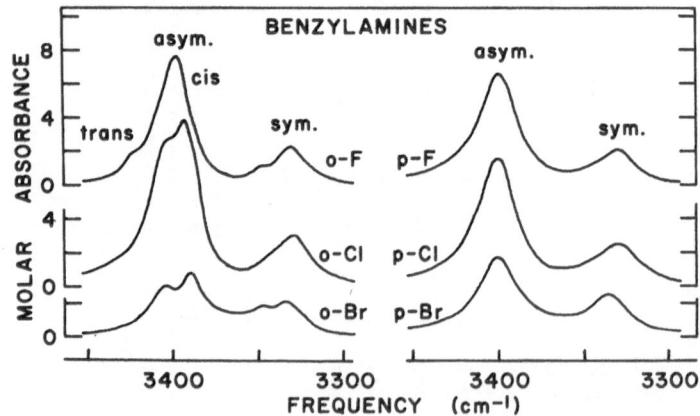

Fig. 3. Halobenzylamine spectra in dilute C$_2$Cl$_4$ solution, showing cis—trans isomerism of the NH$_2$ group in the o-halo compounds.

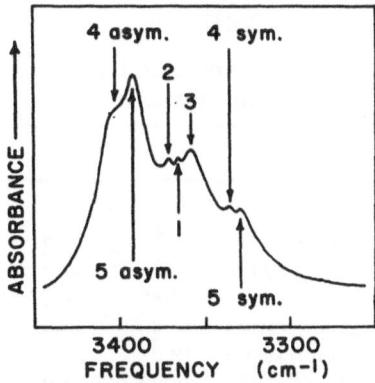

Fig. 4. Spectrum of partially-deuterated o-chlorobenzylamine in dilute C_2Cl_4 solution, showing seven distinct NH stretching bands (apparent maxima at 3403, 3392.5, 3372, 3365.5, 3359, 3336.5, and 3330 cm^{-1}). Bands are numbered to correspond with conformers in equilibria (3) and (4).

For the o-haloanilines the separation of the "free" NH and NH···X bands for the NHD group (Δ in Table I) increases in the order X = F<Cl<Br<I, indicating that the NH···X interaction in the five-membered chelate ring depends primarily on the size of the halogen atom. In the ND stretching region, the corresponding band separation (Δ' in Table I) confirms this conclusion.*

For the mono-deutero-o-halobenzamides, there is no "free" NH band, and the Δ and Δ' values arise because of differences in the NH···O=C$\lesssim$ and NH···X interactions [f_o and f_x, respectively, in equilibrium (2)]. Since both Δ and Δ' for o-fluorobenzamide exceed the corresponding values in benzamide, this can be taken to mean that the N–H bonds in the former compound are more nonequivalent than in the latter, i.e., $f_x \gg f_o$. A comparison of the NH$_2$, NHD, and ND$_2$ frequencies in benzamide and the four p-halobenzamides shows that the electronic effects of the substituent on the NH$_2$ and ND$_2$ frequencies are virtually zero, and very small on the NHD frequencies (affecting these indirectly through the charge on the carbonyl oxygen atom). Since the electronic effects of o- and p-substituents through the aromatic ring are nearly identical, f_o is taken to be roughly constant in the benzamides under consideration. In going

*Hambly and O'Grady [15] have also placed the intramolecular NH···X hydrogen-bond strengths in o-haloanilines in this order on the basis of deviations from the equation $\nu_s = 0.7033 \nu_{as} + 948.2$ cm^{-1}. The extent of splitting of the first overtone of the NH$_2$ symmetric stretching vibration in o-haloanilines further indicates that NH···F interaction is weakest [18].

TABLE III

Comparative Intensities and Band Widths (in cm^{-1}) for
Halobenzamides and Haloanilines in Dilute CCl_4 Solution*

Compound	X =	F		Cl		Br		I
o-$XC_6H_4CONH_2$	A_{as}/A_s	1.4	>	1.3	>	1.2	>	0.9
	$\Delta\nu_{1/2}^a$ asym	8.9	<	10.5	<	12.6	≈	12.5
	$\Delta\nu_{1/2}^a$ sym	7.7	<	10.0	<	10.5	<	13.5
p-$XC_6H_4CONH_2$	A_{as}/A_s	0.80	<	0.88	≈	0.84	<	1.16
	$\Delta\nu_{1/2}^a$ asym	15.9		15.9		14.1		17.1
	$\Delta\nu_{1/2}^a$ sym	12.5		13.5		12.2		11.7
o-$XC_6H_4NH_2$	A_{as}/A_s	0.98	<	1.02	<	1.17	<	1.42
	$\Delta\nu_{1/2}^a$ asym	31.7	>	31.5	>	31	>	25.7
	$\Delta\nu_{1/2}^a$ sym	22.4	>	22	=	22	>	18.9
p-$XC_6H_4NH_2$	A_{as}/A_s	1.01		0.83		0.80		0.85
	$\Delta\nu_{1/2}^a$ asym	44.5	>	38.5	=	38.5	>	33.9
	$\Delta\nu_{1/2}^a$ sym	31.5	>	26.5	>	25.5	>	23.1

*Aniline data taken from [1]. See experimental section for details concerning
determination of integrated intensities and half-band widths for amides.

from o-fluorobenzamide to o-chlorobenzamide, both Δ and Δ'
decrease by 12.5 cm^{-1} to values significantly lower than the
corresponding ones in benzamide; and, as the halogen is suc-
cessively changed to Br and I, the separations begin to increase
again. These data can be accommodated by the suggestion
that $f_x > f_o$ for o-fluorobenzamide, but that $f_x < f_o$ for o-chlorobenzamide, and that f_x falls still further for Br and I (the
greatest drop in electronegativity occurs between F and Cl).
This permits Δ and Δ' to pass through minima. The lowest
frequency NH band of the NHD group in o-fluorobenzamide is
therefore assigned to NH···F, whereas it is assigned to
NH···O=C< in the other three halobenzamides (similarly for
the ND bands, see Table I). Although some small ambiguity
remains in this treatment, since the NH frequency of the NHD
group may be affected by the ND···X or ND···O=C< hydrogen
bond (and vice versa),* the order of intramolecular hydrogen
bonding to covalently-bonded halogen atoms also appears to
follow the electronegativity sequence in this system.

*For example, Moritz [11] has discussed the raising of the "free" NH frequency in
mono-deutero-o-nitroaniline as a consequence of delocalization of electrons away
from the nitrogen atom by the ND···O_2N bonding, although recently Nyquist [19]
has considered an increase in the electron density on the nitrogen atom due to
intramolecular NH···X hydrogen bonding.

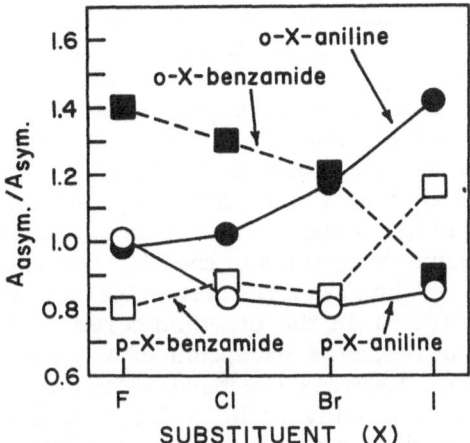

Fig. 5. Effect of orientation of substituent on the relative asymmetric and symmetric NH_2 intensities in haloanilines and halobenzamides.

The integrated NH_2 band intensities (A_s and A_{as}) provide further evidence on the order of NH···X hydrogen–bond strengths in o–haloanilines and o–halobenzamides. According to the vibrational mechanism previously developed [1], a relative increase in A_{as} may be expected on intramolecular hydrogen–bond formation of an ortho–substituent to an NH_2 group. The ortho–substituent is more nearly in line with the direction of the transition moment in ν_{as} than in ν_s, and hence the former mode is intensified. Table III gives the (A_{as}/A_s) ratios for o–haloanilines and o–halobenzamides, together with the reference data for the corresponding p–compounds. For the o–haloanilines, (A_{as}/A_s) increases markedly in the order X=F<Cl <Br<I, whereas a much smaller effect in the opposite direction is observed for p–haloanilines (Fig. 5). Since the latter is indicative of the electronic effect of the substituents through the aromatic ring, the change in the ratio for o–haloanilines must mean that the NH···X interaction increases in the order X=F<Cl<Br<I. The NH···F interaction is very weak since the (A_{as}/A_s) ratios for o– and p–fluoroaniline are identical.

For o–halobenzamides the (A_{as}/A_s) ratio increases in the order X=I<Br<Cl<F. A much smaller effect in the opposite direction is observed for the p–halobenzamides (Fig. 5). These results are consistent with increasing NH···X hydrogen–bond strength in the order X=I<Br<Cl<F in the benzamide series.

Since the intramolecular NH$\cdots$X hydrogen bonds are relatively weak in the five-membered chelate ring formed in o-haloanilines, the $\Delta\nu_{1/2}^a$ values for both modes follow the same trend as in the p-haloanilines; i.e., the electronic effect of the halogen through the aromatic bond system dominates, and the widths decrease in the order X = F>Cl>Br>I. However, the o-haloaniline band widths are uniformly smaller than the corresponding widths in the p-haloanilines, and this may, in part, be due to some restriction of the NH$_2$ motion on hydrogen bonding. This is much more noticeable for the stronger NH$\cdots$X bonds formed in the six-membered chelate ring in o-halobenzamides (with the exception of $\Delta\nu_{1/2}^a$ for ν_s in o- and p-iodobenzamide). Whereas the band widths for the p-halobenzamides vary relatively little, a large regular decrease in $\Delta\nu_{1/2}^a$ for both modes is noted for the o-halobenzamides in the order X = I>Br>Cl>F, which again corresponds to band narrowing as the NH$\cdots$X interaction increases.

Although it would be of interest to deduce the relative NH$\cdots$X and ND$\cdots$X hydrogen-bond strengths in these systems, it is doubtful whether this is possible from these data alone. As Moritz has shown [12], the relative frequency shifts $(\Delta/\nu_{NH \text{"free"}})$ and $(\Delta'/\nu_{ND \text{"free"}})$ are equal within experimental error for a substantial range of Δ and Δ', although there is a suggestion from our work that the relative displacement for the N-D bond is slightly smaller in the o-haloanilines. In any event, the frequency shift is not necessarily a valid direct measure of the strength of the hydrogen bond. Over the range 15—70° Moritz also found no temperature dependence in the relative intensities of the NHD bands in o-nitroaniline in CCl$_4$, which suggests similar thermodynamic stability for the two isomers. No consistent conclusions can be drawn from the relative intensities of the NH and ND bands of the NHD group in the compounds investigated because of the enhancement of the extinction coefficient on hydrogen-bond formation. Thus, for deuterated benzamide and the p-halo derivatives, the NH$\cdots$O$=$C$<$ and ND$\cdots$O$=$C$<$ bands of the NHD group, respectively, are more intense than the accompanying NH "free" and ND "free" bands, with relatively little change in the ratio. On the one hand, this suggests that more NH is hydrogen-bonded, and on the other hand, that more ND is bonded to the carbonyl (or that both are hydrogen-bonded to the same extent, with intensification of the absorption of the bonded group).

TABLE IV

Intensity Ratio for NHD Bands (NH···O=C$\lneq$/NH···X)

	Protium	Deuterium
o-fluorobenzamide	0.98	0.90
o-chlorobenzamide	1.17	1.85
o-bromobenzamide	1.61	2.32
o-iodobenzamide	(1.5)	(2.3)
benzamide*	1.55	1.57

*Here the comparable ratio is (NH···O=C$\lneq$/NH "free").

For the o-halobenzamides, with both the proton and the deuteron of the NHD group bonded at all times, the situation is even more complex. All that can be said is that the relative intensities appear to be consistent with the band assignments made, namely that the intensity of the NH···F band exceeds that of the NH···O=C$\lneq$ band (and the ND···F band that of the ND···O=C$\lneq$ band) in o-fluorobenzamide ($f_x > f_o$), whereas in the other three o-halobenzamides ($f_x < f_o$) the relative intensities are reversed. The trends in the ratios on the basis of this assignment are also fairly consistent (Table IV). The increase in these ratios down the columns may give an indication of the decreasing NH···X and ND···X interaction.

In the deuterated o-haloanilines where the NH···X interaction is much weaker than in the benzamides, the hydrogen bonds appear to be stronger than the corresponding deuterium bonds. For o-Cl, o-Br, and o-I substituents, the (ND···X/ND "free") absorbance ratio increases in the order Cl<Br<I, with all values well below unity. Since enhancement of the ND···X band, because of hydrogen bonding, means that the real ratios would be even lower than the apparent ratios, this can only be interpreted to mean that the isomer with the ND "free" bond is favored over the isomer with the ND···X bond. Confirmation of this in the NH stretching region is relatively ambiguous since, although the (NH···X/NH "free") ratio exceeds unity for o-Cl and o-Br substituents and is approximately unity for o-I, the numerator would be enhanced by hydrogen bonding. For o-fluoroaniline, the results are inconclusive because the NHD bands are very close together, and there may be component band width variations which give a misleading contour to the unresolved doublets. The apparent weakening of the intra-

molecular hydrogen bond to halogen atoms on deuterium substitution can be accounted for in terms of larger vibrational amplitudes for the NH bond and a longer equilibrium bond length, both of which would favor NH$\cdots$X interaction over ND$\cdots$X interaction in a rigid intramolecular system involving a five-membered chelate ring.

On the basis of relative frequency shifts of the NHD group in o-nitroaniline, Bagratishvili et al. [13] have concluded that the intramolecular NH$\cdots$O hydrogen-bond strength is about 0.6 kcal/mole, whereas that of the ND$\cdots$O hydrogen bond is only 0.4 kcal/mole.

ACKNOWLEDGMENTS

The financial support of the National Research Council of Canada is gratefully acknowledged. Miss Myra Petts assisted with some of the experimental work.

REFERENCES

1. P. J. Krueger, Can. J. Chem., 40, 2300 (1962).
2. H. H. Jaffe, J. Am. Chem. Soc., 79, 2373 (1957).
3. P. von R. Schleyer and R. West, J. Am. Chem. Soc., 81, 3164 (1959).
4. R. West, D. L. Powell, L. S. Whatley, M. K. T. Lee, and P. von R. Schleyer, J. Am. Chem. Soc., 84, 3221 (1962).
5. P. J. Krueger and H. D. Mettee, Can. J. Chem., 42, 326 (1964).
6. O. R. Wulf, V. Liddel, and S. B. Hendricks, J. Am. Chem. Soc., 58, 2287 (1936).
7. L. Pauling, J. Am. Chem. Soc., 58, 94 (1936).
8. A. W. Baker and W. W. Kaeding, J. Am. Chem. Soc., 81, 5904 (1959).
9. H. Bourassa-Bataille, P. Sauvageau, and C. Sandorfy, Can J. Chem., 41, 2240 (1963).
10. E. A. Allan and L. W. Reeves, J. Phys. Chem., 67, 591 (1963).
11. A. G. Moritz, Spectrochim. Acta, 16, 1176 (1960).
12. A. G. Moritz, Spectrochim. Acta, 18, 671 (1962).
13. G. D. Bagratishvili, G. V. Tsitsishvili, and K. A. Bezhashvili, Zhur. Fiz. Khim., 36, 2036 (1962).
14. A. N. Hambly and B. V. O'Grady, Chem. and Ind. (London) 15, 459 (1962).
15. A. N. Hambly and B. V. O'Grady, Australian J. Chem., 15, 626 (1962).
16. P. J. Krueger, Appl. Opt., 1, 443 (1962).
17. S. Califano and R. Moccia, Gazz. chim. ital., 87, 805 (1957).
18. P. J. Krueger, Can. J. Chem., 42, 201 (1964).
19. R. A. Nyquist, Spectrochim. Acta, 19, 1595 (1963).

Ultraviolet and Visible Spectroscopy

A New Molybdenum-Blue Method for Silicon in Steel

Uno T. Hill

Inland Steel Company
East Chicago, Indiana

A new rapid spectrophotometric molybdenum-blue method accurate to ±0.001% silicon in steel is described. A single determination may be carried out in fifteen minutes or less, directly in the spectrophotometric tube, and the method is suited for the routine analysis of a large number of samples.

The steel sample is dissolved in nitric acid in a calibrated dissolving flask. The sample is diluted to volume and an aliquot is placed in a photometric cuvette. Two colorimetric reagents are added, and the optical density is determined against a blank after six minutes developing time.

INTRODUCTION

Samples of steel are frequently submitted to the steel mill laboratory in physical forms or in concentrations of elements which are not suitable for spectrographic analysis. For such samples, other methods of analysis must be employed. In a previous paper from this laboratory, a photometric method for silicon in steel based on the yellow silico-molybdate complex was reported. Recent metallurgical requirements have increased the need for a more sensitive method in the lower concentration ranges capable of about ±0.001% accuracy. In the previous paper, it was indicated that the yellow silico-molybdate color may be reduced to the blue form if the ferric iron is complexed with a fluoride in the presence of ferrous sulfate and sulfuric acid [1]. The molybdenum-blue method offers advantages over the previous method in color stability, speed, and accuracy. Because of the slow rate of dissolution of some steels in sulfuric acid, it was decided to investigate the possibility of employing nitric acid to speed up the dissolution of the sample.

215

EXPERIMENTAL

Equipment Used

The Bausch & Lomb Spectronic 20 spectrophotometer was employed in the routine determinations of silicon in steel. The absorption spectra were determined with the Beckman DU spectrophotometer. Both the blue- and the red-sensitive photocells and the $1/2$- and 1-in. cuvettes were used on the B & L Spectronic 20 for routine analysis.

Color Formation

It was anticipated that the nitrate ion introduced in the dissolving acid would cause blocking of color formation and instability of the color in a reduced solution, and methods for its removal were at first investigated. It was found, however, that excellent color stabilities were possible with sample dissolved in nitric acid, and, for this reason, its removal was not necessary.

When 3N HNO_3 alone was employed to dissolve the steel sample, it was found that the method was limited for the determination of silicon to a maximum of about 0.5%. The 3N HNO_3 used in dissolving the sample at boiling temperatures polymerized silicic acid at concentrations higher than 0.5%, and only a fraction of the theoretically possible silico-molybdate complex formed on addition of ammonium molybdate. The polymerization of silicic acid was prevented by reducing the concentration of nitric acid by the substitution of sulfuric acid. The use of sulfuric acid alone is not practical, since it may require several hours to dissolve some samples of steel with it. Only a small amount of nitric acid is necessary in combination with sulfuric acid to dissolve most steel samples in a few minutes.

Absorption Spectra

The absorption spectra of the silico-molybdenum-blue complex in a steel sample are shown in Fig. 1 at 27 and 31°C. It may be seen that the complex is temperature-sensitive at wavelengths below 950 μ. At 830 μ the loss in absorptivity is 55% for a 10°C increase in temperature from 27 to 37°C. The absorptivity at these lower wavelengths is restored, however, when the sample is cooled to its original temperature. At wave-

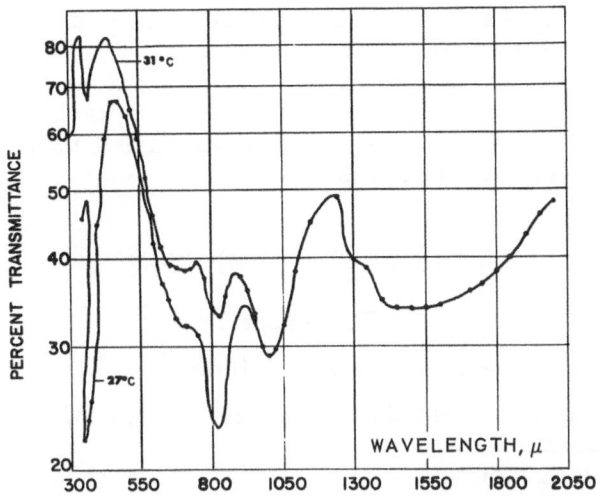

Fig. 1. Absorption spectra of molybdenum-blue silicon complex at 27 and 31°C.

lengths of 950 μ and higher, the effect of temperature on absorptivity is negligible in the usual range of room temperature fluctuations.

Calibration Curves

Calibration curves may be constructed anywhere between wavelengths shown on the absorption spectra. In practice, only the wavelengths at 600 and 950 μ were used. The background absorption due to ferrous iron increases from 98% transmittancy at 600 μ to about 70% at 970 μ. Due to this background color, it may be more desirable to measure the color at 600 μ, in spite of the changes in absorptivity with temperature. If constant temperature cannot be maintained, small changes in the wavelength setting against a silicon standard may be made in order to retain the photometric readings on an established calibration curve. The absorptivity of the silicomolybdate complex is also a function of the ferrous sulfate concentration. The absorptivity increases when the concentration of ferrous sulfate is increased between 10—64 g/liter. If the complex is measured at 950 μ, it is advisable to maintain the low concentration of ferrous sulfate because of its contribution to the background color. At 600 μ, the absorbance of ferrous sulfate is negligible, and color measurements may be made against water or a reagent blank.

Sample sizes may be varied from 0.10 g to as high as 1 g. At the 1-g level, a slight turbidity may form if the developed color is allowed to age; thus, color measurements must be made within fifteen minutes or less.

MATERIALS AND METHOD

Equipment

The equipment required is as follows:

1. 250-ml, wide-mouth Erlenmeyer flask with a 250-ml calibration mark.
2. 50-ml graduate.
3. 2—5 ml pipets.
4. 1—5 ml plastic (Nalgene) pipet.
5. 1—2 ml pipet.
6. spectrophotometer.

Reagents

The reagents required are as follows:

1. 3N nitric—sulfuric acid. Dilute 41 ml of HNO_3, sp. gr. 1.42, and 66.7 ml H_2SO_4, sp. gr. 1.84, to 1 liter.
2. Ammonium molybdate 6.0%. Dissolve 60 g ammonium molybdate in water and dilute to 1 liter.
3. Ferrous sulfate—sodium fluoride—sulfuric acid solution. Combine 32 g $FeSO_4 \cdot 7H_2O$ and 7.68 g NaF in a plastic container. Then, add 432 ml 10N H_2SO_4 and dilute to 1 liter. Store in the plastic container.

Procedure

To a 0.2500-g sample in a 250-ml Erlenmeyer flask, add 50 ml 3N nitric—sulfuric acid solution (1., above), and place on hot plate. Boil out the oxides of nitrogen, cool and dilute to 250 ml, and mix. Pipet two 2-ml aliquots into two, separate, 1-in. colorimeter tubes, one aliquot to serve as a blank. To the sample aliquot, add 2 ml ammonium molybdate solution (2., above), and, to the blank, add 5 ml ferrous sulfate—sodium fluoride—sulfuric acid solution (3., above).

After 7 min or longer, add rapidly 5 ml ferrous sulfate—sodium fluoride—sulfuric acid solution to the sample. This solution should be added rapidly from an automatic pipet and mixed immediately. To the blank, add 2 ml ammonium molybdate and mix.

Obtain the percent transmittancy reading on the instrument at 600 or 950 μ. From a previously prepared calibration curve based on National Bureau of Standards steel samples, obtain the percent silicon. Only one blank is needed for a series of determinations of low-alloy and carbon steel of the same specification.

DISCUSSION

It will be noted that the brownish coloration due to carbides in the sample solution has not been destroyed in the method. Since all of the components present in the sample aliquot are also present in the blank, the destruction of the color contributed by carbon is not necessary. In order to speed up the analysis of a large number of samples, it may be time-saving to limit the number of blanks. In this event, the carbon color is best destroyed to equalize the background absorption by the addition of an oxidizing agent, such as ammonium persulfate or potassium permanganate. Of the two, potassium permanganate is preferred, since less time is taken on the hot plate, minimizing the danger of silicic acid polymerization. Generally, a 2-ml addition of a saturated potassium permanganate solution is sufficient, the excess being removed by the dropwise addition of a 10% sodium nitrite solution after a brief boiling period.

If ammonium persulfate is used to oxidize the carbon color, 5 ml of a 5% ammonium persulfate solution is added, followed by a 4-min boiling period to make certain all of the excess persulfate is destroyed.

RESULTS

Table I shows the results of silicon determination on a wide variety of steels and cast irons. It may be seen that the agreement between the method and the Bureau of Standards values in the lower concentration ranges is within $\pm 0.001\%$.

TABLE I

Analysis of Standard Samples

NBS standard sample		Silicon present, %	Silicon found, %	Difference, %
55 e	Ingot Iron	0.001	0.001	0.000
8 i	Bessemer	0.020	0.021	+0.001
12 g	Basic Open Hearth	0.187	0.187	0.000
14 d	Basic Open Hearth	0.126	0.126	0.000
19 f	Acid Open Hearth	0.204	0.204	0.000
20 f	Acid Open Hearth	0.299	0.297	−0.002
32 d	Nickel Chromium	0.301	0.302	+0.001
65 d	Basic Electric	0.370	0.370	0.000
72 f	Chromium Molybdenum	0.256	0.258	+0.002
73 b	Stainless 13% Chromium	0.437	0.440	+0.003
125 A	Silicon Steel	3.32	3.36	+0.040
4 i	Cast Iron	1.45	1.45	0.000
5 k	Cast Iron	2.08	2.05	−0.03
7 g	Cast Iron	2.41	2.40	−0.01
	Inland High Silicon Cast Iron	4.50	4.50	0.00

The method has been used in routine analysis for a number of years with excellent agreement with spectrographic and gravimetric methods of analysis.

REFERENCE

1. U. T. Hill, Anal. Chem. 21, 589 (1949).

Theory and Applications of Diffuse Reflectance Spectroscopy

Audrey L. Companion

Illinois Institute of Technology
Chicago, Illinois

At present there are at least 100 papers in the literature in which the diffuse reflectance method is used without question to obtain approximate absorption spectra of solids in the visible and near UV, or in which attempts are made to overcome the recognized limitations of the method, namely, that diffuse reflectance spectra are not always identical to transmission spectra. Some of the theories proposed to account for the difference will be discussed, those which purportedly permit one to compute absorption coefficients from diffuse reflectance data: the Kubelka—Munk dilution model, Johnson's plate model, and the Melamed model.

INTRODUCTION

The number of papers in the literature dealing with diffuse reflectance measurements is rapidly increasing as more and more investigators (in particular inorganic chemists) employ this technique to obtain an approximate visible-ultraviolet absorption spectrum of a powdered solid. The governing principle of the method is that, of the monochromatic light impinging on the surface of the powder, some is absorbed, some undergoes specular or mirror reflection at the surface, and some is scattered and after multiple reflections returns to the surface. The total reflected light is usually measured relative to a nonabsorbing standard (MgO, MgCO$_3$, CaF$_2$, NaF, BaSO$_4$) with one of the two main types of spectrophotometer reflectance attachments illustrated in Figs. 1 and 2.

With the ring attachment, monochromatic light emerging from the slit is directed by a plane mirror to the surface of the solid, and diffusely-reflected light between angles of

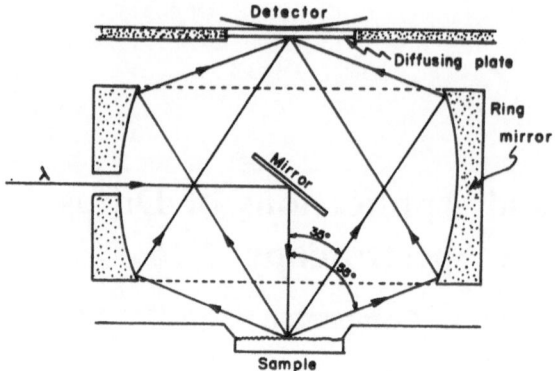

Fig. 1. Schematic diagram of ring-type reflectance attachment.

approximately 35–55° is gathered by an elliptical ring mirror and focused onto a quartz diffusing plate in front of the detector. Separate measurements must be made with sample and standard in position.

The measurements described in this paper were performed with the ring attachment to the Cary Model 14 PMR spectrophotometer.

In the integrating sphere method, chopped monochromatic light falls alternately on sample and standard, is diffusely reflected by the MgO coating on the inner surface of the sphere, and eventually reaches the detector. For nonpowdered samples or standards (such as vitrolite glass), the specular component of reflection can be removed by placing an absorbing light trap in the position indicated.

Sometimes a simple plot of optical density (where

$$OD = \log[R_{(MgO)}/R_{(sample)}]$$

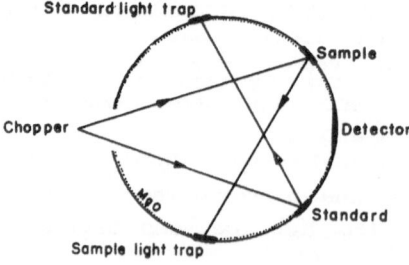

Fig. 2. Schematic diagram of integrating sphere .

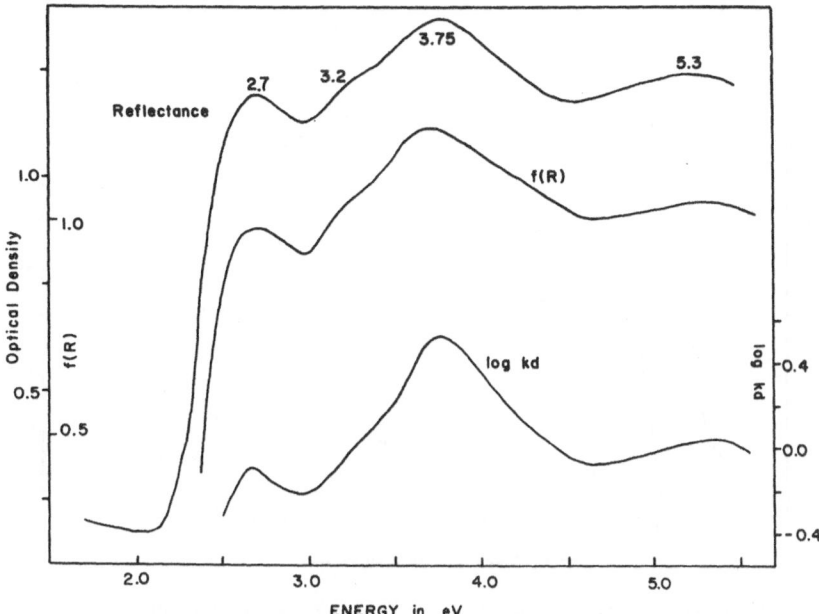

Fig. 3. Top—reflectance spectrum of bulk V_2O_5 plotted as optical density; center—Kubelka—Munk remission function $f(R)$ for bulk V_2O_5; bottom—log (kd) for bulk V_2O_5 as computed from the Melamed model.

and R is diffuse reflectance) directly yields positions of absorption peaks in excellent agreement with those obtained by transmission measurements. For example, our reflectance study of bulk NiO powder indicated maxima at 1.74, 1.95, 2.15, 2.65, 3.00, and 3.29 eV before a cutoff at 3.64 eV, while the transmission measurements of Newman and Chrenko [1] gave peaks at 1.13, 1.75, 2.15, 2.75, 2.95, 3.25, and 3.50 eV before a cutoff at 4.0 eV. Unfortunately though, the reflectance peaks are more often shifted in energy and distorted. For example, Fig. 3 shows the diffuse reflectance spectrum of bulk V_2O_5 (plotted as *OD*), while Fig. 4 shows the transmission spectrum of a polycrystalline sandwich of V_2O_5 between two thin quartz plates.* The two curves are somewhat similar in shape, but the

*Both samples were prepared from recrystallized ammonium vanadate decomposed in oxygen, and both have the same X-ray diffraction pattern. For the sandwich, V_2O_5 was slowly fused between the quartz plates, yielding a layer of nonhomogeneous thickness. Consequently, the *OD* curves obtained with tungsten and hydrogen sources do not meet, since the beams differ slightly in size and shape, illuminating different areas of the sandwich. Since we were interested only in peak position, no attempt was made to control the area illuminated.

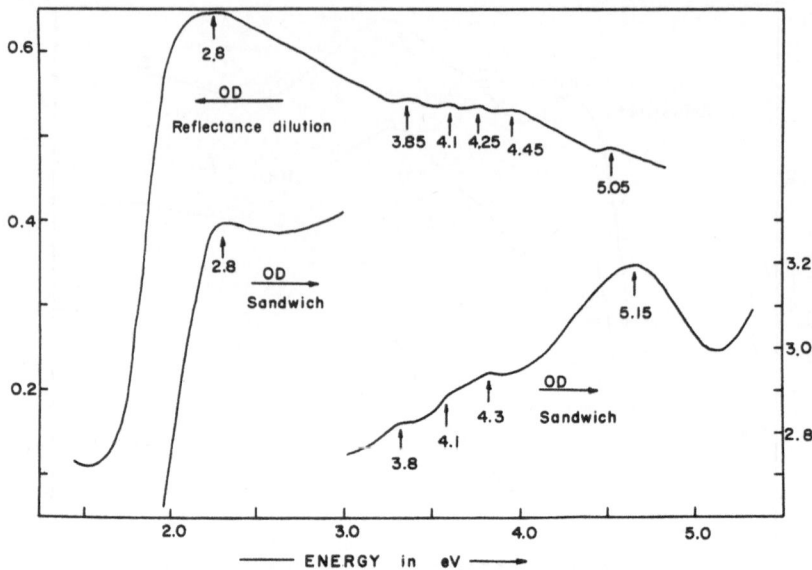

Fig. 4. Top—reflectance spectrum of a 5.7 mol. % V_2O_5 dilution in a MgO matrix; bottom—transmission spectra of polycrystalline V_2O_5 sandwich.

shoulder observed at 3.2 eV in the reflectance spectrum is absent in the sandwich, and other peaks are shifted considerably with respect to one another.

In an attempt toward explanation of those differences in the spectra of V_2O_5 and other compounds, we examined three theoretical models with which purportedly one may compute absolute absorption coefficients from diffuse reflectance data. The main purpose of this paper is to survey these and judge their effectiveness.

Before examining these, we will define symbols of frequent appearance in the discussion to follow. The relative diffuse reflectance R is the ratio of the reflected radiance J of a powdered sample to that of a perfectly-reflecting, perfectly-diffusing standard examined under the same conditions.

$$R = J_{(\text{sample})}/J_{(\text{standard})} \tag{1}$$

The surface or specular reflectivity r is the ratio of reflected to incident radiation I_0 at a particular surface, which, for metallic reflection at normal incidence, is related by Fresnel's formula to the real part of the index of refraction n and the absorption index κ_0.

$$r = J/I_0 = \frac{(n-1)^2 + \kappa_0^2}{(n+1)^2 + \kappa_0^2} \qquad (2)$$

For dielectrics, κ_0 is zero. The absorption index in turn is related to the absorption coefficient k as follows:

$$k = 4\pi \kappa_0 / \lambda \qquad (3)$$

where λ is the wavelength of the light and k itself is defined through the exponential decrease in intensity of light transmitted through an absorbing layer of thickness d.

$$I = I_0 e^{-kd} \cdot \qquad (4)$$

For a more complete discussion of reflection notations see Gibson [2].

THE JOHNSON MODEL

In the reflectance model proposed by P. D. Johnson [3], the powdered sample is approximated by a stack of parallel plates with thickness of the particle size d and characterized by a constant index of refraction and surface reflectivity (Fig. 5). All light incident on the plates is assumed reflected or absorbed. The reflected light is the sum of rays passing through p particles (where p varies from 0 to ∞) by means of reflection or refraction at the crystal—air interfaces. The r value recommended is 1.5 times that computed from Fresnel's formula for normal incidence for a dielectric, 1.5 being an arbitrary correction for incident angles other than normal.

Multiple reflections and loss through scattering are accounted for semiempirically by means of an adjustable parameter y in the expression for the sum of all reflected rays

$$I = I_0 r + 2r I_0 \sum_{p=1}^{\infty} (1-r)^{yp} e^{-2pkd} \qquad (5)$$

Since the above summation may be expressed as a geometric

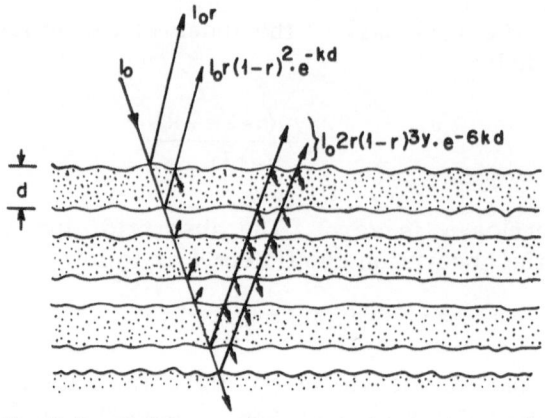

Fig. 5. Parallel plate reflectance model of Johnson [3].

series, equation (5) may be written as

$$I = I_0 \left\{ 2r \left[\frac{e^{y \ln(1-r) - 2kd}}{1 - e^{y \ln(1-r) - 2kd}} \right] + r \right\} \tag{6}$$

With the definition $R = I/I_0$ (an absolute reflectance) and expansion of the exponential term, we have Johnson's equation relating reflectance to particle size and absorption coefficient:

$$R = r \left[\frac{2(1-r)^y \, e^{-kd}}{2kd - y \ln(1-r)} + 1 \right] \tag{7}$$

Figure 5 predicts that even with no multiple reflection or scattering within a layer, y should be 4. However, even for a nonabsorber $(k = 0)$, such as MgO, values of R computed with equation (7) are much less than one when $y = 4$. Johnson suggests a choice of y such that $R = 1$ when $k = 0$ (through the equation obtained from equation (7) under these conditions),

$$\frac{(1-r)^{y-1}}{y} = \frac{1}{2} + \frac{1}{4}r + \frac{1}{6}r^2 + \frac{1}{8}r^3 + \dots \tag{8}$$

and suggests that, for refractive indices under 1.5, a value of $y = 2$ is satisfactory. The smaller y artificially decreases loss due to multiple reflections and scattering.

Thus, for calculation of k, knowledge of R, d, and n is necessary, where n is assumed constant over the entire wavelength range considered.

Johnson's calculations show that the reflectance as computed by equation (7) is quite insensitive to changes in k for large kd (due either to strong absorption or large particle size) and that R converges to the surface reflectivity r as k approaches infinity. On the other hand, for low values of k and consequently high values of R, equation (7) is unsatisfactory because of the importance of multiple reflections and the arbitrary fit of the parameter y. Johnson feels that his equation is most acceptable for reflectance values lying between 0.1 and 0.8.

In our application of equation (7) to bulk V_2O_5, we assumed an index of refraction of approximately 2.4 with a corresponding dielectric reflectivity r of 0.17. According to Johnson's model, for this value of r, no OD above 0.59 should be observed. Since most of the structure in our reflectance curve occurs above this OD, the Johnson y correction for scattering and multiple reflection loss was judged inadequate. No further calculations were performed with equation (7) after the more-promising Melamed model appeared early in 1963, a model which explicitly includes multiple reflection.

THE MELAMED MODEL

Melamed [4] considers an assembly of particles of uniform size but arbitrary shape, represented in Fig. 6 as spheres of diameter d, and sums in a manner similar to that of Johnson rays transmitted, reflected, and scattered through the assembly. His final equation* relates diffuse reflectance R, an averaged reflection coefficient for externally incident radiation r, the particle transmission T, and a probability factor x that light emerging from a particle within the assembly is scattered toward a particle nearer the surface.

$$R = 2xr + \frac{x(1-2xr)\,T\,(1-rR)}{(1-rR)-(1-x)(1-r)\,TR} \tag{9}$$

---

*Unfortunately, the final equation in the paper and some of the statements leading to it contain typographical errors.

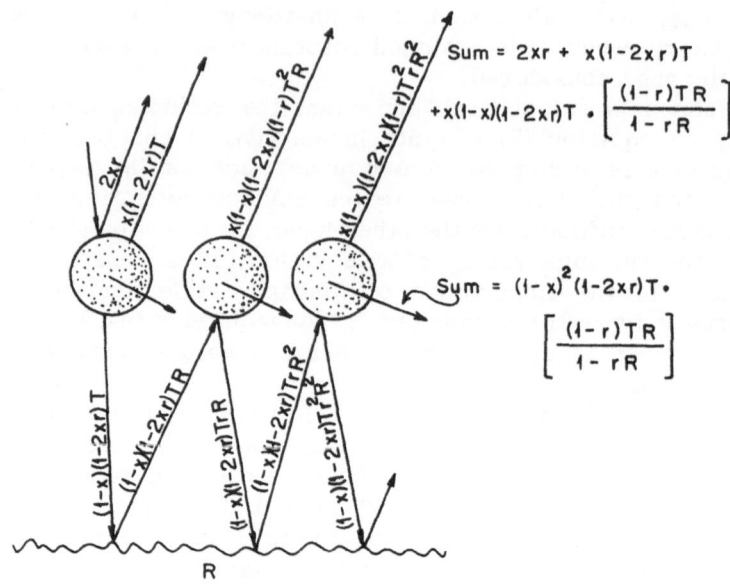

Fig. 6. Reflectance model of Melamed [4].

T itself depends on an internal reflection coefficient r_i and a quantity M, a function of kd, representing the total radiation reaching the surface of a particle after one pass through it:

$$T = \frac{(1 - r_i) M}{(1 - r_i\, M)} \tag{10}$$

$$M = \frac{2}{(kd)^2} \left[1 - (1 + kd)\, e^{-kd} \right] \tag{11}$$

External and internal reflection coefficients are computed using a combination of Lambert's cosine law and the Fresnel relation for specular reflection, and, from the graphs shown in the paper, these coefficients are obviously not functions of κ_0. The probability factor x is rationalized as a function of the particle's transmission T and a constant upwards-scattering factor x_u dependent upon particle shape. For an assembly of close-packed spheres, $x_u = 0.284$.

Unlike Johnson, Melamed claims summation over all rays in the assembly. For $kd = 0$, one can easily show that $R = 1$.

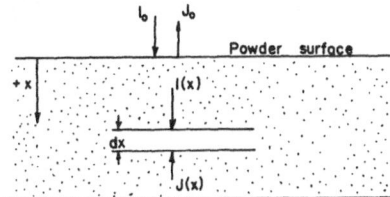

Fig. 7. Kubelka—Munk reflectance model, as described by Schreyer [5].

In our application of the Melamed equations to V_2O_5, we assumed a constant n of 2.4, and interpolated tables provided privately by the author were used to obtain external and internal reflection coefficients. The remaining calculations were programmed for the Univac 1105 computer to yield kd values corresponding to the R values of our experimental reflectance curve. Again, to account for the high OD regions, it was necessary to reduce the x_u factor to approximately 0.1, indicating that, due to the random shapes of our assembly of crystallites, considerably more shadowing occurs than in an assembly of close-packed spheres. Figure 3 shows the curve resulting from the calculations. No explanation for the distortion observed appears here.

THE KUBELKA—MUNK THEORY

The Kubelka—Munk model for diffuse reflectance is discussed by Schreyer [5] in the paper in which he introduces his own adaption of the original theory. We consider a powder slice of thickness dx lying x units below the surface, illuminated from above by light of intensity $I(x)$, which is less than the incident intensity I_0 (Fig. 7). The particles in the slice are assumed to be of equal dimensions and optically homogeneous. The radiation penetrating the slice is diminished in intensity due to absorption $kIdx$ and scattering $SIdx$, where S is a scattering coefficient, assumed to be wavelength-independent. The radiation returning to the surface, of intensity $J(x)$, suffers a similar loss on passing through the slice. Specular reflectance is assumed negligible. If, as in the original presentation, only two directions are considered (the positive and negative perpendicular to the surface), then the scattered light lost by $I(x)$ contributes to $J(x)$ and vice versa, and two differential equations result:

$$dI = -(k + S)\,Idx + SJdx \qquad (12a)$$

$$dJ = +(k + S) J dx - SI dx \qquad (12b)$$

These have the general solutions

$$I = A(1-b) e^{ax} + B(1+b) e^{-ax} \qquad (13a)$$

$$J = A(1+b) e^{ax} + B(1-b) e^{-ax} \qquad (13b)$$

where $a = \sqrt{k(k+2S)}$ and $b = \sqrt{k/(k+2S)}$. A and B are constants determined through the boundary conditions $I = I_0$ for $x = 0$ and $I = J = 0$ for infinite x. Thus,

$$A = 0 \qquad (14a)$$

$$B = I_0/(1+b) \qquad (14b)$$

$$I - I_0 c^{-ax} \qquad (14c)$$

$$J = \frac{(1-b)}{(1+b)} I_0 e^{-ax} \qquad (14d)$$

At the surface, where $x = 0$ and $I = I_0$, J is given by

$$J_{x=0} = \frac{(1-b)}{(1+b)} I_0 = \left(\frac{k+S - \sqrt{k^2 + 2kS}}{S} \right) I_0 \qquad (15)$$

For a nonabsorbing substance, i.e., the standard, $k(\lambda) = 0$ and $J_{x=0, k=0} = I_0$. Since the diffuse reflectance is the ratio of returned intensities of sample and standard for the same irradiating conditions,

$$R = \frac{J_{x=0} (\text{sample})}{J_{x=0, \ k=0}} = \frac{k+S - \sqrt{k^2 + 2kS}}{S} \qquad (16)$$

where k and S refer to the sample. By definition, the Kubelka — Munk remission function $f(R)$ is the ratio of absorption to scattering coefficients k/S, and by rearrangement of equation (16) can be expressed in terms of the measured diffuse reflectance R.

$$f(r) = k/S = (R-1)^2/2R \qquad (17)$$

As calculated from equation (17), $f(R)$ should vary directly as the absorption coefficient k, provided that S is indeed independent of wavelength. Figure 3 shows the results obtained for

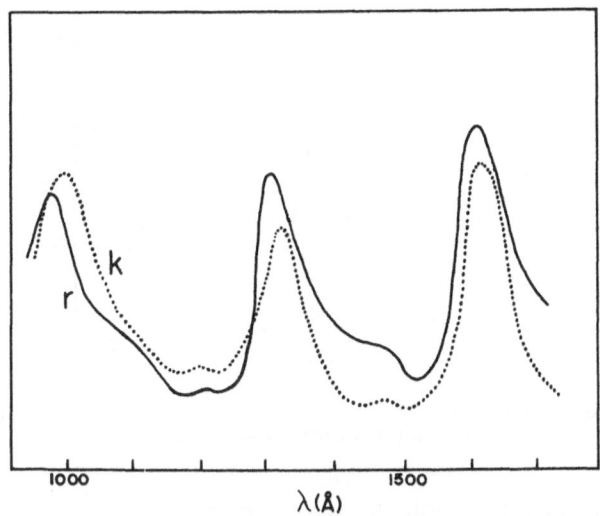

Fig. 8. Absorption and reflectivity of KCl, after Hartman, Nelson, and Siegfried [6].

bulk V_2O_5. Once again, it is disappointing that no explanation of distortion or peak shift is apparent.

VARIATION OF SPECULAR REFLECTIVITY

We believe that the failure of all these models proposed thus far may be attributed to the assumption that the index of refraction, and thus specular reflectivity, are constant for all wavelengths. That this assumption is improper may be illustrated first by the work of Hartman, Nelson, and Siegfried [6], who examined the thin-film absorption and single-crystal reflectivity of KCl (Fig. 8). Note that the measured reflectivity has distinct maxima in the vicinity of absorption maxima, an effect bound to distort the position of a diffuse reflectance peak, since r contributes to the reflected intensity most when k is subtracting from it.

The variation in n, κ, and r for a BaO crystal is illustrated by the work of Jahoda [7], who, with measurements of single-crystal reflectivity, was able to compute by a network analysis the simultaneous changes in n and κ. Note that all three of these optical constants vary rapidly in the vicinity of an absorption peak (Fig. 9).

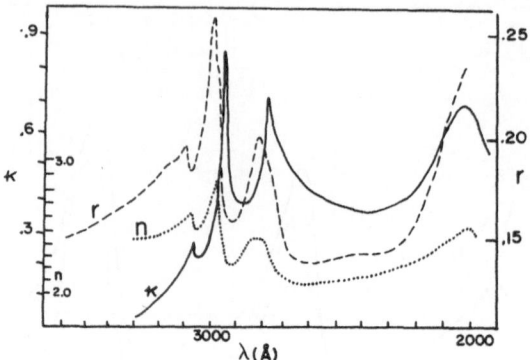

Fig. 9. Single-crystal reflectivity of BaO and computed values of refractive index
and absorption index, after Jahoda [7].

These results indicate that, for proper interpretation of diffuse reflectance results, the variation in specular reflectance with wavelength must be included in the theoretical models (which, through a complicated step, could be introduced into the Melamed or Johnson theories). However, measurement of r would in general require single crystals, and, if such were available, in most cases one would prefer transmission to diffuse reflectance measurements. Moreover, as shown by Jahoda, measurement of single-crystal reflectivity is in itself sufficient to permit calculation of absorption coefficients.

The probable answer to the problem of specular reflectance may lie in an experimental approach advocated primarily by Kortum [8] and later users of the Kubelka—Munk theory.

EFFECT OF DILUTION

Many investigators have observed that when a strongly absorbing sample is physically mixed, i.e., by grinding or milling, with a nonabsorbing matrix, such as the standard, the measured diffuse reflectance spectrum more closely approaches that observed in transmission. Presumably, the specular reflectance of the dilution then approximates that of the standard, and its influence on the measured spectrum is decreased. Figure 4 illustrates one of many dilutions examined for V_2O_5, and, although the curve obtained is not completely satisfactory, it does show that the dilution spectra and sandwich spectra are in somewhat better agreement. The shoulder at 3.2 eV disappears, the peaks near 2.8 eV move into coincidence, weak multiple absorptions appear in the range 3.8 — 4.5 eV, and the last peak near 5.1 eV, though diminished in intensity, shifts to a position closer to that observed in transmission. The dilution principle seems definitely valid experimentally, although its theoretical explanation is still nebulous. Advocates of the Kubelka—Munk theory include the specular reflectance in the scattering terms and assume that, in dilutions, the scattering coefficient of the sample S [equation (16)] is effectively replaced by that of the standard, and that consequently a plot of $f(R)$ versus wavelength properly represents the variation in absorption coefficient. In view of the approximations invoked in the derivation of the remission function and of the fact that a simple OD curve parallels the variation of an $f(R)$ curve (Fig. 3), we doubt the advantages of $f(R)$ calculations. A proper explanation of the dilution effect possibly may be found in the application of Melamed's model to mixtures. Such an investigation is under way in this laboratory.

SUMMARY

1. The diffuse reflectance models proposed by Johnson and Melamed, although elucidating the reflectance process, do not yield absolute absorption coefficients for bulk samples, primarily because of the neglect of the variation in the index of refraction and surface reflectivity near an absorption peak. Correction of these models requires additional experimental data on variation of surface reflectivity for single crystals, which data per se permit calculation of absorption coefficients, making diffuse reflectance measurements unnecessary.

2. Computed Kubelka—Munk remission functions $f(R)$ also do not explicitly include variation in r, maximize at the same wavelengths as simple OD's, and consequently seem of dubious value.

3. Reflectance curves of strongly absorbing materials in nonabsorbing matrices do approach transmission results more closely than reflectance curves of bulk solids and seem worthy of further investigation.

ACKNOWLEDGMENTS

The author is grateful to the donors to the American Chemical Society Petroleum Research Fund for support of this work. Mr. Myron Komarynsky assisted in preparation of the V_2O_5 sandwiches, and Mr. Efrem Zaret carried out many of the measurements and calculations described.

REFERENCES

1. R. Newman and R. M. Chrenko, Phys. Rev. 114, 1507 (1959).
2. K. S. Gibson, Natl. Bur. Standards (U.S.) Circ. 484, Sept. (1949).
3. P. D. Johnson, J. Opt. Soc. Am. 42, 978 (1952).
4. N. T. Melamed, J. Appl. Phys. 34, 560 (1963).
5. G. Schreyer, Z. physik. Chem. N. F. 12, 359 (1957).
6. P. L. Hartman, J. R. Nelson, and J. G. Siegfried, Phys. Rev. 105, 123 (1957).
7. F. C. Jahoda, Phys. Rev. 107, 1261 (1957).
8. G. Kortum, W. Braun, and G. Herzog, Angew. Chem. (Int. Ed.) 2, 333 (1963).

Adaptation of an Inexpensive Ultraviolet-Visible Spectrophotometer for Enzyme Kinetic Work*

David L. Heyse

University of Chicago
Department of Biochemistry
Chicago, Illinois

A Perkin-Elmer Model 202 spectrophotometer was modified to permit sensitive recording of absorbance changes with time by installing a retransmitting slidewire internally. The slidewire, which may be of a standard commercial type, was supplied with an exciting voltage from a battery-powered external circuit consisting of a fixed resistor and two ten-turn potentiometers to permit the use of a variable scale expansion. The output of the slidewire was connected directly to a self-balancing recording potentiometer of a type in common use. This adaptation makes possible an amplification of pen response by a factor of more than four without exceeding the capabilities of the spectrophotometer. It does not interfere with the normal function of the instrument in recording spectra. The modified instrument has been used in the kinetic analysis of several enzyme-catalyzed reactions.

INTRODUCTION

It is the purpose of this paper to describe some equipment which has been used effectively in the biochemistry laboratory of Dr. John Westley at the University of Chicago for enzyme kinetic studies. These studies involved following the changes of optical absorbance as an enzyme-catalyzed reaction proceeded. The reactions were, in many cases, rather fast and the absorbance changes inconveniently small.

The Perkin-Elmer 202 ultraviolet-visible recording spectrophotometer is a convenient and useful instrument in a

*This work was supported by funds made available from a National Science Foundation research grant.

biochemistry laboratory, but, as supplied by the manufacturer, it lacks a means for recording the change in absorbance with time. A time-drive accessory offered by the manufacturer does not have the flexibility or the sensitivity often needed, with the result that small but significant changes with time are difficult to quantitate. Nevertheless, the instrument has a meaningful photometric output corresponding to such changes. This information is available at the quadrant output shaft and can be put to use.

In the adaptation to be described, the 202 is combined with a self-balancing potentiometric recorder of a general-purpose type, such as the Varian G-11-A and the Stoelting Servographic. The result is a more usable record in which the pen movements are magnified by a factor of three, or even five, in terms of the full-scale travel of each instrument. The recorders mentioned are inexpensive. Furthermore, they have a sufficient number of laboratory applications, so that the cost of the recorder need not be entirely charged to the spectrophotometer adaptation.

ADAPTATIONS

An examination of the interior of the spectrophotometer indicates that the manufacturer may have originally intended to provide a kinetic adaptation of this kind, but no such device was made available to us. It would be ideal indeed if the instrument manufacturer could make available promptly and at moderate cost such modifications and adaptations of his equipment that different classes of users require. This will never be the case, due to the high cost of engineering time, expediting effort, the difficulties in production scheduling involving changes or additions to a product, and, finally, due to the manufacturer's hesitancy in releasing a new item because his reputation may be at stake. The user, then, must rely on his own resources, as in the present instance.

Figure 1 illustrates the nature of the problem faced. It shows a comparison of results obtained by this adaptation with those obtained using the manufactured accessory, which arrived only after a crude version of this equipment had been assembled.

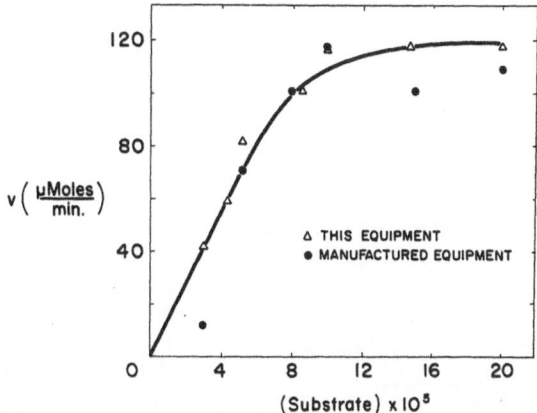

Fig. 1. Comparison of present equipment and manufactured equipment in evaluating initial velocity of the reaction catalyzed by ascorbic acid oxidase.

The curve is drawn to fit the data obtained with this adaptation (the triangles) and furnishes a usable result of the set of experiments depicted. The circles are the data from the same set of experiments recorded at the same time with the manufactured accessory. (Both recording devices were functioning at the same time and without any interaction, as the adaptation does not interfere with the normal use of the instrument.)

It can be seen that the circle-points are not in violent conflict with the graph, but that it would be almost impossible to draw the curve from them. These simultaneous records were not prepared for this presentation, but rather in the normal course of research, as enzymologists in this laboratory felt the need to establish confidence in the modified equipment. These data were taken by Dr. Brenda Gerwin in the course of some studies on the kinetics of ascorbic acid oxidase. Plotted here is the variation of initial velocity with substrate concentration. The results of this research will appear elsewhere. This testing procedure was repeated many times before entirely abandoning the use of the manufactured accessory.

Figure 2 shows one of the simultaneous records mentioned. The upper part of the slide is the record produced by the manufactured accessory. The pen is attached to the spectro-

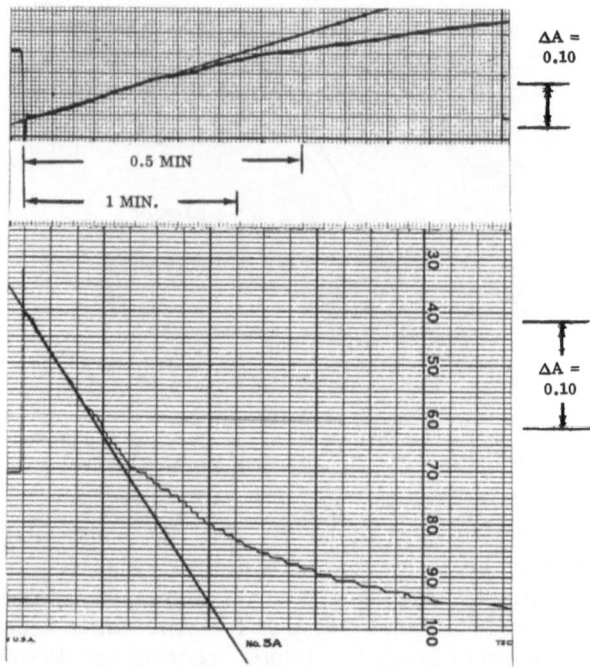

Fig. 2. Examples of the original records constituting primary data for Fig. 1.
The wavelength of the these determinations was 267 mμ.

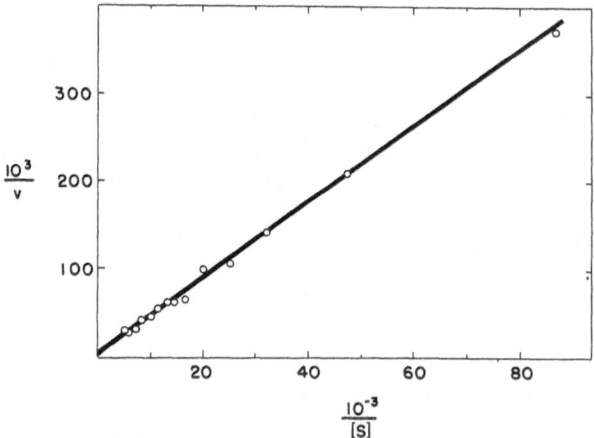

Fig. 3. Conventional Lineweaver—Burk plot of typical enzyme kinetic data. Initial
velocity, v; ascorbate concentration, S.

photometer indicator and writes on a drum substituted for the drum driven by the wavelength drive. The lower part of the slide is the record produced by the external recorder of the set-up being described here. The straight lines are construction lines drawn on the graph afterward.

The upper display does not appear to be such a poor record, but it does fail to supply the needed information when handled by conventional methods, as evidenced by Fig. 1. Each point on Fig. 1 is furnished by one such complete record. The annoying little steps on the lower record will be mentioned later.

Figure 3, taken directly from Dr. Gerwin's study of ascorbic acid oxidase, is a conventional Lineweaver–Burk plot. It is a highly demanding way of testing the instrumentation and experimental procedure from the point of view of detecting both systematic and random errors. As is shown here, the present instrument satisfies this practical statistical test, defining closely the expected straight line and permitting the evaluation of the kinetic constants from the slope and intercept of that line. At this point, if the data are defective, it will be evident

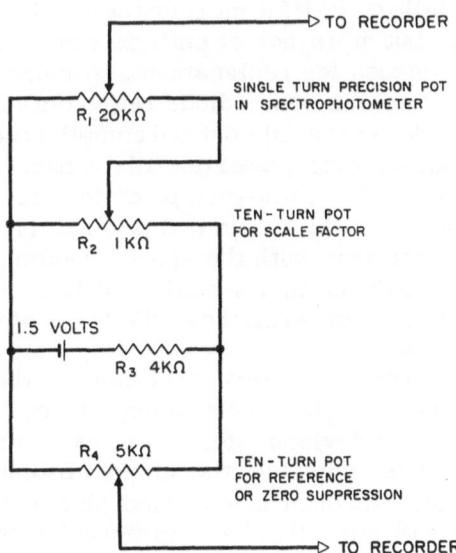

Fig. 4. Circuit diagram of the modification described.

from the shape of this plot. This constitutes further validation of the modified instrument in practical use.

Figure 4 shows the electrical circuit. R_1, the 20,000 -Ω, single-turn, precision potentiometer at the top of the figure, is the output potentiometer, and is the only additional component to be mounted within this instrument. R_2, the 1,000-Ω, ten-turn potentiometer, adjusts the voltage across the slidewire and so establishes the amount of scale amplification obtained. R_3, the 4,000-Ω fixed resistor, drops the baterry voltage to an appropriate level so that R_2 can be used in its full range. It, or a portion of it, might be made adjustable to compensate for the gradual weakening of the battery with age. However, R_2 and R_4 have a sufficient adjustment range to accomplish this without introducing that complication. Other fixed resistors might well have been used in the circuit, but were also omitted to retain simplicity.

There is a wide range of choice for the $1\frac{1}{2}$-V battery. Excluded is a single flashlight cell because it would drift too rapidly. Suggested are a No. 6 dry cell, or a Burgess 4FH lantern battery, which should have a useful life in the circuit of about 2500 hr. A switch here is neither necessary nor desirable. It would result in annoying drifting each time it was turned on. A Mallory RM42R mercury cell with a switch would function nicely, but it is not of sufficient general availability when suddenly needed for replacement. A more sophisticated user would insist upon a zener diode regulated supply, but disadvantages include increased cost and complication, and further possibility of introducing powerline disturbances into the recording. R_4, a 5,000 -Ω, ten-turn potentiometer, supplies the reference voltage for the output potentiometer, and is used to set the recorder to zero with the spectrophotometer indicator at any desired position on its scale. It is required because recorders do not, in general, have the large amount of zero-suppression needed.

Figure 5 shows the interconnection of the spectrophotometer, the control box containing all of the electrical components of the previous figure, except the slidewire, and the recorder. It is advisable, though not absolutely essential, to run the cables through a grounded shield as shown. The control box would normally be a grounded metal enclosure. One might use a plain knob on the zero-suppression control,

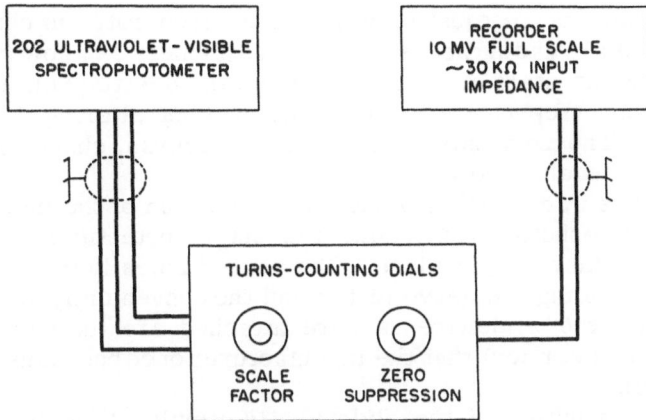

Fig. 5. Block diagram showing cables running through grounded-shield, inter-connecting units.

but the scale factor adjustment should certainly be a turns-counting type of dial having 1,000 divisions for full travel. This is needed for returning to any preselected scale expansion.

When initial adjustment is made, the zero control is manipulated intermittently to keep the recorder on scale, while the indicator on the spectrophotometer is cautiously displaced by hand through some selected fraction of the scale. The scale-

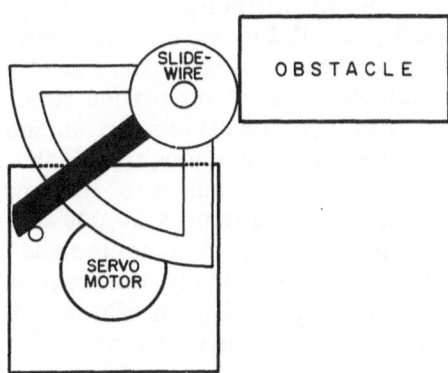

Fig. 6. Interior of the spectrophotometer in the area of interest. The output potentiometer is here labeled slidewire.

factor dial is readjusted until the desired ratio is obtained, and the dial reading noted. The zero adjusment can now be used to place the start of the recording at any desired relationship to the spectrophotometer indicator without altering the scale factor. The zero adjustment will, of course, change for any change in scale factor.

Figure 6, a simplified sketch of a portion of the interior of the spectrophotometer, shows the output potentiometer (here labeled slidewire) in place. The unused holes in the oversize bracket holding the servo motor and the conveniently long shaft to which the slidewire is here attached are the indications (mentioned earlier) that the manufacturer once had plans in this direction.

The mounting of the slidewire is simple, if one has a unit small enough. It is not necessary to drill any holes, uncover any sensitive parts, or otherwise commit any acts of violence on the delicate instrument. It suffices merely to affix a side arm to the potentiometer housing and attach the potentiometer to the shaft extending from the quadrant with a shaft coupling. The side arm is held lightly, by a weak spring, against the unthreaded shank of a bolt passed through one of the holes provided by the manufacturer, the bolt being held in place by two nuts. Finally, a three-wire shielded cable is attached and lead out through any of the many openings available, and the installation is completed.

I now call attention to the object labeled obstacle. This is the occasion for some difficulty in selecting the slidewire. Those most readily available from electronic supply houses cannot be used. If one has a single-turn, precision potentiometer, $\frac{3}{4}$ in. in diameter and of low torque requirements, there is no trouble. For the first version, such a potentiometer could not be obtained quickly enough. As the equipment was urgently needed, a suitable one was improvised using the resistance element removed from a 20,000-Ω Beckman type G potentiometer by mounting it in a homemade housing. This procedure cannot be recommended! Later, it was learned that the Maurey Instrument Company of Chicago has available a reasonably priced, servo-type potentiometer, $\frac{3}{4}$ in. in diameter, which could be used here.

I now wish to discuss those annoying little steps pointed out in Fig. 2, because they concern the selection of the potentiometer. There are 300 increments of resistance in the slidewire for a full scale movement of the indicator. This results in 100 steps on the record at a 3:1 magnification. These are annoying, but never serious. The 25,000 -Ω Maurey potentiometer (their No. 75-M91) would furnish 250 such increments, slightly more annoying but still not serious. This defect will place some limit on the use of the highest magnifications. (Our instrument is usually used at 3:1, and occasionally at 5:1.)

A number of potentiometer manufacturers are able to produce infinite resolution potentiometers of a suitable size, which would eliminate these discontinuities, but the prices and delivery times are not attractive. The Beckman Cermet potentiometer should be ideal here, but cost is considerably greater and the availability has been questionable. The New England Instrument Company also produces a suitable infinite resolution potentiometer.

An infinite resolution potentiometer would make the use of a much less sensitive recorder possible by reducing the resistance of this part, but the other resistance values would have to be reduced and the power supply capabilities increased.

The resistance value of the slidewire can vary over quite a wide range without changing the rest of the circuit, as long as it remains high with respect to the other circuit values.

CONCLUSION

A Perkin-Elmer Model 202 spectrophotometer was modified to permit sensitive recording of absorbance changes with time. The adaptation is simple, effective, and economical enough to be worthy of duplication by anyone faced with a similar problem (at least until the manufacturer decides to incorporate a similar adaptation in his equipment*). There are presumably many instruments already in use to which this application might well be considered. It is not offered as an instrumental innovation— the principles are old and well-known. What has been shown here is the increased utility to be obtained from a good spectrophotometer by these simple means.

---

*Subsequent to the presentation of this paper, the author has been advised by a representative of the Perkin-Elmer Corporation that the manufacturer does now offer for purchase an adaptation of the kind described here.

ACKNOWLEDGMENTS

Dr. Brenda Gerwin's patient testing of the equipment is greatly appreciated and is one of the reasons use of the present modification can be suggested to others.

Investigation of Cathodo-Luminescence
with the Petrographic Microscope

Paul Weiblen

Department of Geology
University of Minnesota
Minneapolis, Minnesota

An attachment can be constructed for any electron microscope which allows bombardment of a solid sample surface by the electron beam and simultaneous observation with a petrographic microscope. Cathodo-luminescence is found to be sufficiently specific in multiphase systems to permit identification of major phases, recognition and determination of the distribution of fine-grained and interstitial phases, and investigation of textural features, such as zoning, exsolution, and grain boundary relations. In combination with a photomultiplier tube, the technique could be used to determine the relative amounts of phases in a sample.

INTRODUCTION

Cathodo-luminescence, the emission of light from a substance excited by an electron beam, has been under investigation since the advent of cathode-ray devices. In solids, its cause has been found to be related to impurity atoms and structural defects [2]. However, the details of the unique luminescence of particular solids are understood for only a limited number of substances.

The development of the electron microprobe has led to the recognition of the use of cathodo-luminescence as a qualitative analytical tool. The microprobe incorporates two features required to observe cathodo-luminescence—an electron beam (in this case of variable energy, intensity, and focus) and an optical viewing system. In addition, the facility provided by the microprobe for elemental analysis of micron-size volumes affords a new technique for investigating the cause of cathodo-luminescence [3].

Investigation of cathodo-luminescence effects in a number of rock-forming minerals with the petrographic microscope has shown that luminescence is a useful petrologic tool [5]. The effects noted by Stenstrom and Smith [5] and those discussed below show that cathodo-luminescence is sufficiently specific in multiphase systems to permit identification of major phases, recognition and determination of the distribution of fine-grained and interstitial phases, and investigation of textural features, such as zoning, exsolution, and grain boundary relations. Despite the lack of complete understanding of the causes of cathodo-luminescence and without the aid of X-ray analysis, determination of cathodo-luminescent effects with a petrographic microscope has been found to be a valuable qualitative analytical technique. As more data become available on the correlation of chemical composition with cathodo-luminescence, the value of this technique will increase as an aid to microscopic investigation and microprobe analysis.

INSTRUMENTAL ADAPTATIONS

A relatively simple adaptation can be made for most electron microscopes which makes it possible to view cathodo-luminescent effects with an ordinary petrographic microscope. Figure 1 is a schematic diagram of the adaptation. Using the RCA EMU 2 electron microscope, it is necessary only to remove the electron gun assembly and auxiliary lens from the column of the microscope, attach these to a suitable stand, and construct a sample holder of appropriate dimensions to allow excitation of the sample by the electron beam and simultaneous observation with a conventional petrographic microscope.

The effects described below were observed with an adaptation made for an RCA EMU 2 microscope. The adaptation permits focusing of an approximately 1- μA, 50-kV electron beam to a 1-mm spot. One-inch diameter, 30-μ thick, polished, thin sections were viewed with a petrographic microscope with facilities for both transmitted and reflected light. Magnifications of 50, 100, and 400 were used. Sample preparation is discussed by Cadwell and Weiblen [1]. Unless otherwise

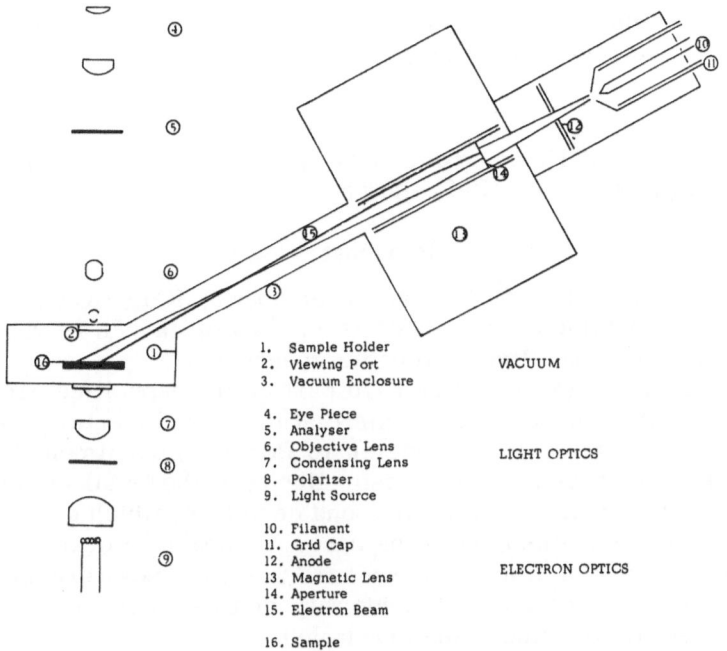

1. Sample Holder	
2. Viewing Port	VACUUM
3. Vacuum Enclosure	

4. Eye Piece	
5. Analyser	
6. Objective Lens	
7. Condensing Lens	LIGHT OPTICS
8. Polarizer	
9. Light Source	

10. Filament	
11. Grid Cap	
12. Anode	
13. Magnetic Lens	ELECTRON OPTICS
14. Aperture	
15. Electron Beam	

16. Sample

Fig. 1. Diagrammatic sketch of the instrumentation required to observe cathodo-luminescence of solids with a petrographic microscope.

indicated, samples are from the Thomson Formation of northern Minnesota.*

RESULTS

Oxides

Stenstrom and Smith [5] have noted that quartz (SiO_2) generally luminesces blue, but recrystallized borders of fused quartz grains in quartzite luminesce red. This was found to be true also for samples of a quartzite from southwestern Minnesota. Quartz in a completely recrystallized metamorphic rock from the Front Range of Colorado luminesces uniformly dull red. This is an indication that luminescence may be useful in tracing the extent of some solid-state reactions.

Synthetic corundum (Al_2O_3) of unknown trace element com-

*A sequence of Precambrian metasedimentary rocks. See Schwartz [4] for a description of mineralogy and lithology.

position luminesces a faint red.   Rutile (TiO_2), magnetite (Fe_3O_4), and hematite (Fe_2O_3) do not luminesce.

Sulfides

All the sulfides examined—pyrite (FeS_2), chalcopyrite ($CuFeS_2$), pyrrhotite (FeS)—do not luminesce.

Phosphates

The amount and distribution of apatite [$Ca_5(F,OH)(PO_4)_3$] needles contained in a basalt from California can be readily determined by the light yellow luminescence.

A conodont (a tooth-like hard part of an unspecified organism which can be used in correlation of sedimentary rocks ranging in age from Ordovician to Permian) specimen shows zonal luminescence. The incisor-like tip of the fossil luminesces light blue and the root-like end deep blue, which grades to longer wavelengths toward the base. Comparison of the luminescence with similarly shaped hard parts of known organisms may aid in establishing the identity of this fossil, as well as increase its usefulness in correlation.

Detrital apatite grains only a few microns in diameter in samples from the Thomson Formation luminesce bright yellow. Their distribution can be studied at magnifications as low as $50\times$.   This is also true for other accessory minerals such as zircon, tourmaline, and epidote.

Silicates

Varieties of olivine [$(Mg, Fe)_2SiO_4$] with as low as 8% FeO do not luminesce.

Clinopyroxenes [$(Ca, Mg, Fe) (Al, Fe) AlSiO_6$] containing 8 to 34% FeO from a variety of igneous rocks luminesce faint blue.   With FeO content between 2 and 3%, the luminescence changes to yellow-orange and the light is polarized.

Enstatite (Mg_2SiO_3) from the meteorite Abee luminesces bright blue.

Epidote [$Ca_2(Al, Fe)_3(SiO_4)_3(OH)$], which contains approximately 80% of the $Ca_2Al_3Si_3O_{12}(OH)$ end member, luminesces light green.

Iron-rich varieties of garnet [$(Ca, Mg, Mn, Fe)_3 (Al, Cr, Fe)_2 (SiO_4)_3$] do not luminesce.

Iron-rich amphiboles [$Ca_2(Mg, Fe)_4, Al(OH)_2, AlSi_7O_{22}$] show no luminescence.

Zircon ($ZrSiO_4$) crystals in glass shards from a volcanic tuff from California luminesce an intense blue, and zoning is clearly displayed.

Cordierite [$Mg_2Al_3(AlSi_5O_{18})$] from a completely recrystallized gneiss from the Front Range of Colorado luminesces bright blue and is readily distinguishable from quartz (red) and feldspar (light blue-green) in the same rock. Cordierite can be distinguished from these minerals with the petrographic microscope generally only by determining its biaxial character and optic sign; this is a laborious procedure if studies are to be made of both the amount and distribution of the mineral.

Carbonates

Several investigators have begun studies of the carbonates [3,5]. Long [3] has found that luminescence in calcite ($CaCO_3$) is probably due to manganese substitution and that iron acts as a quencher. The effects noted above for pyroxene, olivine, garnet, and amphibole suggest that iron also quenches the luminescence in these silicates.

Carbonate concretions, which probably formed during compaction of the sediments in the Thomson Formation, luminesce uniformly light orange. Grain boundaries and textural features, such as cone-in-cone structure which are visible in transmitted light, are not apparent in the luminescence. On the other hand, calcite intergrown with dolomite in nodules that formed from hydrothermal solutions show bright bands of orange luminescence parallel to cleavage and twinning directions. This type of luminescence indicates substitution of manganese in particular positions in the calcite structure [3]. This provides a means of distinguishing the two types of nodules, which may be of similar external form. It also suggests that diffusion of manganese possibly occurs in the calcite structure under specific conditions. If the cause of these effects can be determined, luminescence may provide information on the mode of formation of a particular occurrence of calcite.

Other Effects

The effects noted in the sample containing cordierite suggest that cathodo-luminescence can be used in determining relative amounts of different phases in a system. Some methods—physical separation by mechanical, density, or magnetic techniques, or point counts with a microscope—are often time-consuming, inaccurate, or impossible. In samples in which the different phases luminesce with sufficiently different wavelengths, it should be possible to determine the amount of a phase present by detecting with a photomultiplier tube the fraction of the surface viewed that luminesces within a restricted wavelength range.

Observation of cathodo-luminescence with the petrographic microscope also affords a means for determining decay times. A sample can be moved across the beam at a known speed, and decay times can be obtained by measuring the length of the tail of the beam.

Polarization of the luminescence was noted in several minerals. This effect is particularly striking for twinned feldspar. Extinction angles measured by luminescence correspond to those measured with transmitted, polarized light. However, all grains examined in a single 1-in. diameter sample did not display polarized effects, although all grains in the sample show twinning in transmitted light. This suggests that polarization of the luminescence may depend on orientation of the mineral with respect to the electron beam. Thus, it might be possible to obtain crystal structure information from cathodo-luminescence effects.

CONCLUSION

The few examples described above and this brief discussion can possibly serve only to suggest the usefulness of cathodo-luminescence in a variety of problems in petrology, ceramics, phase equilibria studies, and analytical problems in organic and inorganic chemistry. Continued compilation of cathodo-luminescent effects, as well as further investigation of their cause, is needed.

REFERENCES

1. D. Cadwell, and P. W. Weiblen, Diamond disk preparation of polished thin sections for electron microprobe analysis, (to be published).
2. C. C. Klick, and J. H. Schulman, Luminescence in solids, in: Solid State Physics, Vol. 5, Academic Press, New York (1957), pp. 97–172.

3. J. V. P. Long, The application of the electronprobe microanalyser to metallurgy and mineralogy, in Pattee, Cosslett, and Engstrom (eds.): X-Ray Optics and X-Ray Microanalysis, Academic Press, New York (1963), pp. 279-295.
4. G. M. Schwartz, Correlation and metamorphism of the Thomson Formation, Bull. Geol. Soc. Am. 53, 1001-1020 (1942).
5. R. C. Stenstrom, and J. V. Smith, Electron excited luminescence as a petrologic tool, Geol. Soc. Am. Spec. Papers 76, 158 (1963).

Gas Chromatography

The Detection of Submicrogram Quantities of Carcinogenic Polynuclear Hydrocarbons Using Electron Capture

William Lijinsky and Irving Domsky

Division of Oncology
The Chicago Medical School
Institute for Medical Research
Chicago, Illinois

The ionization detector of a Barber–Colman Model 20 instrument has been converted to an electron-capture detector by applying a potential of 10.5 V from batteries and reversing the polarity of the electrometer.

As little as 0.1μg of the carcinogen benzo(a)pyrene gives a good response under these conditions when chromatographed on a column of glass microbeads (60—80 mesh) coated with 0.25% SE-30 silicone. The retention time at 230° is 8 min in an 8 ft by 1/4 in. column. The minimum detectable amount of benzopyrene in the argon ionization detector was 1 μg.

This method has proved suitable for the rapid detection of benzo-(a)pyrene and other polynuclear hydrocarbons added to materials such as dairy wax at a concentration of 0.1 ppm. Analytical methods previously employed required a week or longer to achieve this sensitivity. A 10-g sample of wax, to which was added 1 μg of benzopyrene, was dissolved in hexane, extracted with three portions of dimethylsulfoxide, a few ml of benzene added, followed by water, and the benzene extract was evaporated to dryness. The residue was diluted to 50 μ liter with iso-octane. A 25 μ liter sample of this solution (corresponding to 0.5 μg of BP) gave a good recorder response. It is hoped that this sensitivity can be further increased and that the method can be applied to the analysis of mixtures of polynuclear hydrocarbons.

INTRODUCTION

The problem of detecting very small quantities of poly-nuclear hydrocarbons has become of great concern in recent years because among this group of compounds are found the most widespread of known chemical carcinogens. Since passage of the Food, Drug, and Cosmetic Act of 1958, containing the

Delaney Amendment requiring proof of the absence of chemical
carcinogens from any food additive, intended or incidental, it
has become imperative to find rapid methods of screening a
wide variety of materials for the presence (or absence) of
these carcinogens. This is necessary both for industry, which
must comply with the law, and for the government agencies
which must enforce the law.

Methods have been developed for the detection of as little
as 1 part of such compounds in 100 million parts of a petroleum
wax [1]. These methods, involving extensive paper chromatography analysis combined with ultraviolet spectrometry, are
necessarily very lengthy. (A single sample takes more than
a week.)

Shorter methods have been developed, more suitable for
routine examination of samples, to limit the possible concentration of such carcinogenic hydrocarbons as benzo(a)pyrene to
1 part in 10 million, based on maximum total absorption within
certain wavelength ranges of a suitably prepared extract of
the material [2,3]. However, even these short methods take
1 or 2 days per sample. They suffer from the great disadvantage of being completely nonspecific, since noncarcin-
ogens, as well as completely unrelated compounds, might
contribute to the absorption. Therefore, materials completely
free of carcinogens might be rejected on this entirely arbitrary
basis.

EXPERIMENTAL

This very unsatisfactory state of affairs has led us to search
for a more specific rapid method of detecting carcinogenic
hydrocarbons at very low concentrations. Gas chromatography
seemed to be the answer. When it was found that the very high-
boiling polynuclear hydrocarbons (many of them boil above
500°C), such as benzo(a)pyrene and dibenz(a,h)anthracene (both
of which are carcinogenic), have sufficient vapor pressure to
be gas-chromatographed at temperatures below 220°C, the way
seemed open.

Our first experiments were carried out on a Barber-Colman
Model 20 instrument, using a strontium-90 ionization detector
with argon as carrier gas. As previously reported [4], one
deficiency of the argon ionization detector in working with this
group of compounds is the wide range of sensitivity among the

different hydrocarbons. Thus, the sensitivity to pyrene, for example, is two orders of magnitude greater than the sensitivity to benzo(a)pyrene and the other high-boiling polynuclear hydrocarbons. Unfortunately, all of the carcinogenic hydrocarbons are high-boiling-point compounds; hence, this system is least sensitive to those compounds that are of greatest interest and happen to be those that are usually present in smallest concentration. Using temperature programming, it was possible to increase the sensitivity to the high-boiling compounds to some small extent [5], but the minimum detectable quantity of benzo(a)pyrene was still of the order of 1 μg. This would mean that to obtain an overall sensitivity of 1 part per 100 million, the extract of 100 g of a material would have to be chromatographed in a single small volume of solvent. This proved impractical.

The recent introduction of the electron-capture detector seemed to provide a way out of the dilemma, since it was likely that this detector would be much more sensitive to polynuclear compounds than is the ionization detector. The electron-capture detector has the additional advantage of being insensitive to aliphatic and alicyclic compounds, which are always present in the materials of concern to us and which are detected in the ionization detector; this is possibly one of the reasons for the inadequacies of the ionization detector for analysis of the commercial material we have studied, as compared with mixtures of pure compounds. We, therefore, modified the ionization detector of the Barber-Colman Model 20 instrument for electron-capture detection [6] by reducing the cell potential to the range of 0 — 50 V using batteries and using nitrogen as carrier gas. The results were not as good as we had expected. Although the sensitivity was much more uniform from one polynuclear hydrocarbon to another, the sensitivity to the compounds of greatest interest was not very much greater than using the ionization detector. The limit of detectability of benzo(a)pyrene was 0.25 — 0.5 μg. Even using on-column injection and larger volumes of more dilute solutions to minimize losses during injection of the sample, the smallest detectable amount of benzopyrene was $\frac{1}{6}$ μg at an optimum potential of 10.5 V and with maximum gain and minimum attenuation of the electrometer. In addition, the peaks were broad and poorly resolved, although this might have been due to our use of an 8-ft copper column.

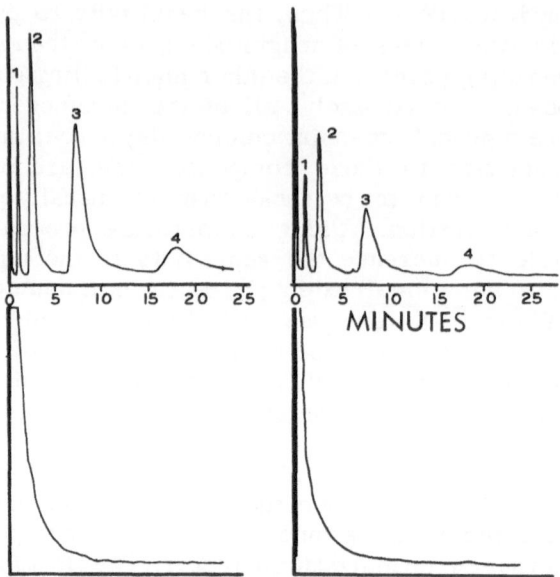

Fig. 1. Chromatogram of polynuclear hydrocarbons on 0.25% SE-30 on glass beads; 6 ft × 1/4 in. glass column; nitrogen flow 100 ml/min; split 1:1. Column temperature 200°C, detector temperature 207°C. Upper tracings from electron-capture detector, lower tracings from hydrogen-flame detector. 1 = pyrene, 2 = benz(a)anthracene, 3 = benzo(a)pyrene, 4 = dibenz(a,h)anthracene. Lefthand chromatograms are of 0.05 μg of each component in 5 μliter of solution; righthand chromatograms are 0.02 μg of each component in 2 μliter of solution.

Since this was obviously unsatisfactory, we obtained a Packard gas chromatograph equipped with electron-capture and flame-ionization detectors with two recorders. The columns were of glass, 6 ft long and packed with our usual support, 60—80 mesh glass beads coated with 0.25% SE-30 silicone. On-column injection was used. The gas flow was split 1:1, half of the sample going to each of the two detectors. A typical chromatogram of a mixture of pyrene, benz(a)anthracene, benzo(a)pyrene, and dibenz(a,h)anthracene is shown in Fig. 1. The chromatogram on the left is of 0.05 μg of each hydrocarbon in 5 μliter of solution; the right-hand chromatogram is of 0.02 μg of each hydrocarbon, or 0.01 μg in each detector. The flame-ionization detector showed no response whatever. This sensitivity of the electron-capture detector was much higher than we had been able to obtain previously, 0.01 μg giving a reasonable response in the case of benzopyrene and a slight re-

sponse in the case of dibenzanthracene. At a column temperature of 200°C, the retention times were considerably shorter
with the glass column than with the copper column—1 min for
pyrene, 3 for benzanthracene, $7\frac{1}{2}$ for benzopyrene, and 18 for
dibenzanthracene. The prospects seemed excellent for developing a sensitive method for determining polynuclear hydrocarbons in mixtures containing them in minute traces. Some
studies of petroleum waxes will illustrate this.

A 20-g sample of a wax was dissolved in 100 ml of hot
iso-octane which was shaken vigorously with 3 × 50 ml of warm
dimethylsulfoxide, which preferentially extracts polynuclear
compounds. The combined DMSO extracts were back extracted
with 20 ml of iso-octane to remove entrained aliphatic
material.   To the clear DMSO solution was added 50 ml of
benzene, followed by 600 — 700 ml of water. The upper benzene
layer was shaken twice with small volumes of water to remove
DMSO and the benzene removed in a stream of nitrogen. The
residue, weighing a few milligrams, was dissolved in 0.2 ml of
iso-octane and 20 μliter of this solution used for injection.
The extraction was repeated in exactly the same way after addition of 2 μg of benzopyrene or dibenzanthracene. Chromatograms of the two solutions are shown in Fig. 2. The same

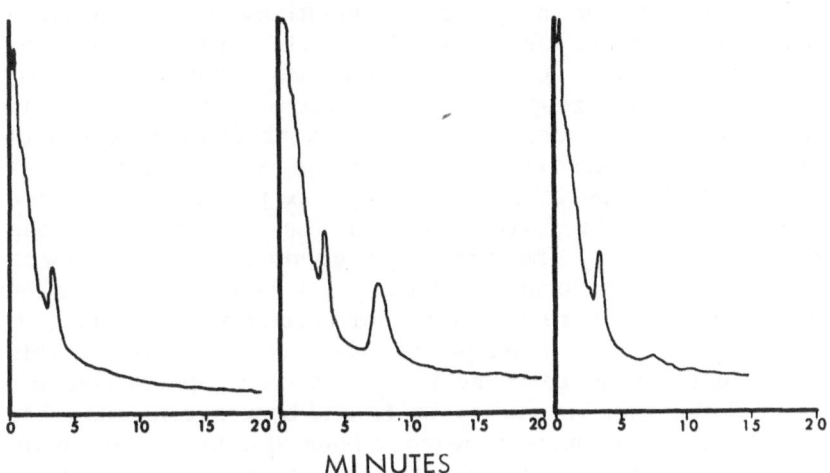

MINUTES

Fig. 2. Extract of 20 g of paraffin wax in 0.1 ml of iso-octane. 1 μ liter sample.
Lefthand chromatogram is wax alone; middle chromatogram with 0.1 ppm benzo(a)
pyrene; righthand chromatogram with 0.02 ppm benzo(a)pyrene.

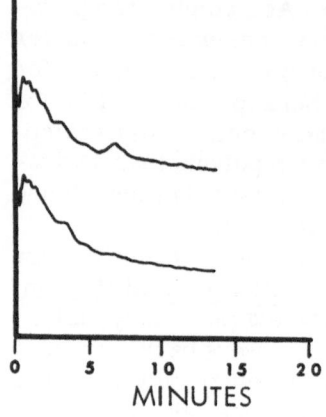

Fig. 3. Extract of 20 g of microcrystalline wax in 0.1 ml of iso-octane. 1 μliter sample. Lower chromatogram is wax alone; upper chromatogram with 0.1 ppm benzo(a)pyrene.

MINUTES

experiment was carried out with a microcrystalline wax and the results are shown in Fig. 3. To assure the adequacy of the extraction of the added polynuclear hydrocarbons, a further chromatogram was run of a sample of the wax extract to which the same proportion of benzopyrene was added subsequently; the response was very nearly the same.

Finally, some wax samples known to contain polynuclear hydrocarbons (as shown by the lengthy paper chromatographic method) were carried through the same extraction procedure.

It appears that the practical sensitivity of this procedure for the detection of benzopyrene in a wax is about 0.2 ppm. The presence of some compounds other than polynuclear hydrocarbons giving a response in the electron-capture detector in the extract of the microcrystalline wax makes the detection of very low concentrations of benzopyrene in this wax more difficult than in a paraffin wax. It might be possible to increase the sensitivity by further refining the extract before gas chromatography. The time of analysis is 2 — 3 hr, which compares very favorably with the 1 — 2 days required for the very inadequate screening tests for carcinogens now used. It seems certain that a sample of wax extract prepared in this way that gives no response in the electron-capture detector after 7 min can be said to be free of benzopyrene and other carcinogenic polynuclear hydrocarbons at a level of 0.1 ppm.

Extracts of various other materials suspected of containing polynuclear hydrocarbons and analyzed by paper chromatography have been gas-chromatographed in our instrument and have shown that the system described is suitable for rapid

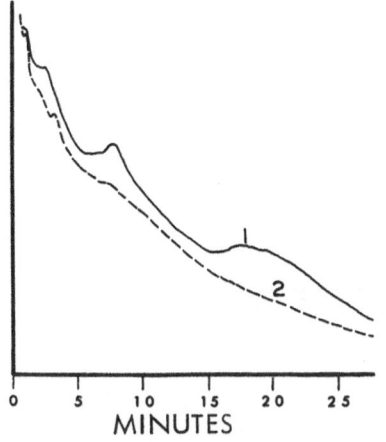

Fig. 4. Extract of charcoal-broiled steak in iso-octane, showing presence of benzo(a)pyrene. 1 μliter sample.

MINUTES

qualitative analysis of these materials for carcinogenic poly-nuclear hydrocarbons.

Figure 4 shows chromatograms of an extract of two dif-ferent batches of charcoal-broiled steak, one home-cooked and one commercial. The mixture of polynuclear hydro-carbons present is shown well, including benzo(a)pyrene, which has been demonstrated by paper chromatography [7].

Figure 5 is a chromatogram of an extract of coal tar, showing clearly the presence of benzo(a)pyrene. Figure 6 is of the same extract, but programmed from 100—200°C, which reveals the presence of many lower-boiling polynuclear hydro-carbons that are not seen in isothermal operation at 200°C.

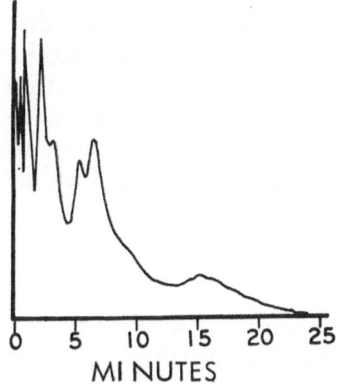

Fig. 5. Coal tar solution in iso-octane/benzene, 5.6 mg/ml, 1 μ liter sample. Benz(a)anthracene at 3 min; benzo(a)pyrene at 7 min; benzo(g,h,i) perylene at 16 min. Column temp., 198°C.

MINUTES

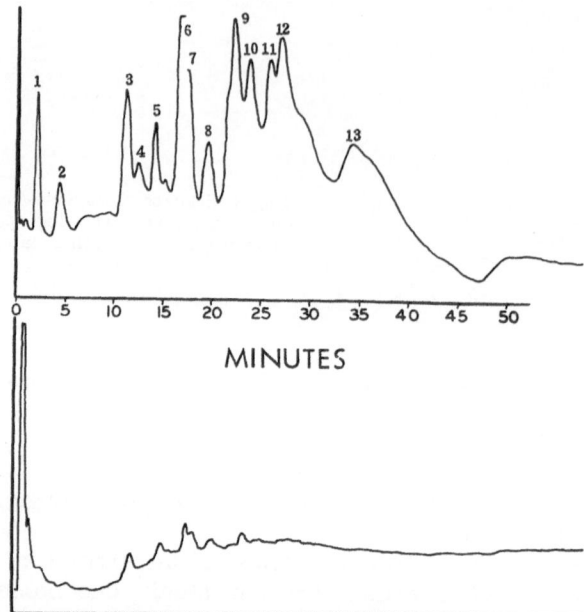

Fig. 6. Same solution of coal tar, 1 μliter sample, programmed from 100—200°C at 3½ °/min. 1 = naphthalene, 2 = methylnaphthalene, 3 = phenanthrene/anthracene, 4 = ?, 5 = carbazole, 6 = fluoranthene, 7 = pyrene, 8 = benz(a)anthracene/chrysene, 9 = benzo(a)pyrene/benzo(b)chrysene, 10 = perylene/benzo(k)fluoranthene, 11 = ?, 12 = benzo(g,h,i)perylene/dibenz(a,h)anthracene, 13 = ?.   Upper chromatogram—electron-capture response; lower chromatogram—hydrogen-flame response.

CONCLUSION

Having reached a sensitivy of 0.05 ppm for a carcinogenic polynuclear hydrocarbon, the question might be asked why the sensitivity cannot be increased still further. It undoubtedly can be, but we shall soon reach a level at which these ubiquitous hydrocarbon carcinogens will be detected in some concentration everywhere. Then, in spite of the wording of the law which sets a zero tolerance for chemical carcinogens, we shall have to face the problem of assessing the biological significance of these minute concentrations.

ACKNOWLEDGMENTS

This investigation was supported by Public Health Service Research Grant No. CA-05170-04 from the National Cancer Institute.

REFERENCES

1. W. Lijinsky, Anal. Chem. 32, 684 (1960).
2. E.O. Haenni, F. L. Joe, J.W. Howard, and R. L. Leibel, J. Assoc. Offic. Agr. Chemists 45, 59 (1962).
3. W. Lijinsky, I. Domsky, and C.R. Raha, J. Assoc. Offic. Agr. Chemists 46, 725 (1963).
4. W. Lijinsky, I. Domsky, G. Mason, H.Y. Ramahi, and T. Safavi, Anal. Chem. 35, 952 (1963).
5. W. Lijinsky, and G. Mason, J. Gas Chromatog. 1, 12 (1963).
6. W.D. Ross, Anal. Chem. 35, 1596 (1963).
7. W. Lijinsky and P. Shubik, Science 145, 53 (1964).

Chromatographic Analysis of Evolved Contaminants From Spacecraft Materials

Norman T. Gonnella

Honeywell, Aero Division
Minneapolis, Minnesota

The organic materials used in the inhabited area of manned spacecraft were examined for their atmospheric contaminating potential. A sample of each spacecraft material was sealed in a closed, environmentally-controlled chamber and the chamber gas from each sample was subjected to the following tests: (1) odor analysis, (2) carbon monoxide and carbon dioxide analysis, (3) total hydrocarbon content by flame ionization, and (4) qualitative and quantitative analysis of organic contaminants by gas chromatography using three different packed columns.

The first three tests screened the materials selected for possible use in the spacecraft. Only the materials meeting the requirements of these tests were subjected to complete qualitative and quantitative analysis. Forty contaminating compounds were separated from the spacecraft materials. These contaminating compounds were identified as alcohols, aldehydes, ketones, hydrocarbons, halogenated hydrocarbons, and aromatics.

INTRODUCTION

Honeywell for the past few years has been a participant in the Apollo Project. The Apollo spacecraft provides a closed environment and must have atmospheric control over the duration of its extended space mission. This control must maintain a proper oxygen balance in this atmosphere and keep the concentration of toxic contaminants at or below noninjurious levels.

Controlling the atmospheric contaminants in the spacecraft is accomplished by two methods: The first method is by using an effective air purification system; the second method is by eliminating the source of the contaminants. The latter involves selecting specific materials required in construction

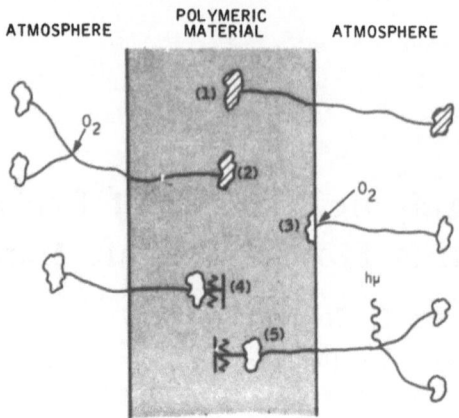

Fig. 1. Probable gassing processes.

and operation of the spacecraft to achieve low toxic gas concentrations. While both of these methods are used in the Apollo spacecraft, Honeywell has been concerned with the material selection problem. This selection had to be made from a lengthy list of materials and alternates submitted by development engineers. The list contained numerous polymeric materials, such as plastics, elastomers, silicones, and organic lubricants. Such materials are potential sources of atmospheric contamination by a variety of physical and chemical processes. The more probable of these processes are depicted in Fig. 1.

Process (1) illustrates the case where a species is physically entrapped and diffuses unchanged through the polymeric material into the gas state. Process (2) is the same process followed by a chemical reaction in the gas state between the diffusing species and the oxygen of the atmosphere, resulting in new products. Process (3) represents a two-phase or surface reaction between the polymer and the oxygen of the atmosphere, again resulting in a new product. Process (4) illustrates the case where a chemical reaction with the polymer generates the species which then diffuses out into the gas state to become a contaminant. Process (5) is the same as process (4) except that the species decomposes due to light (photolysis), thus resulting in the formation of new contaminant.

In addition to the large number of materials to be tested, the problem of atmospheric control through material selection

TABLE I

Maximum Allowable Concentrations

Compound	ppm
Acrolein	0.5
Carbon disulfide	20.0
Benzene	35.0
n-Butanol	50.0
Methyl isobutyl ketone	100.0
Trichloroethylene	100.0
Toluene	200.0
Xylene	200.0
Acetone	500.0

*These values are recommended by the American Conference of Government Industrial Hygienists.

is further complicated by the concentration in parts per million of the contaminants evolved into the enclosed atmosphere. In the ordinary terrestrial atmosphere, the concentration of these evolved contaminants presents no appreciable problem, but, in the spacecraft, this concentration becomes significant because the environment is closed and many of these evolved organic contaminants are toxic even in low concentration. This is illustrated in Table I which lists the maximum allowable concentrations for certain organic compounds.

It is to be noted that the concentrations given as the maximum allowable toxicant concentration must be significantly lower. Analysis for these low concentrations requires the high sensitivity available in the ionization detector in the gas chromatograph.

Still more aspects of the problem require consideration. Evaluation of contaminants from a material is dependent on several factors, such as past history and the environment to which it is subjected. Complete analysis of the contaminants evolved from a material under all conditions is a difficult and time-consuming task, especially for the large number of materials to be tested. Due to the Apollo time schedule, time limitation was the governing consideration in determining the methods and extensiveness of the attack on the problem of controlling the Apollo atmosphere.

The solution consisted of four basic steps in the analysis of the materials: preparation of gas samples from the material specimens, a pre-screening system for rapid elimination of the more objectionable materials, a final analysis by gas chromatography, and normalization of data.

Preparation of Gas Samples from Material Specimens

The first item in the gas analysis was the preparation of gas samples containing contaminants from material specimens. This required designing and constructing a sealed-chamber gassing apparatus. This apparatus, illustrated in Figs. 2 and 3 consists of an all-glass reactor which is sealed with a base plate

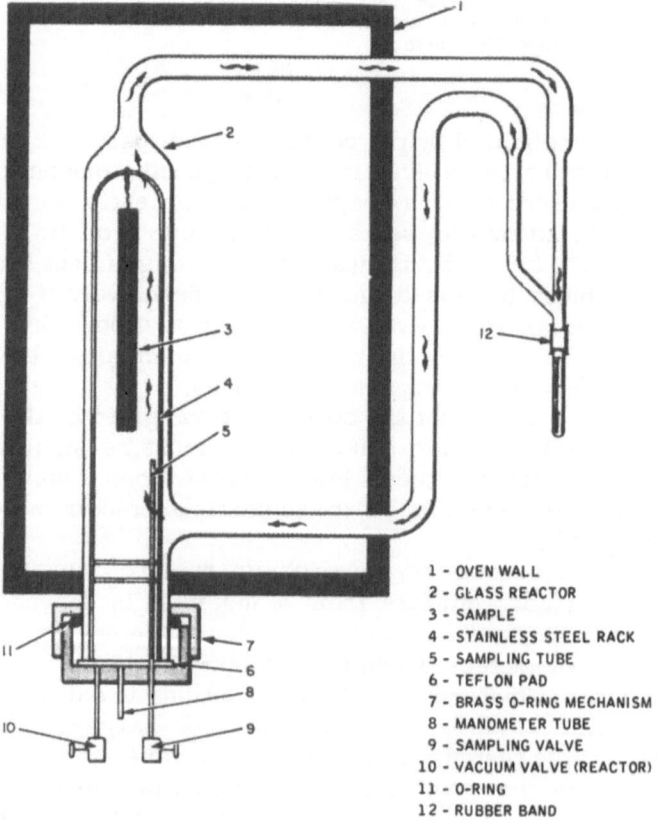

1 - OVEN WALL
2 - GLASS REACTOR
3 - SAMPLE
4 - STAINLESS STEEL RACK
5 - SAMPLING TUBE
6 - TEFLON PAD
7 - BRASS O-RING MECHANISM
8 - MANOMETER TUBE
9 - SAMPLING VALVE
10 - VACUUM VALVE (REACTOR)
11 - O-RING
12 - RUBBER BAND

Fig. 2. Gassing apparatus.

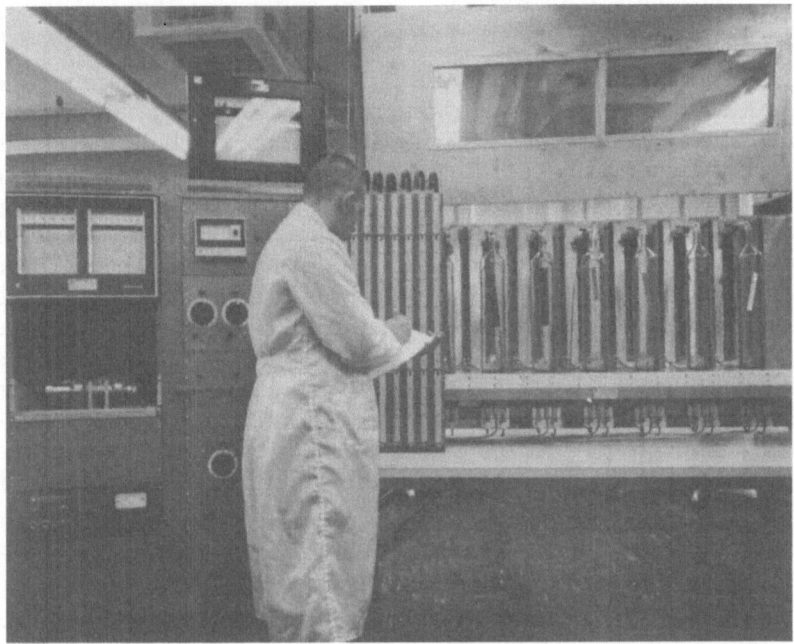

Fig. 3. Gassing equipment.

containing an O-ring type seal. The gases within the reactor are controlled and sampled by special valving in this base plate. The major portion of the reactor is contained in an oven which will maintain an elevated temperature. The remaining portion of the reactor penetrates the oven wall to form an external room-temperature trap. In actual use, several reactors were combined in the oven for simultaneous use.

The next step was the development of a standard set of conditions for evolution of the contaminants with consideration for the time limitation. These conditions were as follows:

1. Samples were prepared according to specifications for the use of each one in the spacecraft. These specifications covered composition, curing conditions, cleaning, and the like.
2. Exposure of the samples would be identical to the spacecraft atmosphere, 5 psi pressure of oxygen.
3. Testing temperature was set at 200°F. This is above the maximum temperature to which any material in the

spacecraft is expected to be subjected; thus, a safety margin was provided as well as acceleration of the tests.

In preparation of the actual gas samples containing the evolved contaminants and the subsequent sampling, an orderly procedure was maintained. Test specimens of bulk materials were prepared in the form of $\frac{1}{2}$-in. square bars. Surface finish specimens were coated on $1\frac{1}{4}$-in. aluminum tubing. These specimens were placed in the portion of the reactor inside the oven, then the reactor was evacuated and back-filled with oxygen at 5 psi. The oven was brought to temperature and maintained there for several days. During this time, the gases within the reactor circulated, due to convection, as indicated by arrows in Fig. 2. This circulation resulted in condensation of the less volatile contaminants in the room-temperature trap. At the end of the heat cycle, samples of the gases within the reactor were removed through the valving in the base plate with a 100-ml glass syringe.

Prescreening of Materials

The next step was a quick elimination of obvious undesirable materials to save time. This prescreening consisted of three tests, which established the rejection criteria for the prescreening of all materials. Rejections were based on irritating odor, excess concentration of carbon monoxide, or excessive total hydrocarbon content.

The first test consisted of the gas samples from the reactor being subjected to a panel of three people who rated the odor as undetectable, detectable, objectionable, or irritating. It is interesting to note that the human nose is still one of the more sensitive detectors. Any material which evolved an irritating odor was eliminated from use in the spacecraft, since the irritating odor would interfere with the performance of the astronauts.

The second prescreening test was to determine the carbon monoxide content of each gas sample from the reactor. This was determined by use of an indicator tube. In the cases of high concentration, the determination was verified by gas chromatography. This verification method consisted of using a gas chromatograph equipped with a thermoconductivity detector, molecular sieve column, and helium as a carrier gas. The actual concentration was determined by comparing the peak

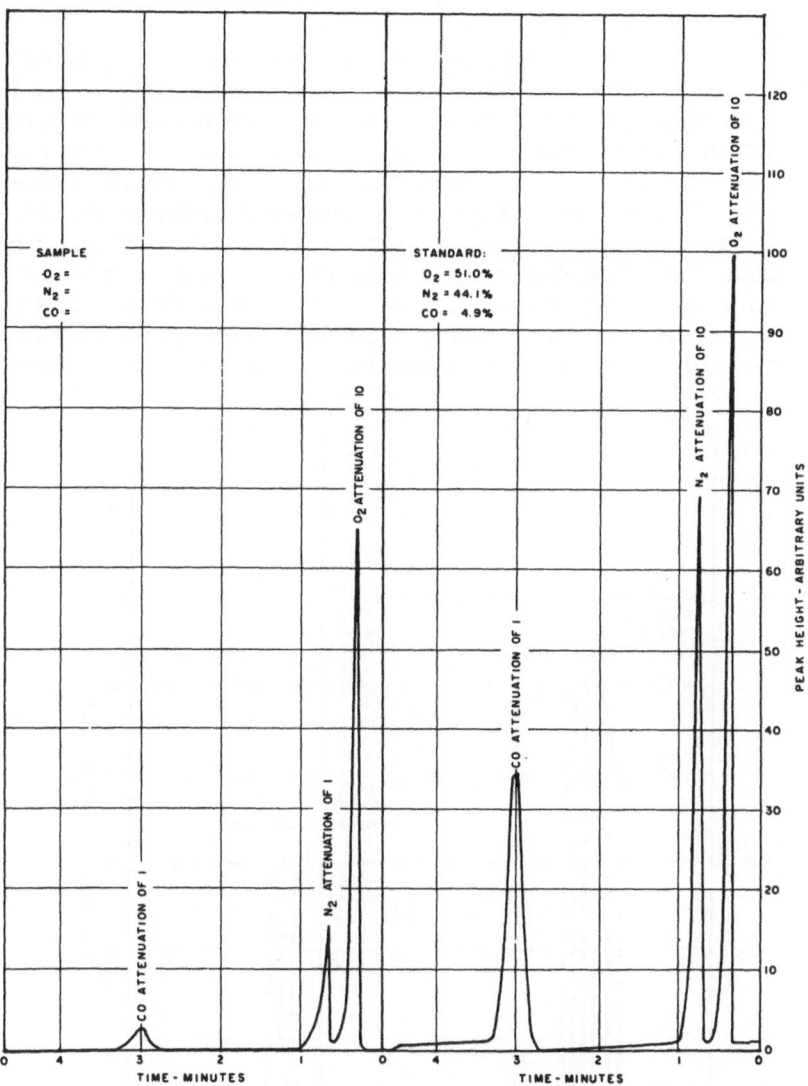

Fig. 4. Example of typical fixed gas chromatogram of standard and sample.

height of the sample with that of a standard obtained from
Matheson Company, as illustrated in Fig. 4. Any material
evolving over 100 ppm of carbon monoxide gas was restricted
from use in the spacecraft.

The final prescreening test was the measuring of total
hydrocarbon content. This was done by using a Wilkens Hy Fi
gas chromatograph with flame ionization. The column consisted
of crushed glass with no liquid phase, which would allow control
of the flow rate without separating the sample into component
peaks. The samples of reactor gases were injected through a
gas sampling valve with a 1-ml volume. The hydrocarbon
content was determined by comparing the peak height of the
sample with that of a pentane standard. The pentane standard
was prepared by evaporating a known volume of pentane in a
known volume of air. The standard sample was run immediately

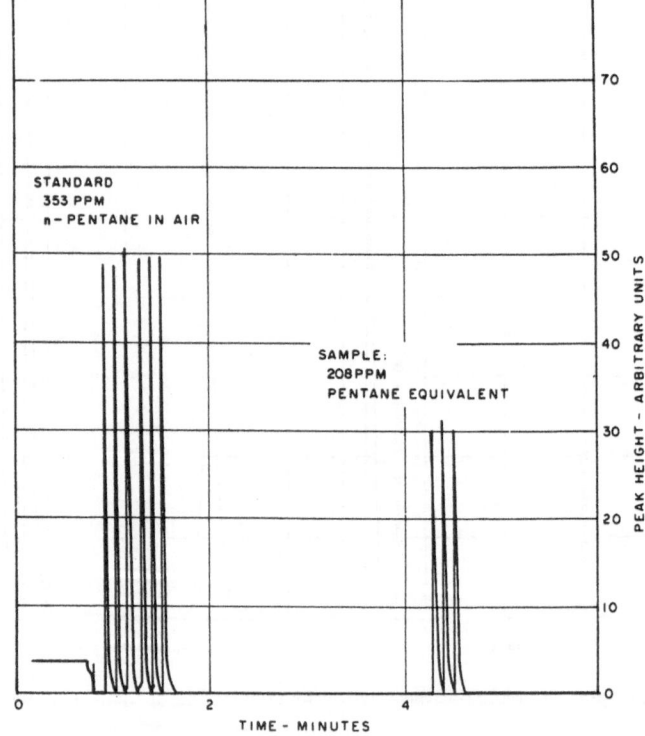

Fig. 5. Example of typical total hydrocarbon content of standard and sample.

prior to the test sample from the reactor, so that direct comparison was possible. The total hydrocarbon content was calibrated in terms of parts per million pentane. Any material having a higher total hydrocarbon content of 250 ppm was restricted from use in the spacecraft. Figure 5 illustrates the comparison of standard and sample.

Final Analysis

The materials which passed the three pre-screening tests were tentatively approved for spacecraft use; however, final approval depended on quantitative and qualitative analysis of the evolved contaminants from each material. The gas chromatograph used for this analysis was a Jarrell-Ash Model 700 gas chromatograph equipped with an argon diode detector. The injection system consists of a gas sampling valve containing a 1-ml sample loop. Although the gas chromatograph was equipped with a stream-splitter, it was closed off to force the entire sample through the detector for greater sensitivity.

All samples from the numerous materials were handled according to the same definite procedure. The steps used in setting up this procedure on the gas chromatograph are as follows: (1) establishment of three columns to cover the spectra of organic compounds, (2) determination of a suitable means of standardizing, so that various runs are comparable, (3) determination of the retention times for various organic compounds of known structure on each of the three columns, and (4) calibration of peak areas in terms of concentration.

The first step in the procedure was to construct columns that would give separation of the various organic compounds. Some difficulty was encountered in making these columns for, in addition to the usual problems of separation, it was found that many columns would absorb oxygen-containing compounds, such as alcohols and ketones, in the concentrations of 25 ppm. This was attributed to absorption by the treated diatomaceous earth support.

After considerable effort, three columns were established. They were Carbowax 1540, di-n-decyl phthalate, and silicone DC200 with a viscosity of 300,000 cs. Each of these columns was packed in $\frac{1}{8}$-in. stainless steel tubing with a liquid phase of about 5%. The DC200 column did not generally show oxygen-containing compounds.

After construction of the three columns, the operating

TABLE II

Column Parameters

Column	Pressure, psi	Flow rate, ml/min	Low temperature, °C	High temperature, °C
Carbowax 1540	18.5	20	60	100
Di-n-decyl phthalate	33.5	15	75	115
Silicone DC200	21.5	20	60	100

conditions were determined for each column and fixed analysis procedure established. Since the material samples in the reactor were maintained at 200°F, the organic contaminants were fairly volatile. A temperature programming procedure was established which required 1 hr, 40 min at a low temperature, 10 min of temperature programming at 4°F/min, and 10 min at the high temperature. The operating conditions for each column are summarized in Table II.

The temperature programming and the high temperature not only forced out the less volatile compounds, but also purged the column for the next run. Following this 1 hr procedure caused the retention times to be reasonably repeatable.

Next, a standardization means was established. In order to compensate for the small variations in retention time due to slight deviations in column temperature and flow rate, an internal standard was employed. This consisted of injecting a known concentration of organic vapor into gas samples, thus creating a peak on the chromatographic chart. By use of a simple proportionality, a correction factor was obtained and used to correct the other peaks on the chromatographic chart. Chlorobenzene was chosen as the organic vapor because it falls in the mid-portion of the chromatographic run. This type of standardization will compensate for variations in temperature and in flow rate from one run to another, but it will not compensate for fluctuations within the same run.

The third step in the procedure was the determination of retention times for the various organic compounds of known structure. This was necessary to calibrate the columns for qualitative analysis. Gas samples of known concentration were made up using apparatus as shown in Fig. 6. The flask

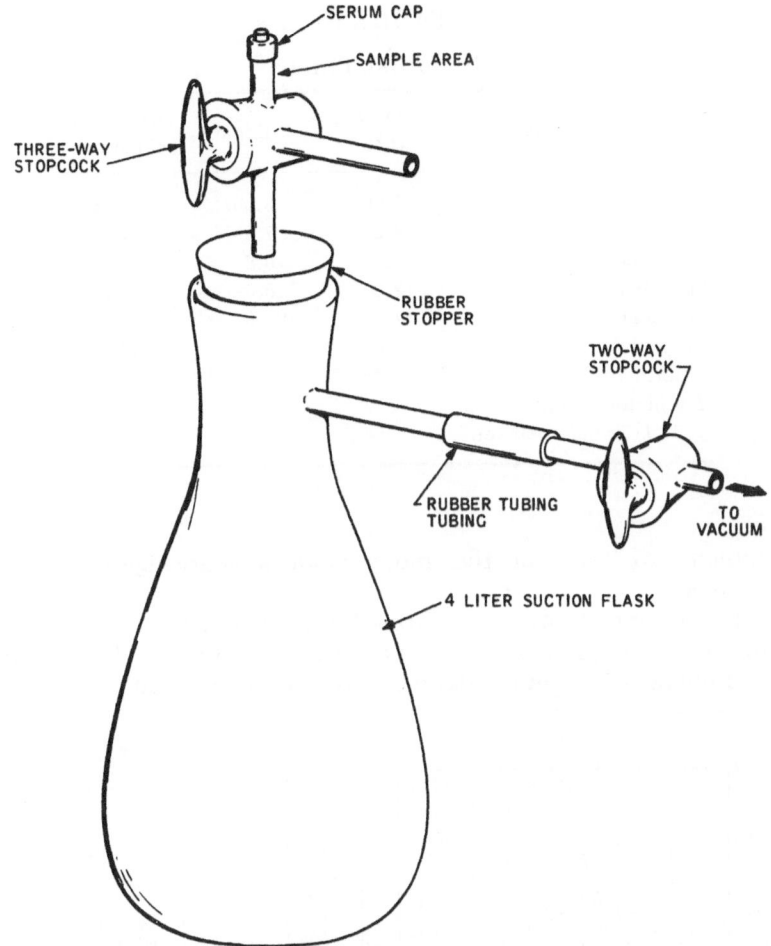

Fig. 6. Gas standards apparatus.

was evacuated except for the sample area which was kept at atmospheric pressure. A known volume of an organic liquid was injected into the sample area with a microliter syringe. The sample area was then opened to the flask and the sample allowed to evaporate in the evacuated flask. The flask was then backfilled with air, thus forming a gas sample of known concentration to be used as a standard. These samples, which were treated with chlorobenzene vapor, were injected into the gas chromatograph and the retention time for each column determined. Table III lists the retention times on the three

TABLE III

Retention Times

Compounds	Carbowax 1540	Silicone	Di-n-decyl phthalate
Benzene	3.06	3.28	6.36
Toluene	6.28	7.80	16.0
Ethyl benzene	11.6	16.8	34.0
m-Xylene	12.8	18.0	39.6
o-Xylene	17.1	21.6	45.8
Cumene	16.3	28.0	47.7
Chloroform	5.48	2.09	5.74
Trichloroethylene	4.41	4.33	8.97
1, 1, 1-Trichloroethane	1.95	2.80	5.15

columns for part of the more than seventy hydrocarbons examined.

The fourth step, the calibration of peak areas in terms of concentration, provided for quantitative analysis. By varying the amount of liquid injected into the standards apparatus,

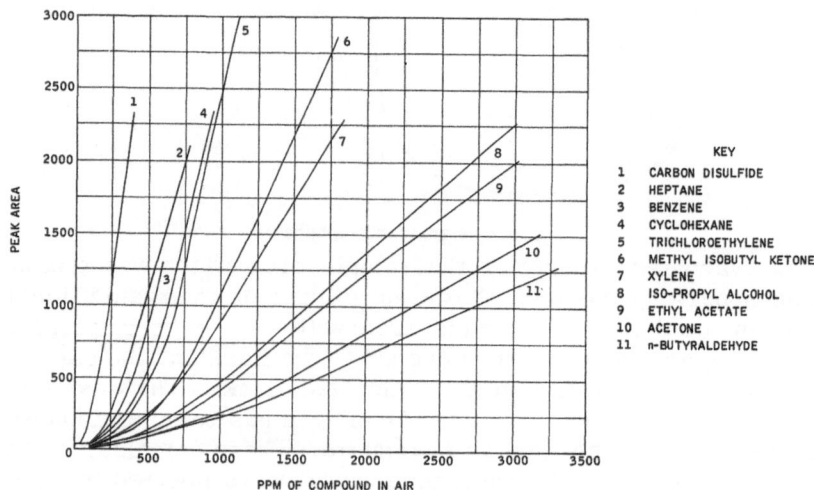

Fig. 7. Chart for peak area versus parts per million of various compounds.

standards of various concentrations were produced. After introduction of these standards into the chromatograph, the areas of the resulting peaks were measured using a planimeter. The known concentration values and the peak area measurements were plotted, as shown in Fig. 7, to provide standards for quantitative measurements.

After calibration of the retention times and areas of the peaks eluted from each column, the chromatograph was ready for analyzing the reactor gas samples from each of the various materials. However, before withdrawing of the gas samples from the reactor, the chlorobenzene internal standard had to be added to the samples. This was accomplished by making a known concentration of chlorobenzene vapor in an inert carrier gas and injecting a known volume of this vapor into the reactor. The gases were allowed to mix for about an hour before a sample was withdrawn and injected into the chromatograph.

After the reactor gas sample had been run through the chromatograph at the proper attenuation, the chromatographic chart was removed from the recorder and pertinent information added. The retention time of each peak on the chart was determined and corrected by use of the chlorobenzene correction factor. The corresponding area of each peak was then measured with a planimeter. The corrected retention times and the areas were recorded directly on each chart. Figure 8 illustrates a typical chromatographic chart. This procedure of injection and chart analysis was repeated on all three of the columns, thus resulting in three sets of corrected retention times and areas.

Identification of the various peaks was accomplished by comparing the corrected retention time of the peak on one column with the retention times of known organic compounds. Since positive identification was required, a corresponding peak of the correct area and retention time had to appear on the other two columns.

The concentration of a contaminant in the reactor was determined from the area of the peak on the chromatographic chart. Such an area can be directly converted into concentration by choosing the proper curve on the concentration versus area plot, typified by Fig. 7. In cases where the peak could not be identified, the methyl isobutyl ketone curve was used, since this was an average curve.

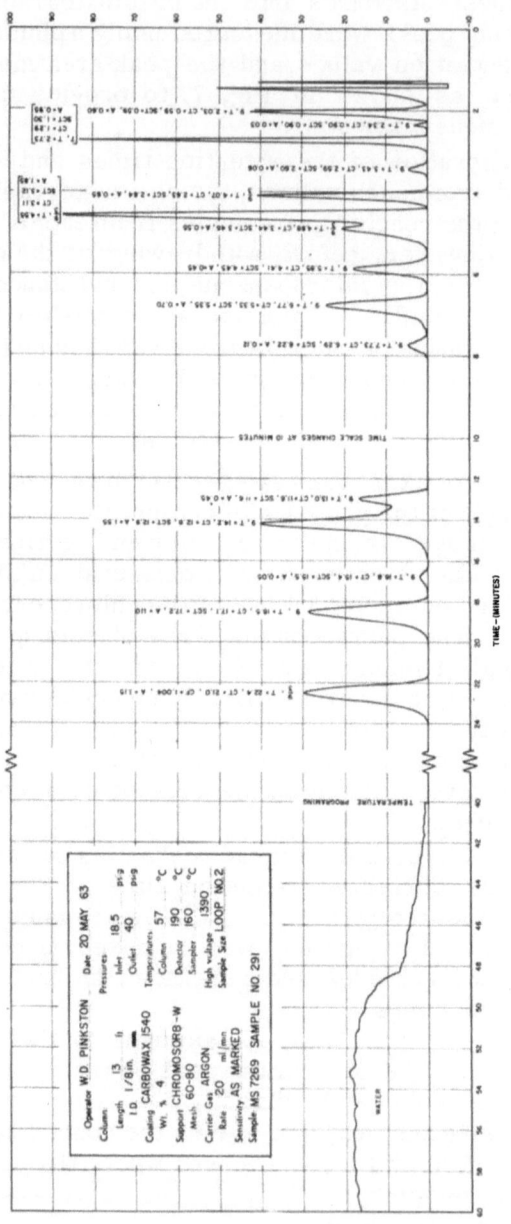

Fig. 8. Example of gas chromatographic chart.

Normalization of Data

In order to make the concentration of a contaminant independent of sample weight and chamber volume, the concept of atmospheric contaminating potential (*ACP*) was developed. This concept is defined as a quantitative measure of the amount of a particular contaminant produced by a given weight of a given material. It is calculated by using the following equation:

$$ACP = \frac{V_t C_t}{f M_t}$$

where V_t is the volume of the test chamber, C_t is the concentration of contaminant evolved from the material in the test chamber in parts per million by volume, f is the gram molecular volume, and M_t is the weight of the material test sample. The normal units for *ACP* are micromoles of contaminant per gram of sample material. It should be realized that each material will have one *ACP* value for every contaminant it evolves.

The *ACP* values are useful in two ways: as criteria for selecting materials for use in manned spacecraft and in estimating space cabin contamination arising from the types and amounts of the various materials used in the spacecraft.

RESULTS AND TESTING

During this material selection program, over 300 gassing tests were carried out on 150 different materials. The contaminants evolved from these materials included 40 different organic compounds listed in Table IV.

These contaminants, as previously stated, result from entrapped solvents and impurities, and chemical reactions. An example of a contaminant resulting from a chemical reaction is carbon monoxide. A surprising number of materials, including thermosetting plastics, adhesives, and finishes, evolve small amounts of carbon monoxide. This probably is a result of surface oxidation or decomposition of organic structures, such as that of a carbonyl group in the presence of ultraviolet light.

TABLE IV

Some Organic Contaminants Evolved from Spacecraft Materials

Amyl acetate	Cumene	Methylene chloride	Propionaldehyde
Acetone	Ethyl alcohol	Methyl chloroform	Pentane
Acetaldehyde	Ethyl benzene	Methyl ethyl ketone	sec-Butyl alcohol
Acrolein	Ethyl acetate	Mesitylene	tert-Butyl alcohol
Acetylene	Heptane	2-Methyl furan	Trichloroethylene
Butyl acetate	Hexane	Methyl isobutyl ketone	Tetrahydrofuran
Benzene	Iso-amyl alcohol	Methyl methacrylate	1, 2, 4-Trimethyl benzene
Carbon disulfide	Iso-butyl alcohol	n-Butyraldehyde	Toluene
Cellusolve acetate	Iso-propyl alcohol	n-Butyl alcohol	o-Xylene
Cyclohexane	Iso-valeraldehyde	n-Valeraldehyde	m- and p-Xylene

The worst group of materials for evolving contamination were the paints and finishes. This is an unhappy circumstance, for large quantities of these are needed to cover exposed surfaces within the spacecraft. The required amounts of finishes and paints thus present a definite toxic hazard. The contaminants from this group result largely from entrapped solvents, as would be expected.

In the case of one particular silicone paint, large quantities of benzene were detected. Upon close examination, it was found that the benzene did not occur as a solvent, but was rather a result of chemical reaction. The source of the benzene appeared to be the phenyl siloxane binder. This was probably due to catalytic cleavage of the phenyl—silicon bond.

Some of the elastomers tested gave interesting results. Many natural and synthetic rubbers vulcanized with sulfur evolved considerable quantities of carbon disulfide. While some RTV silicone rubbers were found to give off alcohol, these alcohols were a result of catalytic action during the curing reaction.

The thermoplastics—polyethylene, Teflon, Kel F, and nylon—evolved extremely small amounts of contamination. This was a result of handling and soiling of the surfaces of these materials rather than a property inherent in the thermoplastics, themselves.

ACKNOWLEDGMENT

The investigation reported in this paper was conducted as part of a program funded by the Space and Information Systems Division of North American Aviation, Inc. The program is related to the Apollo manned lunar landing effort.

A Novel Method for Collecting Samples for Infrared Identification

Lillian Churchill

Universal Oil Products Company
Des Plaines, Illinois

The selection of the best technique and apparatus for collecting individual components from gas chromatographic separations depends on the type of compound, its concentration in the sample, and the separation available. A new technique is described for concentrating an impurity in the difficult case of a sample having an impurity of less than 10%, which impurity elutes just before the major peak (> 80%) with almost baseline resolution at this concentration, and the retention time of which is several minutes. Concentration of this minor component is done on the collector which, after several trappings, also becomes the sample introduction instrument. Following the concentration step, the minor component may be adequately resolved on the analytical column and collected from it by such means as deposition on a salt plate or condensation in a microcell. This concentration technique allows the analytical column used to accomplish the initial separation also, to be used instead of a preparative column in the separation of minor components for subsequent identification by infrared examination. Much time is saved in comparison with that required for the conventional, more cumbersome preparative column technique, and the same result is accomplished.

INTRODUCTION

Several methods—among them direct trapping into micro infrared cells, direct deposition on a salt plate, and condensation into specially-designed traps—are available for collecting individual peaks, or components, after they have been separated by gas chromatographic techniques. The selection of the best trapping technique and apparatus depends on the type of compound to be collected, its boiling point, the quantity required, and the separation available on the gas chromatographic column.

The direct conventional methods do not work well if the quantity of component to be collected is a very small portion of the total sample and so closely adjacent to the elution peak of a major component that it is difficult to resolve when large samples are chromatographed. Thus, an improved technique is highly desirable.

The usual maximum amount of an injected sample on an analytical column ranges from 2 to 20 μl, for example. Even if 50 μl were injected, an impurity of less than 10% concentration with complete (100%) collection represents only 5 μl of material. This 5 μl of component is diluted by a large volume of carrier gas when eluted and, therefore, is difficult to trap by condensation. Loss of resolution on analytical columns occurs rapidly if the sample size is further increased in an attempt to secure the usual volume of 50 μl of pure minor component needed to provide adequate microsamples for identification with infrared, mass spectrometer, and nuclear magnetic resonance spectrometry.

An example of such a problem is a sample of N-cyclohexylaniline containing an impurity of 3.7% which required separation and identification. In this sample, the impurity was eluted just before the major peak with almost baseline resolution on the analytical column.

Although preparative gas chromatography can be considered in many such cases, preparative columns often do not have the same resolution that is obtainable from analytical columns, and also require higher temperatures or have longer retention times. The time spent in establishing conditions and making the columns for a preparative run is often several days. Therefore, an attempt was made to develop a technique which could use the analytical column and the conditions originally employed, tolerate a somewhat larger sample, and ultimately provide a sample free from the major peak.

A novel concentration technique was developed which employs a collector tube to concentrate the unknown by trapping the component and some of the interfering adjacent peaks on a solid support. The concentration tube containing the trapped material then becomes the sample introduction instrument. The trapped portion of the original sample is rechromatographed, and the material, now more concentrated but still well resolved, can be collected by the usual devices, such as deposition on salt plates. The total time elapsed for concentrating and rechromatographing usually is much less than the preparative scale methods, and often a purer sample will be obtained. Whenever larger quantities (>100μl) of the material are needed, however, conventional preparative techniques should be considered.

EXPERIMENTAL AND RESULTS

This method requires a chromatograph that has an easily accessible column oven and a column that can be changed easily. The collection tubes are made from the same size tubing as the column and are provided with the proper fittings to enable them to be attached in front of the column. A special adapter or means of connection to the exit port is also necessary. The tube is filled with either dry Chromosorb, some preconditioned packing material (the same packing as in the column), or some other solid, inert support which has high surface area. Certain materials, such as alumina and silica gel, should not be used as a trapping medium when the analytical column packing is of the Chromosorb type. The affinity of alumina and silica gel for many hydrocarbons differs so greatly from that of Chromosorb that serious interference with the final trapping may occur. The collectors may range from $1\frac{1}{2}$—6 in. in length. Gas chromatographic conditions used are the same as for the initial analytical run where the presence of the impurity was first noted, if possible, but not all gas chromatography units have the same features.

Cooling the collection tube with dry ice during the trapping period provides more complete collection.

The number of collections is dependent on the concentration of the material to be collected and the portion trapped each time. Usually four collections are made. After these collections, the column in the chromatograph is cooled below the starting temperature, bridge current is shut off, and the collector is connected to the front of the column. The bridge current is turned on, the temperature is raised to the starting temperature and then programmed or raised to the isothermal temperature used to approximate the same conditions used for the original separation. If the original analysis used temperature programming, the retention time will exhibit almost no change from the one originally obtained.

If the sample was chromatographed using isothermal

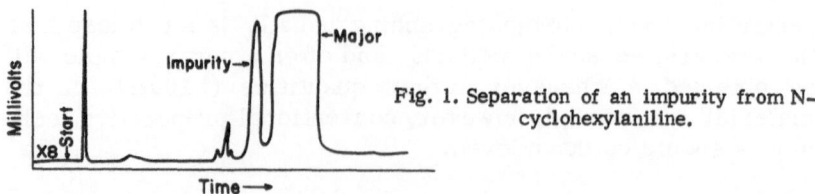

Fig. 1. Separation of an impurity from N-cyclohexylaniline.

temperatures, the retention time will increase and better resolution in some cases will result, because of the nonlinear temperature programming which occurs before the isothermal temperature is reached. The component which is now in greater concentration and which will exhibit the original resolution (Fig. 1) from the major peak is collected on a salt plate and the infrared spectrum obtained.

A sample of N-cyclohexylaniline containing an unidentified peak of 3.7% concentration was tried with this collection method. The impurity peak which preceded the major peak was fairly well resolved (almost to the baseline) when a 10 μl sample was used, but, when 100 μl was injected, the chromatogram showed poor resolution (Fig. 2). The impurity and some of the overlapping major peak was collected, using a 100 μl volume of sample, and the infrared spectrum was obtained.

Although a cleaner spectrum was secured, the same peaks

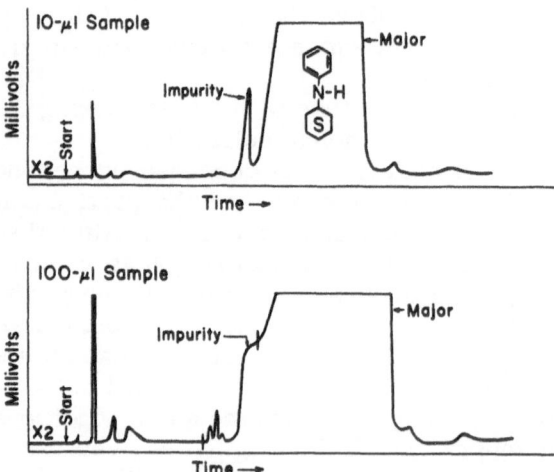

Fig. 2. Comparison of 10- and 100-μl sample sizes in the separation of an impurity from N-cyclohexylaniline.

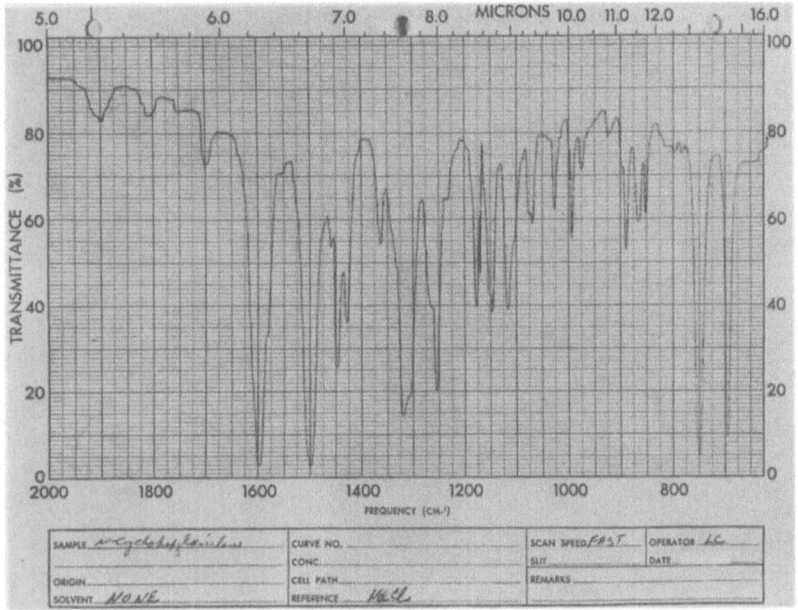

Fig. 3. Infrared spectrum of N–cyclohexylaniline.

as the major component were still shown. The material on the salt plate was washed off, and the solution obtained was rechromatographed. A large quantity of the major peak was found in the sample. Figure 3 is an infrared spectrum of the pure major component which was trapped. The experiment was repeated, but an additional heater was added to the exit line and the exit line was flushed with solvent before the final collection was made, in order to take extreme care to keep the exit line free from contamination.

The material collected this time gave a spectrum which was free from the adsorption bands of the major component. A chromatographic check of the material trapped on the plate revealed it to be quite free of the major peak.

In the above manner, a sufficient amount of material was collected to obtain the following information: The compound was shown to contain a strong carbonyl band at 1712 cm^{-1}, have dicyclic character, and have bands at about 1445 and 1125 cm^{-1} (Fig. 4). The spectrum resembles that of 2-cyclohexylcyclohexanone, but some of the minor bands do not correspond exactly. Mass spectrometry gave a molecular weight

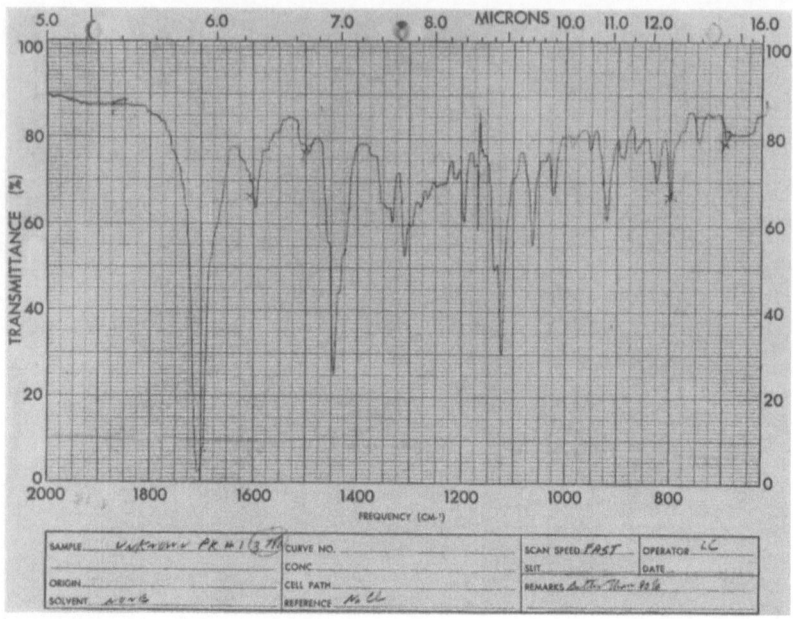

Fig. 4. Infrared spectrum of an impurity separated from N-cyclohexylaniline.

of 178. This information was deemed adequate, without need to specifically identify the impurity, because it indicated that the impurity could be removed by chemical means. The spectra were obtained on a Perkin-Elmer 237 grating infrared spectrophotometer.

Determination of Traces of Glycols by Gas Chromatography

Abram Davis, Arthur Roaldi,
and Lewis E. Tufts

Research Center
Hooker Chemical Corp.
Niagara Falls, New York

Fractional parts per million of ethylene glycol, propylene glycol, and diethylene glycol in water are determined by gas chromatography using a flame-ionization detector, a mixture of helium and steam as the carrier gas, and teflon as the column support. A prior distillation to remove most of the water and other more volatile solvents reduces the limit of detection to 0.02 ppm glycol. Traces of glycols in water-immiscible organic solvents are determined by extracting the glycols into water before analysis.

INTRODUCTION

No suitable procedure was found in the literature for the analysis of glycol in water by gas chromatography at the parts-per-million level. Nadeau and Oaks [1] analyzed glycol samples containing up to 90% water using a thermal-conductivity detector. Bennett, Dal Nogare, Safranski, and Lewis [2] were able to detect 2 ppm methanol in water by amplifying the signal from a thermistor detector. The glycols of interest in the present problem elute later than water. Hence, these must be detected in the troublesome tail of the water peak.

A flame-ionization detector was used because of its high sensitivity and relative insensitivity for water [3,4]. Teflon column support, together with a flame-ionization detector, greatly reduced the effects of water on the glycol analysis. Actually, a negative peak for water is observed when a high concentration of water vapor passes through the flame detector. Possibly, this is due to a change in flame temperature.

289

Preliminary experiments with dry helium carrier gas, teflon column support, and a flame-ionization detector provided sufficient reduction in tailing of the water peak to yield detection limits of 100 ppm of the glycols. However, subsequent injections of pure water produced glycol peaks indicating a "memory" or "repeater effect" [5-7].

Several special carrier gases have been used to reduce memory effect and tailing. Examples are ammonia [8], carbon dioxide [9], formic acid [5], and water vapor [10]. Large water injections have also been used to remove material absorbed on GC columns.

It was hoped that the use of steam as a carrier gas would eliminate the memory effect [11]. The use of steam carrier gas from a Wilkins Instrument steam generator did eliminate memory effects and tailing. A major disadvantage of pure steam as the carrier gas was a high noise level, allowing a minimum detectable level of only 10 ppm ethylene glycol.

A mixture of helium and steam as carrier gas greatly improved the signal-to-noise ratio, while still eliminating tailing and memory problems. Column conditioning [13] is also very important when steam is used as a carrier gas. Three to four weeks were needed for good conditioning.

The best results were obtained with a mixture of helium and steam as the carrier gas, poly-m-phenyl ether (6-ring) as the liquid substrate coated on a teflon column support, and a flame-ionization detector. This allowed detection of 0.2 ppm of ethylene glycol in water.

A further increase in sensitivity was achieved by distillation to concentrate the glycols by a factor of ten.

EXPERIMENTAL

Apparatus

A Perkin-Elmer model 154-C gas chromatograph and hydrogen-flame detector was used throughout this investigation. The GC inlet system was changed to allow injection at the column head. The total column effluent was passed through the flame-ionization detector. Figure 1 shows schematically the flow system used for mixing carrier gas and minimizing dead volume.

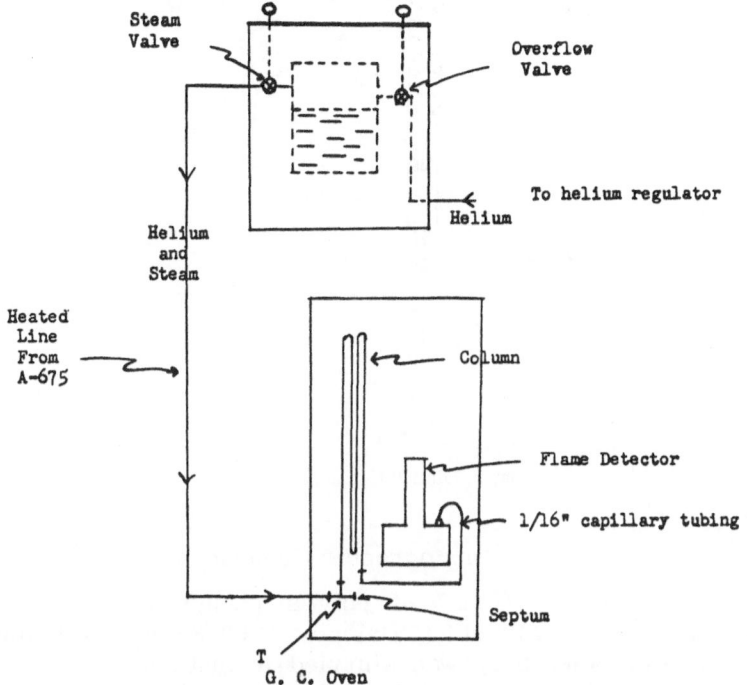

Fig. 1. Schematic diagram of the flow system.

A Wilkins Instrument Corporation steam generator [12], model A-675, was modified to prevent temperature cycling by connecting the heater directly to a variac which supplied continuous power at a low-voltage setting. The water reservoir was further insulated to reduce slight variations in steam temperature due to ambient temperature changes.

Preparation of Samples

The glycols used in preparing standard samples were obtained from Eastman Organic Chemicals Lab. The distilled water used did not show peaks in the glycol region.

The simple distillation apparatus used to concentrate the glycol is shown in Fig. 2. A 25-ml sample of the water glycol was distilled leaving a volume of 2.5 ml. No indication of glycol loss from this distillation was detected. The expected 10:1 increase in sensitivity was obtained.

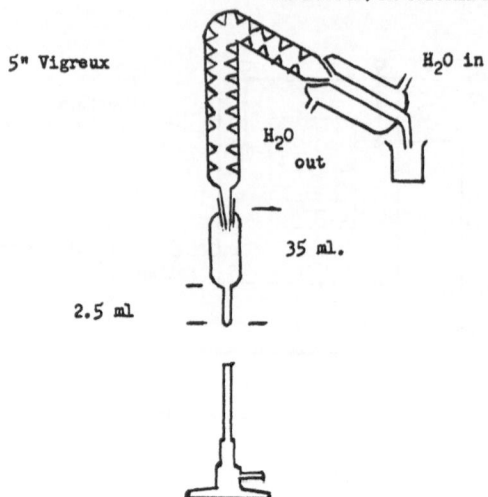

Fig. 2. Distillation apparatus.

Gas Chromatographic Conditions

A mixture of 10 wt.% of poly-m-phenyl ether, 6-ring, (Wilkins Instrument) and 90% Haloport F 60/80 mesh (obtained from F & M Scientific) was slurried in methylene chloride. To prevent agglomeration of the teflon particles, the methylene chloride was stripped off at low temperature in a rotating vacuum evaporator. Two meters of $1/4$-in. stainless steel tubing was packed using care to avoid excessive tamping.

The column was conditioned at 180°C with pure steam at 130°C (flow, ~80 ml/min) for four weeks. Periodically, column background was checked operating the instrument under the conditions listed in Table I.

Helium was mixed with the steam by passing the helium into the water overflow fitting of the steam generator. The helium pressure was controlled by a regulator attached to the helium cylinder and steam pressure by choice of temperature.

For maximum sensitivity, it is necessary to adjust elution time of the desired glycol as soon after the negative water peak as baseline stabilization allows. The peaks broaden considerably as elution time is extended, resulting in a decrease of sensitivity. Therefore, for the most sensitive results, diethylene glycol was run at a higher temperature than ethylene and propylene glycol.

TABLE I

Operating Conditions

	Conditions for ethylene glycol and propylene glycol	Conditions for diethylene glycol
Temperature GC oven, °C	110	150
Steam generator temperature, °C	70	100
with Flow, ml/min	55	~50
Helium pressure, psig	6	6
with Flow, ml/min	~30	~28
Hydrogen for flame detector, psig	30	30
Air for flame detector, psig	40	40
Sample size, μliter	50 or 5	50 or 5

RESULTS AND DISCUSSION

Known solutions were run to determine the experimental error and limit of detectability (Table II). Figure 3 is the chromatogram of 1 ppm propylene glycol not concentrated by distillation. Figure 4 is the chromatogram of 0.1 ppm diethylene glycol concentrated by distillation. Figure 5 is the chromatogram of ethylene glycol 0.02 ppm concentrated by distillation. The background of pure water is dotted in on this chromatogram. Separation of ethylene and propylene glycol was not attempted, as the purpose of this work was to detect traces of glycols, not to identify.

CONCLUSION

The use of steam or a mixture of steam and helium as a carrier gas greatly simplifies the analysis of water samples for trace organics. We have extended the method to determinations of trace glycol in hexane. The glycols were extracted into water—this allows for a 20:1 concentration step—and then the method outlined in this paper was applied. This method of analysis can be applied to solvents immiscible with water, where the components of interest have partition coefficients highly favorable for water extraction.

TABLE II

Determination of Experimental Error and Detection Limits for Known Samples

Sample, ppm	Found, ppm	Percent error	Distillation	Sample size, μ liter	Area, cm^2	Minimum detection limit, ppm
Ethylene glycol:						
2	1.9	5.0	No	50	9.2	0.2
2	2.2	10.0	No	50	11.0	0.2
1	0.9	10.0	Yes	5	4.1	0.2
1	0.8	20.0	Yes	5	3.6	0.2
2	1.9	5.0	Yes	5	9.0	0.2
0.1	0.10	0.0	Yes	50	5.0	0.02
Propylene glycol:						
2	2.0	0.0	No	50	9.4	0.2
2	2.1	5.0	No	50	10.5	0.2
1	1.0	0.0	Yes	5	4.5	0.2
0.4	0.40	0.0	Yes	50	4.6	0.02
Diethylene glycol:						
2.5	2.4	4.0	No	50	7.0	0.4
2.5	2.5	0.0	No	50	7.4	0.4
2.5	2.6	4.0	No	50	7.8	0.4
5	5.0	0.0	No	50	147	0.4
1	0.96	4.0	Yes	50	283	0.04
0.1	0.14	4.0	Yes	50	40	0.04
1	0.9	10.0	Yes	5	27	0.4
1	1.2	20.0	Yes	5	35	0.4

Standard deviation—8%.

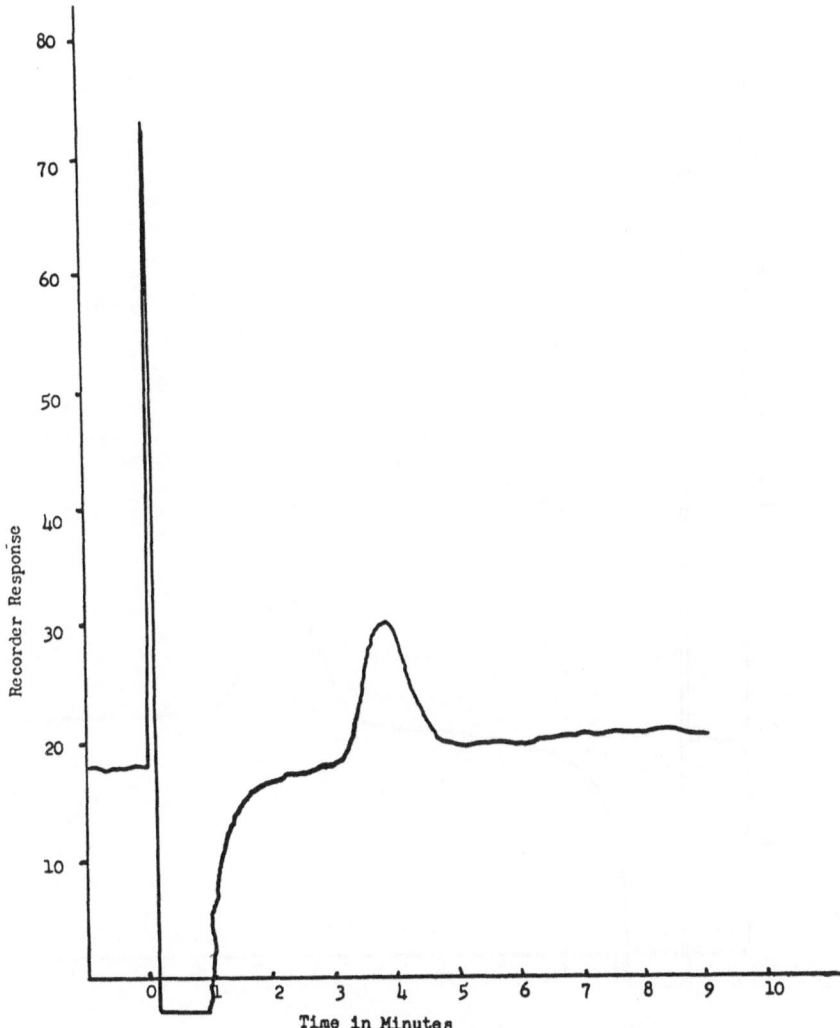

Fig. 3. Chromatogram of 1 ppm propylene glycol, not concentrated by distillation. 50-μliter sample H_2O.

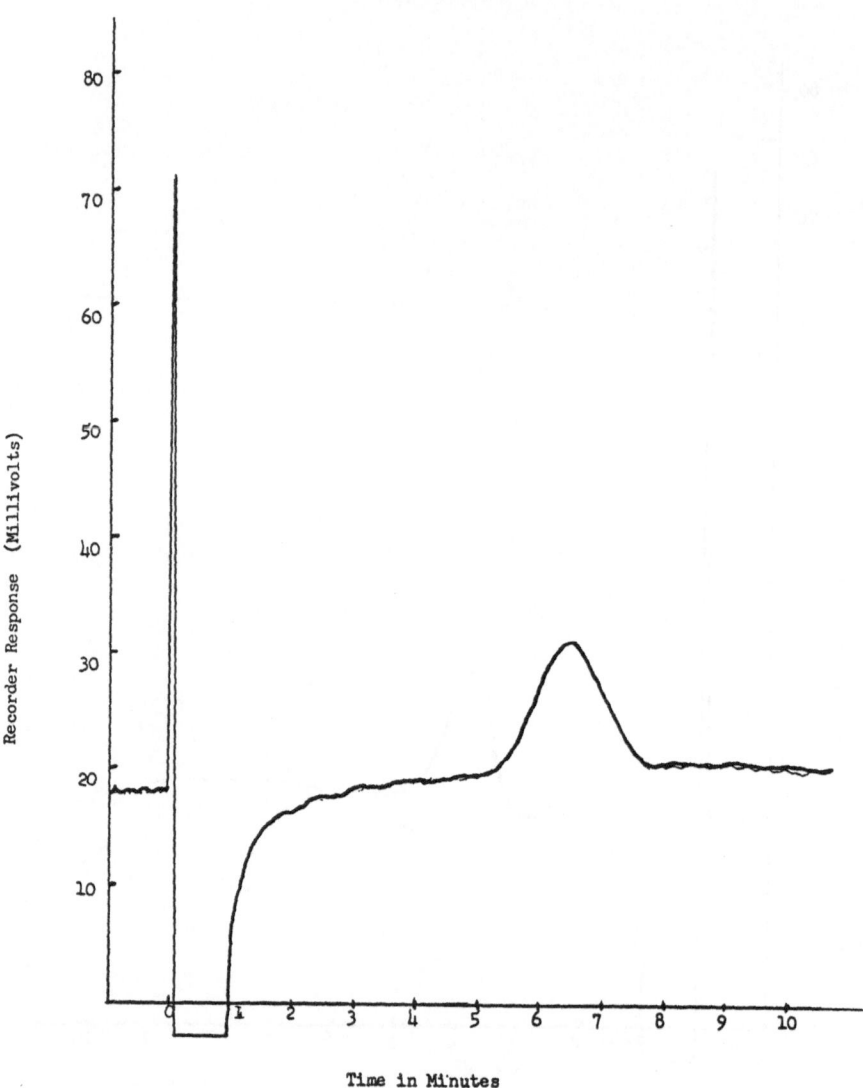

Fig. 4. Chromatogram of 0.1 ppm diethylene glycol concentrated by distillation. 50-μliter sample H_2O.

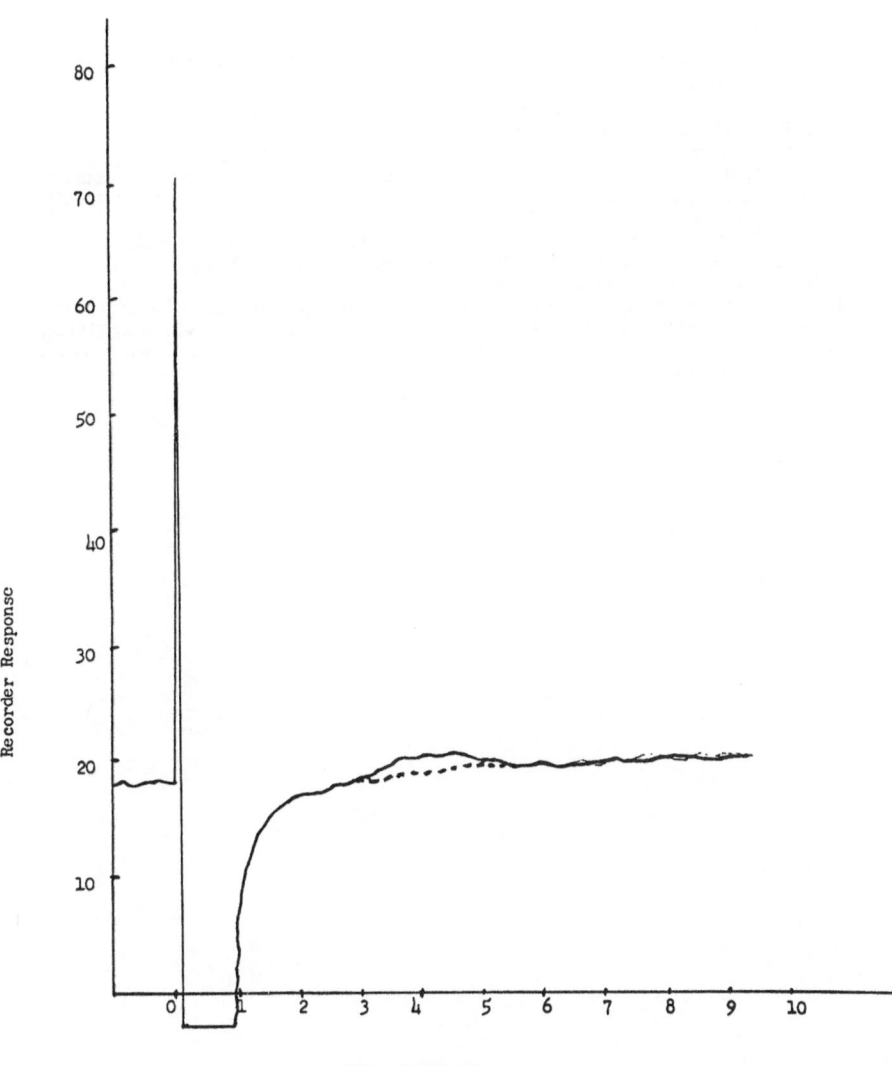

Fig. 5. Chromatogram of 0.02 ppm ethylene glycol concentrated by distillation. 50-μliter sample H_2O.

REFERENCES

1. H. G. Nadeau and D. M. Oaks, Anal. Chem. 32, 1760 (1960).
2. C. E. Bennett, S. Dal Nogare, L. W. Safranski, and C. D. Lewis, Anal. Chem. 30, 898 (1958).
3. E. M. Emery and W. E. Koerner, Anal. Chem. 33, 146 (1961).
4. J. E. Lovelock, Anal. Chem. 33, 162 (1961).
5. R. G. Achman and R. D. Burghen, Anal. Chem. 35, 647 (1963).
6. D. P. Manka, Treatment of columns in gas chromatographic analysis of aqueous solutions, paper presented at the Pittsburgh Analytical Conference, Marcy (1963).
7. E. D. Smith and A. B. Gusnell, Anal. Chem. 34, 438 (1962).
8. H. A. Suroff, A. Karmen, and J. W. Healy, J. Chromatog. 9, 122 (1962).
9. A. Karmen, I. McCafferey, and R. L. Bowman, Nature 193, 575 (1962).
10. H. S. Knight, Anal. Chem. 30, 2030 (1958).
11. Wilkins Instruments and Research Inc., Aerograph Research Notes (1961).
12. Wilkins Instruments and Research Inc., Previews and Reviews, May (1962).
13. Wilkins Instruments and Research Inc., instructions received with A-675 steam generator.

Gas Chromatographic Analysis
of Alpha-Hydroxy Carboxylic Acids

Norman E. Hoffman and Peter J. Conigliaro

Marquette University
Milwaukee, Wisconsin

Dilute solutions of alpha-hydroxy acids in water can be analyzed by oxidation of the acids with periodic acid in the chromatograph. The quantity of aldehyde or ketone produced in oxidation is measured. When a Carbowax 20M-on-firebrick column is used, acid concentration is proportional to aldehyde or ketone peak height.

INTRODUCTION

Recently a gas chromatographic method for analyzing lactic acid in aqueous solution has been developed [1]. The method consists of adding an excess of periodic acid to a sample containing lactic acid and injecting the resulting solution into a gas chromatograph. The periodic acid oxidizes the lactic acid in the chromatograph's injection system, and the acetaldehyde produced is rapidly eluted. The acetaldehyde peak height is linearly related to the concentration of lactic acid in the solution.

$$\overset{\displaystyle OH}{\underset{\displaystyle |}{CH_3\text{-}CH\text{-}COOH}} + HIO_4 \longrightarrow CH_3\text{-}CHO + H_2O + CO_2 + HIO_3 \quad (1)$$

The primary objective of the investigation described herein was to determine whether the method for lactic acid analysis could be applied generally to alpha-hydroxy acids.

$$\underset{\underset{\displaystyle R'}{\displaystyle |}}{\overset{\displaystyle OH}{\underset{\displaystyle |}{R\text{-}C}}}\text{-}COOH + HIO_4 \longrightarrow \overset{\displaystyle O}{\overset{\displaystyle \|}{R\text{-}C}}\text{-}R' + H_2O + CO_2 + HIO_3 \quad (2)$$

The general applicability of the method was not obvious in view of the reported difficulty in oxidizing the alpha-hydroxy acids with periodic acid [2-5] and the possibility of a competitive dehydration reaction.

$$R-CH_2-\underset{\underset{R'}{|}}{\overset{\overset{OH}{|}}{C}}-COOH \longrightarrow R-CH=\underset{\underset{R'}{|}}{C}-COOH + H_2O \qquad (3)$$

These hydroxy acids are not stable at the high temperature required for their elution [6]. Furthermore, even if suitable temperature and column conditions could be found, it is likely that they would tail badly because of the combination of hydroxyl and carboxyl groups in their structure. Therefore, the periodic acid method could offer a simple alternate analytical route to the sometimes time-consuming and troublesome esterification procedure.

In addition to determining the general applicability of the method, other objectives of this work were to determine the relationship between periodic acid—hydroxy acid molar ratio and the extent of oxidation, the relationship between injection volume and response, whether all of the hydroxy acid reacted and all of the ketone or aldehyde produced was eluted, the fate of excess periodic acid, the retention times of the various aldehydes and ketones relative to water, and the extent of adsorption (if any) of the aldehydes and ketones on the column support.

The compounds chosen for investigation were methyllactic acid [R=CH$_3$, R'=CH$_3$ in equation (2)], a-methyl-a-hydroxybutyric acid (R=C$_2$H$_5$, R'=CH$_3$), a-ethyl-a-hydroxybutyric acid (R=C$_2$H$_5$, R'=C$_2$H$_5$), a-hydroxybutyric acid (R=C$_2$H$_5$, R'=H), -hydroxyvaleric acid (R=n-C$_3$H$_7$, R'=H), and mandelic acid (R=C$_6$H$_5$, R'=H).

EXPERIMENTAL

Apparatus

For analyzing acids other than mandelic acid, an Aerograph model 90CS gas chromatograph equipped with a thermal-conductivity detector was used. Helium was the carrier gas. The column was a 5 ft by $\frac{1}{4}$ in. coiled copper tube packed with

15% Carbowax 20M on 60/80 firebrick. The sensitivity range used was from $1/32$ of the maximum to maximum.

For mandelic acid analysis, an Aerograph Hy-FL gas chromatograph equipped with a hydrogen flame-ionization detector was used. Nitrogen was the carrier gas. The column was a 5 ft by $1/8$ in. coiled aluminum tube packed with 15% Carbowax 20M on 60/80 firebrick. The sensitivity range was from $1/640$ to $1/10$ maximum.

Chromatographic Conditions

The chromatographic conditions were chosen to given rapid elution of the aldehydes and ketones, i.e., the maximum peak height possible without interference from the air and water peaks. The conditions employed for the analysis of alpha-hydroxy acids other than mandelic were a carrier gas flow rate of 80 ml/min, a column temperature of 88°C, and an injection block temperature of 200°C. For mandelic acid analysis, the conditions employed were a flow rate of 24 ml/min, a column temperature of 160°C, and an injection block temperature of 238°C. The efficiency of the column used for the analysis of acids other than mandelic was about 1200 theoretical plates.

Materials

Methyllactic acid, mandelic acid, acetone, methyl ethyl ketone, diethyl ketone, butyraldehyde, benzaldehyde, propionaldehyde, a-bromobutyric acid, a-bromovaleric acid, and periodic acid were purchased from laboratory supply houses and were the purest samples found available. a-Methyl-a-hydroxybutyric and a-ethyl-a-hydroxybutyric acids were synthesized by hydrolysis of the cyanohydrins of methyl ethyl ketone and diethyl ketone. The cyanohydrins were prepared by hydrogen cyanide addition to the ketone. The method followed was one described by Young, Dillon, and Lucas [7]. Purification was carried out by sublimation. Fine needlelike crystals of the acids were obtained with the following melting points: 71.5—72.5°C for a-methyl-a-hydroxybutyric acid and 76.5—78°C for a-ethyl-a-hydroxybutyric acid.

a-Hydroxybutyric and a-hydroxyvaleric acids were synthesized by hydrolysis of the corresponding a-bromoacids by aqueous potassium carbonate following the method described by Levenne and Haller [8]. Purification was carried out by sublimation. Needlelike crystals of both acids were obtained with melting points of 41—43°C for α-hydroxybutyric acid and 30—33°C for α-hydroxyvaleric acid.

Preparation of Samples and Procedure

Water solutions of hydroxy acids of varied concentrations were prepared. A solution containing 60 wt.% of paraperiodic acid (4.4 M) was also prepared. A constant volume of periodic acid was added to solutions of alpha-hydroxy acids to give a 4:1 minimum molar ratio of periodic acid to hydroxy acid. The resulting sample solutions were injected directly into the chromatograph. The concentrations of hydroxy acids in the sample solutions ranged from 0.005 to 0.5 M for acids other than mandelic, and 0.00033 to 0.033 M for mandelic acid.

Studies of molar ratios were made with methyllactic acid by adding various amounts of water and 60% periodic acid to solutions of methyllactic acid to give solutions having the same concentration of methyllactic acid, but different concentrations of periodic acid.

For chromatography of ketones and aldehydes with and without periodic acid, water solutions were again used.

A 10 μ liter syringe was used to inject into the chromatograph 1.9—2.2 μ liter of the sample solution of acids other than mandelic and 1.3—1.4 μ liter of mandelic acid sample solution. Since there was found to be a linear relationship between injection volume and response in the range 1.3—2.2 μ liter, the response was divided by the injection volume in all cases to compensate for differences in the actual volume of sample solution injected.

The response obtained from a hydroxy acid sample began to drop when the sample was allowed to stand for longer than one hour. There was a continued, but very erratic, drop in response upon further standing. Analyses were thus performed as soon as possible after addition of periodic acid.

RESULTS

Generality of the Method

All of the alpha-hydroxy acids studied were rapidly oxidized to aldehydes or ketones by periodic acid in the injection block

of the chromatograph.   The aldehydes and ketones produced
were eluted rapidly as relatively symmetric bands.   Figure 1
presents a typical chromatogram; a-methyl-a-hydroxybutyric
acid was analyzed by oxidation to methyl ethyl ketone (MEK).
All gave a linear relationship between peak height of the
aldehyde or ketone produced and concentration of the hydroxy
acid.   For acids other than mandelic, this linear relationship
was found to hold down to a concentration of about 0.008 M in
the sample solutions.   At lower concentrations, the peak height
was too small for good accuracy.   At concentrations below
about 0.03 M, the response became more erratic than at higher
concentrations.   The analysis of mandelic acid gave a linear
relationship between peak height and concentration in the range
0.00033—0.033 M in the sample solution.   The peak height vs.
concentration curves for a-methyl-a-hydroxybutyric acid and
mandelic acid are given in Figs. 2 and 3, respectively.   They
are representative of all the acids studied.   The peak heights
given for these curves are the products of the recorder attenuation and the number of scale divisions on the recorder

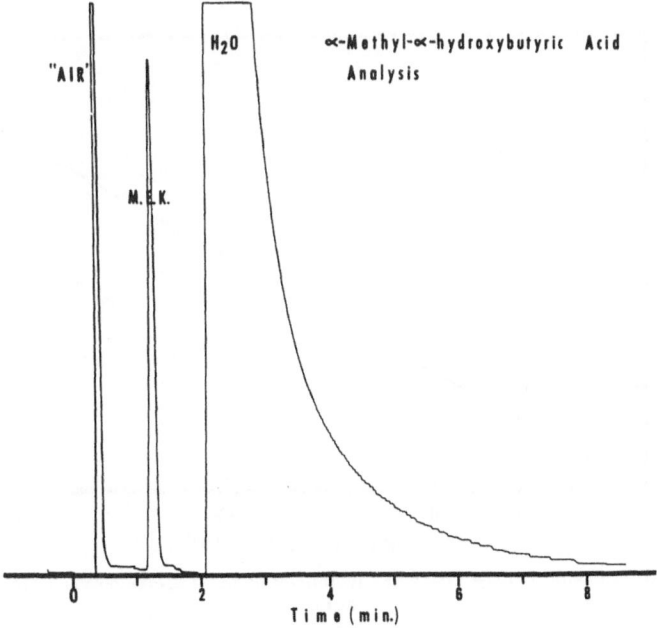

Fig. 1. Typical chromatogram in α-hydroxy acid analysis.

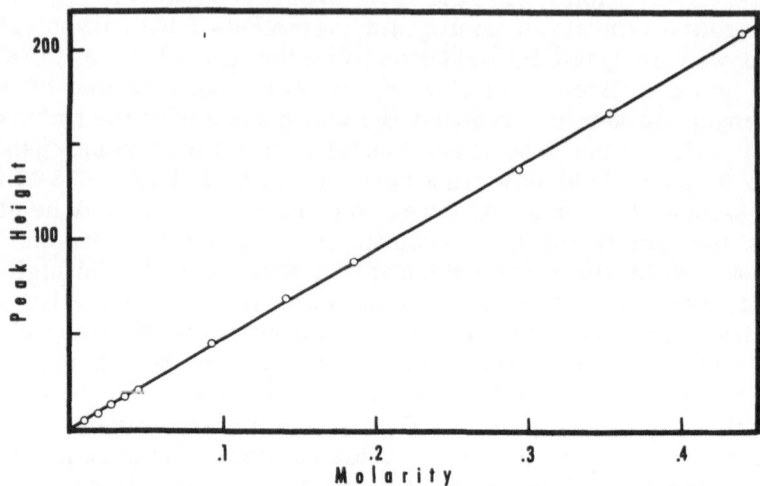

Fig. 2. Relationship between methyl ethyl ketone peak height and concentration of
α-methyl-α-hydroxybutyric acid.

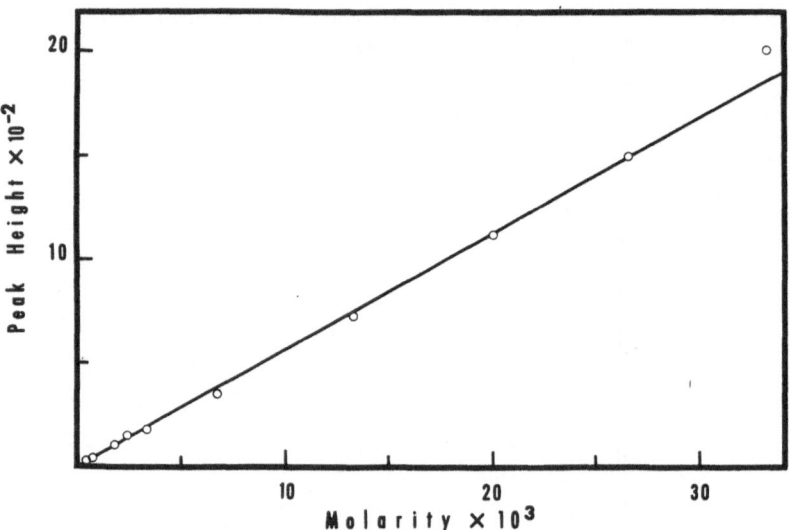

Fig. 3. Relationship between benzaldehyde peak height and concentration of mandelic
acid.

TABLE I

Reproducibility of Alpha-Hydroxy Acid
Analysis at Various Concentrations
(Five Injections)

Methyllactic acid		Mandelic acid	
Concentration (molarity)	Standard deviation, %	Concentration (molarity)	Standard deviation, %
0.428	2.28	0.0333	± 5.8
0.260	2.96	0.0133	± 4.3
0.096	2.12	0.00667	± 4.9
0.0302	5.25	0.00333	± 19
0.0078	6.85	0.000667	± 11.6
		0.000333	± 15.4

chart (100 divisions = full scale) divided by the volume (in microliters) of sample injected. Except for the injection volume studies, all subsequent peak height values were calculated in this way.

Multiple analyses were made of samples of methyllactic acid and mandelic acid at various concentrations to determine the precision of the method. The results are given in Table I. Methyllactic acid, α-methyl-α-hydroxybutyric acid, and α-hydroxyvaleric acid analyzed in solutions containing constant 5:1 molar excesses of periodic acid also gave a linear relationship between peak height and concentration, indicating that the method did not depend on high concentrations of periodic acid in the sample solution.

Injection Volumes

Several different volumes of sample solutions of methyllactic acid at concentrations of 0.285 and 0.038 M were injected into the chromatograph. The results, given as peak height vs. injection volume, are illustrated in Fig. 4. There is a linear relationship between response and injection volume up to about 3.5 μ liter. Above this injection volume, there is a deviation from linearity. A number of consecutive injections of 3 μ liter or more, made 10 min or less apart, tended to build

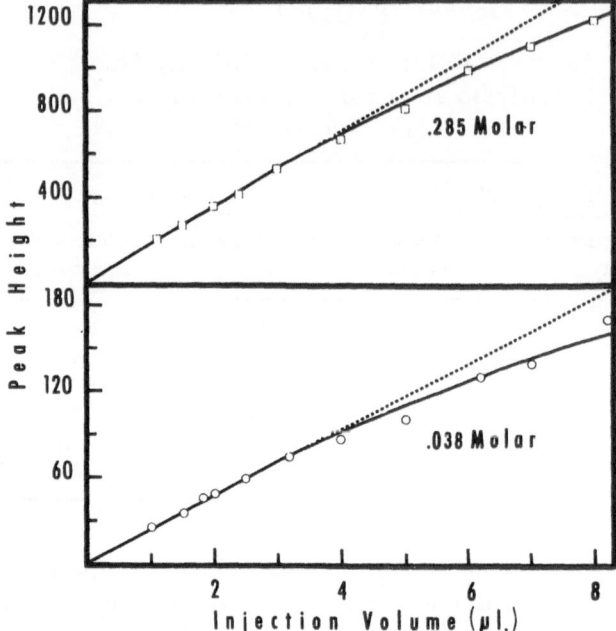

Fig. 4. Relationship between acetone peak height and injected volume of methyl-lactic acid at two concentrations.

up a residue in the injection block which remained one hour or more. This residue was probably iodic acid.

Molar Ratios

The relationship between the molar ratio of periodic acid to alpha-hydroxy acid and peak height was studied with methyllactic acid at concentrations of 0.047, 0.083, and 0.20 M. The results are shown in Fig. 5. In the molar ratio range of 4–80, the peak height was constant for a given concentration of hydroxy acid. Between 2–4, the response dropped slightly with decreasing molar ratios of periodic acid to methyllactic acid. Below 2, the response dropped sharply but erratically as the molar ratio was lowered. The drop was the result of a decrease in peak area; the width of the base did not change.

These results indicate that a minimum ratio of about 4 would be necessary, if linearity between peak height and hydroxy acid concentration was to be achieved. Actually, this work did not establish the upper linearity limit of the periodic—hydroxy acid ratio.   The value of 80 was merely the highest value obtainable in the concentration ranges studied. Previous work with lactic acid would indicate that the value may be as high as 1000.   The analysis of the alpha-hydroxy acids, discussed previously, was found applicable down to a concentration of 0.008 M, at which the molar excess of periodic acid was as high as 250.

Recovery of Aldehydes and Ketones

Table II summarizes the chromatographic recovery of aldehydes and ketones produced by periodic acid oxidation. Prior analysis of aldehyde and ketone solutions gave linear relationships between peak height and concentration similar to the alpha-hydroxy acids studied.   The theoretical peak height per mole per liter of a given aldehyde or ketone was calculated from the slope of a plot of peak height vs. concentration for a given aldehyde or ketone.   By comparing these values with

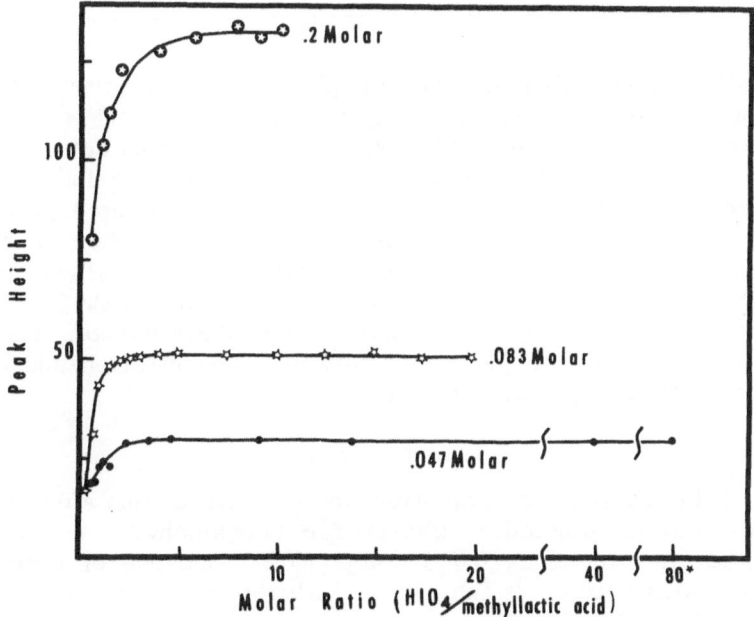

Fig. 5. Relationship between acetone peak height and molar ratio of periodic acid to methyllactic acid at three methyllactic acid concentrations *0.047/2M×2.

TABLE II

Recovery of Aldehydes and Ketones

Compound	Theoretical peak height	Observed peak height	Percent recovery
Methyllactic acid	640	633	99
Acetone HIO_4	640	544	85
α-Methyl-α-hydroxy-butyric acid	471	467	99
Methyl ethyl ketone HIO_4	471	407	87
α-Ethyl-α-hydroxy-butyric acid	390	374	96
Diethyl ketone HIO_4	390	277	71
α-Hydroxybutyric acid	515	410	80
Propionaldehyde HIO_4	515	283	55
α-Hydroxyvaleric acid	532	468	88
Butyraldehyde HIO_4	532	319	60
Mandelic acid	76,700	56,800	74
Benzaldehyde HIO_4	76,700	66,700	87

similar values obtained from the plots of peak height vs. concentration for the corresponding alpha-hydroxy acid (Figs. 2 and 3), the percent of aldehyde or ketone recovery was calculated.

Solutions of aldehydes and ketones containing periodic acid were also analyzed and the percent recovery calculated. There was some loss of aldehyde and ketone observed after the addition of periodic acid. The actual recovery values were extremely erratic and varied over a wide range ($\pm 20\%$). The percent recovery figures are averages of these values. There was no relationship observed between concentration and the amount of aldehyde or ketone lost.

Retention Times

Table III gives the retention times of the various aldehydes and ketones studied. Except for benzaldehyde, water is eluted after these carbonyl compounds. Because of water's wide tailing band, it interferes with the benzaldehyde peak. Mandelic acid, therefore, was analyzed with a chromatograph having the water-transparent flame-ionization detector.

TABLE III

Retention Times

Compound	Time, sec
Propionaldehyde	48
Acetone	54
Butyraldehyde	69
Methyl ethyl ketone	75
Diethyl ketone	110
Water	150

In all samples analyzed with a chromatograph having a thermal-conductivity detector, a large peak was observed in the vicinity of the air peak. This peak was found to increase in area with an increase in the concentration of periodic acid, and was also found when only water and periodic acid were injected. It is highly probable that this peak was mostly oxygen with some carbon dioxide. In some analyses, a "split peak" was actually noted. The carbon dioxide was produced by the oxidation of the alpha-hydroxy acids, while the oxygen was probably produced by the decomposition of periodic acid. Periodic acid is reported to decompose at high temperatures [9].

This peak was large enough so that, at low concentrations of α-hydroxybutyric and methyllactic acids (<0.008) and high molar excesses of periodic acid (>250), it overlapped the propionaldehyde and acetone peaks and made analysis difficult. With the other hydroxy acids studied, retention times of the ketones and aldehydes were sufficiently large so that there were no problems of interference from this peak.

Adsorption

Several replicate injections in rapid succession were made of sample solutions of methyllactic acid, α-hydroxyvaleric acid, and α-methyl-α-hydroxybutyric acid at concentrations of 0.03 and 0.3 M. This produced several consecutive peaks before water was eluted. At a flow rate of 80 ml/min, the peaks produced were of approximately the same height for any given sample solution. At a flow rate of 40 ml/min, however, the peaks were progressively higher in each case, with the most pronounced differences being between the first two peaks. The difference between the first and last peaks was about 15—20% at the 0.03 M concentrations and about 5% at the 0.3 M concentrations. This study indicates that at the 40 ml/min flow rate there is some adsorption in the initial samples injected. At the 80 ml/min flow rate, however, this phenomenon was not observed.

DISCUSSION

The most significant result of this investigation was the establishment of proportionality between peak height and hydroxy acid concentration with a molar excess of periodic acid of at least 4. This result shows that the method can be used for quantitative analysis of the alpha-hydroxy acids. Thus, periodic acid in molar excess of at least 4 can be added to a sample and the resulting sample solution injected into the chromatograph. The height of the elution peak of the carbonyl compound produced in oxidation can be used to determine the concentration of the alpha-hydroxy acid in the sample solution.

By employing a chromatograph with a thermal-conductivity detector, sample solutions of alpha-hydroxy acids of concentrations as low as 0.008 M were analyzed. The standard deviation was found to be about ± 2—3% at the higher concentrations and about ±5—7% at the lower concentrations studied. Chromatographs having more sensitive detection systems could be used for the analysis of samples of much lower concentrations. The analysis of mandelic acid was achieved down to a concentration of 0.00033 M. The standard deviation was found to be higher, however (±4—19%).

The study of injection volumes indicated that unless a constant injection volume can be carefully maintained, injections should be of sufficiently small volume to fall within the range of linearity shown in Fig. 4 in order to maintain proportionality between response and sample volume.

Table II shows that methyllactic and a-methyl-a-hydroxybutyric acids, both of which produce ketones on oxidation, are completely converted to ketones, and the ketones are completely transferred to the column during analysis. Mandelic, a-hydroxybutyric, and a-hydroxyvaleric acids, which produce aldehydes on oxidation, give less than theoretical peak height during analysis. It seems reasonable, in view of the ease of oxidation of aldehydes, to conclude that the acids are completely oxidized, but that not all of the aldehyde produced is transferred to the column. Some is oxidized further. Fortunately, the fraction lost is constant and independent of the hydroxy acid concentration, and a linear peak height vs. concentration curve is obtained.

It is noteworthy that in both cases, when ketones containing periodic acid are analyzed, recovery is less than 100% (Table II). This may be related to the fact that the carbonyl compounds are present initially with an excess of periodic acid. In contrast, by the time they are formed in high concentration by oxidation in hydroxy acid analysis, the excess of oxidizing agent has diminished through decomposition, and the ketones are rapidly eluted as soon as they are formed.

The long retention time of water extends the time of analysis when several α-hydroxy acid samples are to be analyzed consecutively. A considerable time elapses before complete elution of water with its lengthy tail. In most cases (except α-ethyl-α-hydroxybutyric acid), two injections of the sample may be made before water is eluted. Some time may also be saved by increasing the flow rate of carrier gas during the elution of water. With instruments having the water-transparent, hydrogen flame-ionization detector, more frequent injections may be made rapidly, since there is no waiting for the elution of water. The location of the water peak would indicate that certain hydroxy acids, probably in the range of seven to ten carbons, would give ketones and aldehydes that elute with water. For these acids, as was the case of mandelic acid, a flame-ionization detector would have to be used.

The appearance of so much oxygen and the location of its band was of interest. The oxygen made quantitative analysis of lactic acid difficult with a thermal-conductivity detector because oxygen's retention time was so close to that of acetaldehyde. Nevertheless, in the lactic acid analysis method recently developed [1] where the concentrations were such that the acetaldehyde was being eluted along with a large excess of oxygen, no difficulties were encountered. A flame-ionization detector was used.

The decomposition of periodic acid is probably related to the requirement of a minimum ratio of periodic acid to hydroxy acid of 4 in order to obtain the maximum peak height for the eluted carbonyl compound. At high ratios of periodic acid to hydroxy acid, despite some loss of periodic acid to decomposition, there is enough remaining to completely oxidize the hydroxy acid rapidly. At low ratios, as the rapid competitive decomposition proceeds, there is not enough remaining to oxidize the hydroxy acid completely, and the amount of carbonyl compound eluted as a sharp band is lower than it should be.

SUMMARY

Dilute solutions of alpha-hydroxy acids in water can be analyzed by oxidation of the acids with periodic acid in the chromatograph. The quantity of aldehyde or ketone produced in oxidation is measured. When a Carbowax 20M-on-firebrick column is used, acid concentration is proportional to aldehyde or ketone peak height.

REFERENCES

1. N. E. Hoffman, J. J. Barboriak, and H. F. Hardman, Analytical Biochemistry 9, 175 (1964).
2. S. Siggia, Quantitative Organic Analysis via Functional Groups, John Wiley and Sons, Inc., New York (1949), p. 8.
3. P. Fleury and R. Boisson, J. Pharm. Chim. 30, 145 (1939).
4. P. Fleury and J. Courtois, Bull. soc. chim. France, 190 (1948).
5. D. B. Spoison and E. Chargoff, J. Biol. Chem. 164, 433 (1946).
6. N. E. Hoffman, unpublished results.
7. W. G. Young, R. Dillon, and H. J. Lucas, J. Am. Chem. Soc. 51, 2528 (1929).
8. P. A. Levenne and H. L. Haller, J. Biol. Chem. 77, 55 (1928).
9. W. M. Latimer and J. H. Hildebrand, Reference Book of Inorganic Chemistry, The Macmillan Co., New York (1940), p. 177.

The Direct Gas Chromatographic Determination of Low Molecular Weight Fatty Acids in Rumen Fluid

J. B. Martin, Jr.

Abbott Laboratories
North Chicago, Illinois

A gas chromatographic procedure to analyze rumen fluids for the low molecular weight fatty acids through n-valeric acid has been developed. The method requires little pretreatment of the rumen sample. The column developed for the analysis overcomes many difficulties associated with the gas chromatographic analysis of very polar materials in aqueous media.

Low molecular weight fatty acids originate via fermentation of foodstuffs in the rumen of ruminant animals. These fatty acids are the primary source of energy for these animals. The relative concentrations of these fatty acids in the rumen fluid have been related to efficient energy conversion by such animals [2]. The low molecular weight fatty acids found in the rumen are acetic, propionic, isobutyric, n-butyric, isovaleric, and n-valeric, plus smaller amounts of the six-carbon acids, as well as lactic and pyruvic acids.

In various studies concerning the rumen, it was necessary to determine the concentrations of these acids. A simple, rapid, accurate method of analysis was needed to expedite these studies. The method developed in this work was concerned only with the analysis of the monofunctional, monobasic acids containing up to and including five carbons.

Several methods [1,3] have been employed for the determination of these low molecular weight acids. One important method consists of steam distillation of an acidified rumen fluid sample followed by conversion to the sodium salts. These

313

salts are subsequently converted to methyl esters and are analyzed by gas chromatography. This particular method has several drawbacks, including excessive handling of the sample which may lead to possible losses and give results which are, at best, only semiquantitative. A more desirable method would be one which would permit the direct gas chromatographic separation of all components with minimum treatment and handling of the sample. The search for a suitable column to meet these requirements was the major purpose of this work.

The basic approach to the type of analysis desired was reported by Erwin, Marco, and Emery [5]. Their work involved the direct analysis of the acids found in the rumen fluid by gas chromatographic analysis of an acidified rumen sample.

Erwin and co-workers were particularly concerned with the separation of the isoacids from the straight chain isomers. Their choice of partitioning agent was Tween-80, a poly-oxyethylene sorbitan mono-oleate, modified with ortho-phos-phoric acid, according to the suggestion of Metcalfe [6]. Emery and Koerner [4] had previously investigated this par-titioning agent without the phosphoric acid, using hydrogen flame-ionization detection and had obtained symmetrical peaks with the free fatty acids with the desired separation of the isoacids from the next lower normal homolog. However, excessive ghosting was a problem. Ghosting is a phenomenon found when highly polar materials are strongly adsorbed on the column. Subsequent injection of a more polar solvent, such as water, results in a chromatogram of the adsorbed materials, each appearing at the approximate retention time one would find if one injected the adsorbed material onto the column. What apparently happens is that water displaces the adsorbed materials, and this adsorbed material is subsequently chromatographed.

However, with other sources of Tween-80, Erwin and co-workers were not able to effect the separation of the acids. They were able, however, to obtain reproducible results with acidified Tween-80, even though ghosting was still a problem. And, since the ghosting was essentially reproducible, calibra-tions could be made which took this into account.

Figure 1 shows results obtained with a similar column—20% Tween-80, 3% ortho-phosphoric acid on 60—80-mesh, acid-washed Chromasorb W, packed in a $^3/_{16}$-in. ID × 8 ft glass

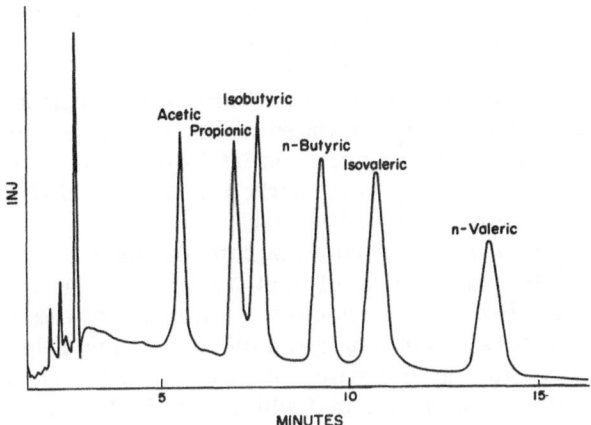

Fig. 1. Separation of fatty acids on Tween-80—H_3PO_4 column at 143°C.

column. The work was performed on a Barber-Colman Model 10 gas chromatograph equipped with a Wilkens flame-ionization detector and a Wilkens Model A500 electrometer. With solutions containing the acid in somewhat greater concentrations than actual rumen samples, the column gave good separations. The separation between propionic acid and isobutyric acid, the most difficult pair to resolve, was adequate for the more concentrated blends.

The chromatogram represents approximately 10 μg per component. However, with rumen fluid where the concentrations of the acids are not all the same—for example, much less isobutyric acid than propionic acid is found in rumen fluid—one has more difficulty with the separation; and, as the solutions become more dilute, water interference presents more of a problem, in spite of flame-ionization detection. Water is eluted very quickly from the column and exhibits a long tail which can run over into the acetic acid peak and cause errors in quantitative interpretation. After considerable conditioning during which time repeated injections of water were made, there was less interference by water. The conditioning consisted of baking the column for one week at 150°C during which time a continuous purge of nitrogen at approximately 40 cc/min was maintained. Each day approximately four 50-μliter portions of water were injected onto the column.

The column was then tested with an actual rumen sample.

The rumen sample was prepared for analysis as follows: 5 ml of rumen fluid was centrifuged and the supernatant liquid removed. This supernatant liquid is then acidified with 1 ml of 25% aqueous meta-phosphoric acid. This mixture is allowed to stand for 20—30 min, centrifuged again to remove coagulated matter, and the resulting supernatant liquid analyzed directly by gas chromatography. This method of preparation was used throughout the remainder of the work.

Figure 2 shows the chromatogram of such a rumen fluid sample. Since the amount of isobutyric acid is very small, this component becomes a shoulder on the side of the propionic acid peak. Several operating variables were investigated with this column, including variation of flow rate and temperature, but no conditions could be found which would allow the accurate determination of very small amounts of isobutyric acid in the rumen fluid. The column's life time was relatively short — a matter of a few weeks — and ghosting and noise were also problems. These faults spurred the search for other partitioning agents.

The second partitioning agent investigated was Carbowax 6000/ortho-phosphoric acid in a 20:3% ratio on acid-washed Chromasorb W, 60-80 mesh. A chromatogram showing the separation of the acids in a blend is shown in Fig. 3. Very good chromatograms were obtained with little or no interference by water. Unfortunately, this partitioning agent does not separate propionic acid and isobutyric acid, but rather elutes them as a single peak. The column was discarded on this basis. No tests were made for ghosting.

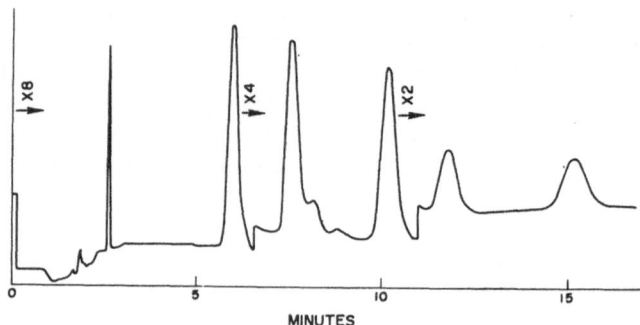

Fig. 2. Separation of rumen fluid on Tween-80—H_3PO_4 column at 140°C.

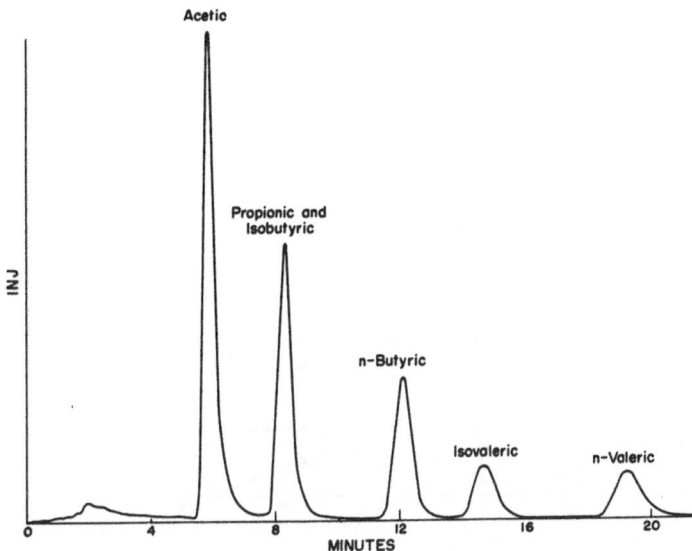

Fig. 3. Chromatogram of fatty acid blend on Carbowax 6000—H_3PO_4 column.

The next phase, Versamid 900, a polyamid, and isophthalic acid on Chromasorb W [7] was found to give an adequate separation of all six acids in aqueous solution. The column contained 20% Versamid 900 and 5% isophthalic acid on 60-80 mesh, acid-washed Chromasorb W. Again, extensive conditioning, this time at 160°C with repeated injections of water, was necessary to obtain a stable column.   Ghosting with this column is a problem, but by using the steam-cleaning technique, i.e., the injection of several large quantities of water to remove adsorbed material, one may minimize this problem. However, with this column, the first injection of a sample after steam cleaning gives low results, presumably due to the adsorption of some of these acids.   After this first injection of the day, samples may be run without noticeable additional adsorption, and calibration curves can be constructed. This column proved adequate with work on the rumen acids, but slowly deteriorated to the point that very broad peaks were obtained. While the isobutyric acid still exhibited some separation from propionic acid, the chromatograms had such broad peaks that they were incalculable.   The Versamid—isophthalic acid column seemed quite stable to temperature (170°C) and gave excellent baselines

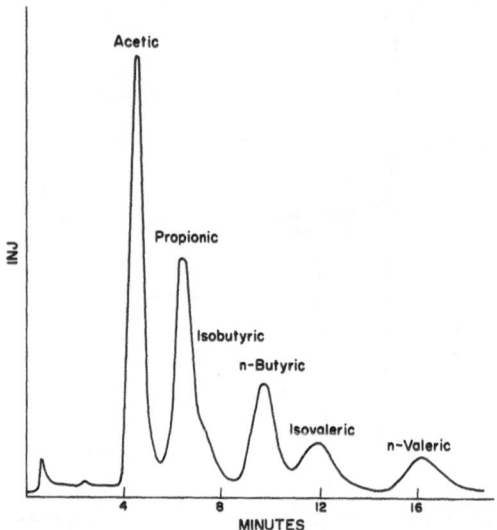

Fig. 4. Separation of fatty acids in rumen sample using old Versamid 900—isophthalic acid column.

with little water interference, even though the separating ability had slowly deteriorated. An example of separations with the old column is shown in Fig. 4.

To extend the work on Versamid 900, a modification was made containing 20% Versamid 900 and 3% ortho-phosphoric acid on 60-70 mesh Gas Chrom P, ABS. This column, even before conditioning, gave remarkable separations of all six acids. Initially, another phenomenon to be designated as "echoing" was found, as shown in Fig. 5. Echoing as used in this paper designates the appearance of secondary peaks after each main peak in the chromatogram. Double peaking could also be used as a term to describe this phenomenon. Also, the water peak interference was considerable before additional conditioning.

Figure 6 shows results of separation after the column had been conditioned for 48 hr with occasional injections of 50 μ liter portions of water. The echoing and water interference disappeared. Separation and efficiency were still quite good; for example, with n-valeric acid, an efficiency of 1800 theoretical plates was obtained.

Since one of the problems with the other columns had been ghosting, this column was carefully investigated for this

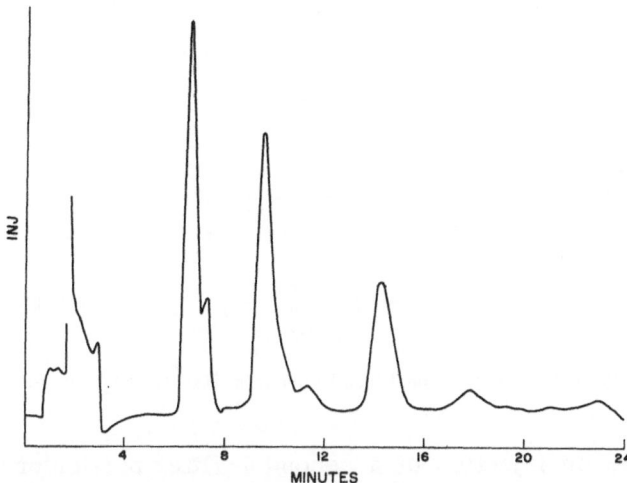

Fig. 5. Echoing phenomenon during separation of fatty acids column: 20% Versamid 900—3% H_3PO_4 at 155°C.

phenomenon. Ghosting proved to be negligible, as shown in Fig. 7. The upper half of this figure shows the chromatogram obtained by injecting 5 µliter of water onto the column just after a blend that contained twice the amount of acids normally found in rumen had been injected. The appearance of slight amounts of acetic and propionic acid is shown here. Acetic and propionic acid are present to the greatest extent in rumen fluid. The bottom half of the figure shows the chromatogram

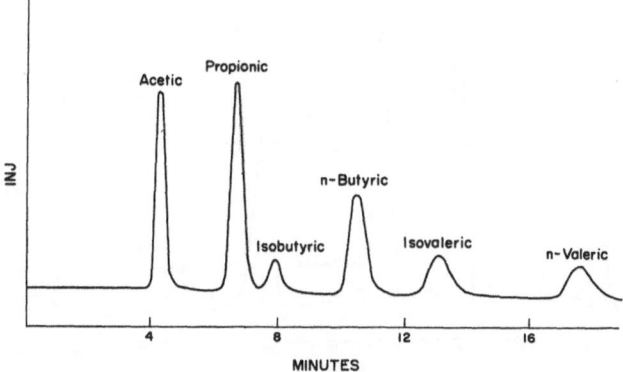

Fig. 6. Separation of fatty acids in blend using conditioned Versamid 900—H_3PO_4 column.

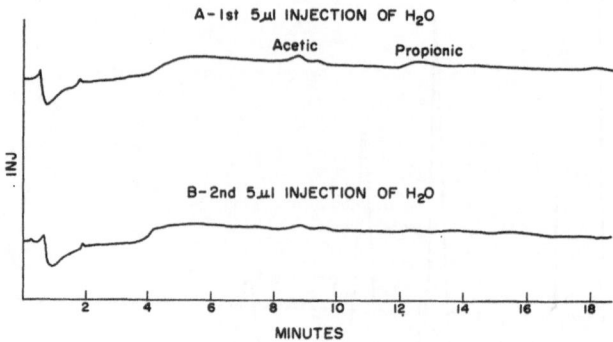

Fig. 7. Ghosting on conditioned Versamid 900—H$_3$PO$_4$ column.

obtained on injection of a second 5-μliter portion of water on the column. Here, the interference has essentially disappeared. Ghosting is therefore considered negligible for our purposes, since these very small peaks do not contribute significantly to error in our analysis, considering that the actual amounts of the acids chromatographed are larger than those found by this ghosting.

Figure 8 is a chromatogram obtained with a rumen acid sample; here the separation between propionic acid and iso-butyric acid is shown clearly. Thus, this partitioning agent

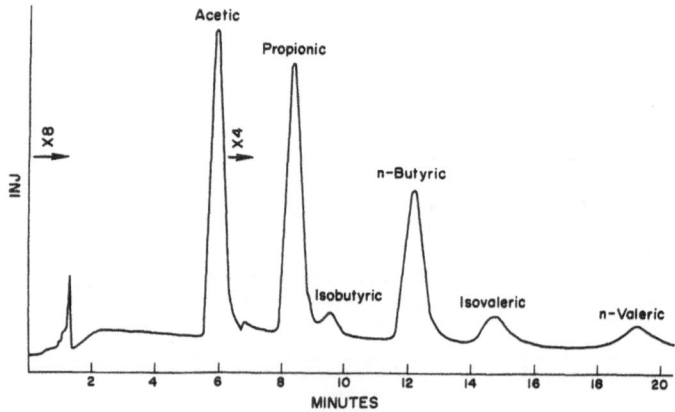

Fig. 8. Chromatogram of rumen acid sample on Versamid 900—H$_3$PO$_4$ column at 155°C.

combination exhibits the desired selectivity with little interference of water, as outlined in our original goals.

On studies related to reproducibility, the following experiment was performed. Six replicates of a rumen sample were analyzed for volatile fatty acids to determine reproducibility. The reproducibility is shown in Table I. In these studies, no internal standard was used, as it was found that using samples as large as 4 or 5 μ liter, one could easily reproduce injections within very small limits of error.

After the method was developed, the work on the Barber-Colman instrument was transferred to a Wilkens A-90-P2 gas chromatograph equipped with hydrogen flame ionization. It was found that two modifications of procedure were necessary. The first procedure involved packing the injection block with glass wool to prevent solids in the rumen fluid from contaminating the column support or baking on the walls of the injection chamber. Secondly, the glass wool has to be removed, the block cleaned out by swabbing with chloroform, and a new plug placed in the chamber every two or three days. The analytical results obtained with these modifications have been comparable to those obtained with the glass column. Table II gives the analytical conditions adopted for this analysis.

It is necessary to allow carrier gas to flow through the column at all times when the column is heated; otherwise, the Versamid 900 may begin to decompose, making replacement of the column mandatory.

In conclusion, a method has been developed which will allow the direct determination of the low molecular weight fatty

TABLE I

Analysis of Rumen Fluid Sample

Acid	Average value* (μg/μliter)	Standard deviation
Acetic acid	5.26	0.017
Propionic acid	1.96	0.085
Isobutyric acid	0.056	0.012
n-Butyric acid	1.45	0.02
Isovaleric acid	0.042	0.002
n-Valeric acid	0.036	0.001

*Five determinations.

TABLE II

Analytical Conditions for Analysis of Fatty Acids in Rumen Fluid

Condition	Description
Instrument	Wilkens A-90-P2 equipped with flame-ionization detection
Temperatures:	
Column	155°C
Flash	170°C
Flow	40 cc/min nitrogen carrier 30 cc/min hydrogen for flame
Column	$^3/_{16}$ in. × 8 ft stainless steel filled with 20% Versamid 900, 3% H_3PO_4 on 60-70 mesh Gas Chrom P, ABS

acids in rumen fluid with minimum pretreatment of the rumen fluid. Analyses require about 15 min instrument time per sample. This method utilizes a gas chromatographic column which gives adequate separation for all the acid isomers up through n-valeric acid, including the difficult pair, propionic acid and isobutyric acid.

REFERENCES

1. E. F. Annison, Biochem. J. 58: 670 (1954).
2. D. G. Armstrong and K. L. Blaxter, Brit. J. Nutrition 11:413 (1957).
3. A. Bensadoun, Ph.D. Dissertation, Cornell University (1960).
4. E. M. Emery and W. E. Koerner, Anal. Chem. 33:146 (1961).
5. E. S. Erwin, G. J. Marco, and E. M. Emery, J. Dairy Sci. 44:1768 (1961).
6. L. D. Metcalfe, Nature 188: 142 (1961).
7. B. R. Baumgardt, University of Wisconsin, Madison, Wisconsin, personal communication (1963).

Simultaneous Gas Chromatography and Radioactivity Analysis: Instrumentation, Calibration, and Application

D. C. Nelson, R. C. Hawes, D. Paull,
and P. C. Ressler, Jr.

Applied Physics Corporation
Monrovia, California

An instrument for simultaneous gas chromatography and radioactivity analysis is described. The ion chamber used for the radioactivity measurement is contained in the same temperature-controlled oven as the normal chromatography detector, and can be heated to over 300°C. Data are given for both programmed and isothermal column operation at high and low temperatures. Sensitivity, linearity of response, response times, and resolution are discussed for the entire system.

Advantage may be taken of the fact that when an ion chamber is connected in series within a gas chromatograph, a given amount of radioactivity in a fraction results in the same ion-chamber signal, regardless of the molecule to which the unstable isotope is attached. The effect of various peak shapes on the signal is treated theoretically and compared to experimental results. A simple generalized method of quantitative analysis is proposed, by which overall accuracy within 1% can be attained, independent of column, mass detector, and other operating conditions.

Selected applications are also discussed.

INTRODUCTION

Not too long after gas chromatography in the conventional sense was well under way in the United States, articles began to appear on the detection of the radioactivity of labeled fractions in addition to the mass detection of the separated components. Two general methods of measuring the activity of a separated component have been used. One is to collect the sample on a coated solid scintillator or in solution, or as the pure material, and then measure the radioactivity with

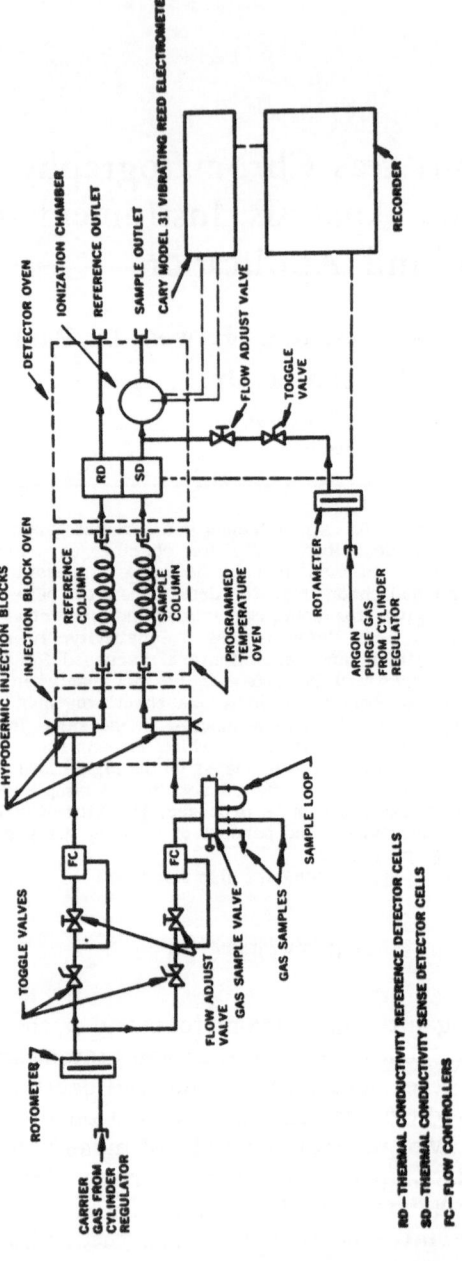

Fig. 1. System flow diagram.

a counter [5, 9, 10, 14]. The second technique, used more often recently, is to measure the activity of the fraction in the following gas phase as it leaves the mass detector.

As early as 1953, Kokes, Tobin, and Emmett [11] passed active fractions containing soft beta emitters by a Geiger counter with a thin window. In 1956, Evans and Willard [6] measured hard beta and gamma particles in a similar manner. However, it was soon realized that the window materials of the detectors make for poor efficiency of measurement for soft beta emitters, such as C^{14} and H^3, the most useful isotopes. Wolfgang and Rowland [18], Wolfgang and Mackay [17[, and James and Piper [8] have investigated proportional counters, with and without windows.

Dobbs [4], Cacace [2], and Karmen and Winkelman [10] have used an ion chamber in series with the GC detector. Although slightly higher sensitivity can be attained by proportional counting, the ion-chamber technique offers the advantages of being simple, easy to install, stable, resistant to poisoning effects (the most serious deficiency of the proportional technique), sensitive to soft beta decay, reproducible, and usable at least to detector temperatures up to 300°C. Samples may be trapped as they emerge from the ion chamber.

EXPERIMENTAL

Gas Chromatograph

The gas chromatograph used for our work is a Cary-Loenco Model 70. A Minneapolis-Honeywell dual channel recorder is used. The gas chromatography channel is 0—1 mV, 1-sec response for full scale, and the radioactivity channel is 0—25 mV full scale regardless of the electrometer range, 1-sec response. The pens are physically separated by a little more than one chart division to keep from colliding. Figure 1 shows the flow diagram for the entire assembly. Separate temperature control is provided for the inlets, detector compartment, and column compartment, where either programmed or isothermal work can be done. All three can be operated to temperatures over 300°C. The detector compartment will accept thermal conductivity, ionization, and radioactivity detectors. After the mass detectors, the reference stream exhausts, and the sample stream is mixed with the diluent gas at a tee, whence it passes into the ion chamber.

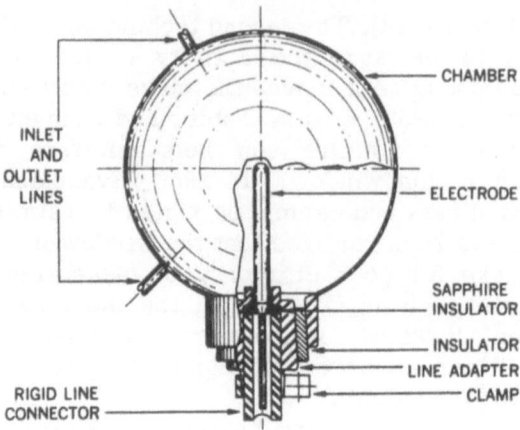

Fig. 2. Cary high-temperature ionization chamber.

Radioactivity Detector

Activity was measured with an ion chamber of 275-cc volume, shown in Fig. 2. As Cacace [2] pointed out, the chamber may be large for higher ionization efficiency without loss of response speed, if a diluent gas is employed. However, smaller chambers have been used [4]. The electrometer used is a Cary Model 31 vibrating reed instrument. The full-scale range normally used is 100 mV when using a 10^{12}-Ω resistor, or 10^{-13} A. The electrometer is preferably critically damped, providing a 1.2-sec, full-scale response speed (98.7% of full scale). The base line noise under these conditions is about 0.5%, and does not change with chamber temperatures at least up to 300°C. We prefer the high-resistance leak method of measurement, which yields peaks similar to mass peaks. The rate of charge method [9], which offers slightly higher sensitivity and results in an integral presentation of charge, is much less convenient to interpret and requires frequent resetting, and the record may run off scale just at the most interesting point.

The chamber wall is negatively charged, with respect to the electrometer ground and probe, with a dry B battery. The probe is attached to a rigid-line connector leading directly to the input of the vibrating reed amplifier. The probe and conductor are both insulated with sapphire. Surrounding the upper end of the stainless steel rigid line is a conical support for the chamber. Between this support and the stainless steel chamber walls is a conically-shaped ceramic insulator. The chamber must be electrically insulated from the rest of the instrument by inserting in the entrance line to and exit line from the chamber a glass-to-Kovar connector. The instrument connected in this manner can easily detect 10^{-15} A.

RESULTS

Response Times

With the usual 275-cc chamber, normal carrier flow rates used in columns of $\frac{1}{4}$ in. or smaller diameter would not sweep the sample through the ion chamber fast enough to permit

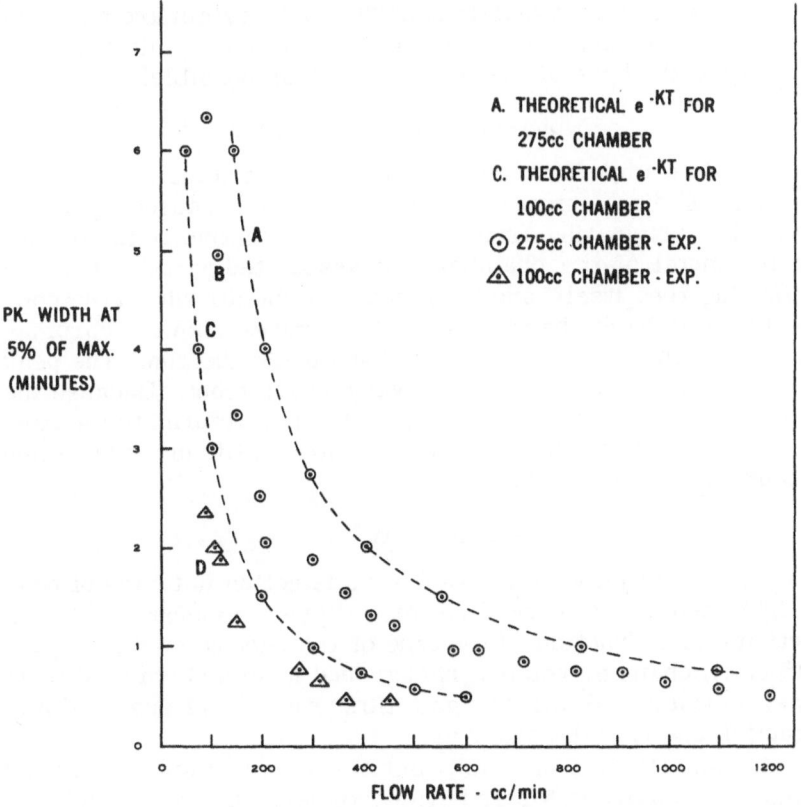

Fig. 3. Flow rate vs. time to decay to 95% of peak height.

resolution of adjacent peaks. Furthermore, helium is not the best gas for ion-pair formation, because of its relatively high ionization energy [13]. For these reasons, argon is added as a diluent gas.

Assuming a "plug" sample and perfect mixing in the ion chamber, response would follow an exponential. Figure 3 shows the peak width at various chamber diluent flow rates, using for width the time at 5% of full peak height. The dashed lines are theoretical.

The shape of the theoretical curve A is followed, but points B show shorter times than predicted. Essentially, the same results were obtained by reversing input and exit lines, or using a 100-cc ion chamber (C and D). The difference is attributed to imperfect mixing. This effect does not affect efficiency and allows a lower diluent flow rate to be used. From Fig. 3 it is apparent that flows of 600—900 cc/min are adequate for the 275-cc chamber. We operated at 900 cc/min normally to narrow the base of the peak as much as possible.

Alpha-Particle Interference

Another important reason for fast electrometer response is ease of discrimination against ionization caused by alpha particles originating from traces of radioactive contaminants in the metal of the chamber and associated parts, within the vibrating reed itself, and from radon in the air within the connecting line from chamber to input. Figure 4 shows a chromatogram with alpha events occuring during the run. The peak always has the characteristic very sharp front. Because the electrometer is critically damped, the pen returns to the base line almost immediately, causing little difficulty even when occurring during a peak.

Sensitivity

The sensitivity of the radiation detection in terms of peak height above the base line for a given amount of injected activity is a function of the type of radioactive emission (C^{14}, H^3, etc.), chamber volume, background level and noise, diluent gas, carrier and diluent gas flow rates, and peak width of fraction entering the chamber.

C^{14} and H^3, both of which emit soft beta particles, are used almost exclusively for radiation studies. Disintegrations of C^{14} have an average energy of 45,000 eV per disintegration,

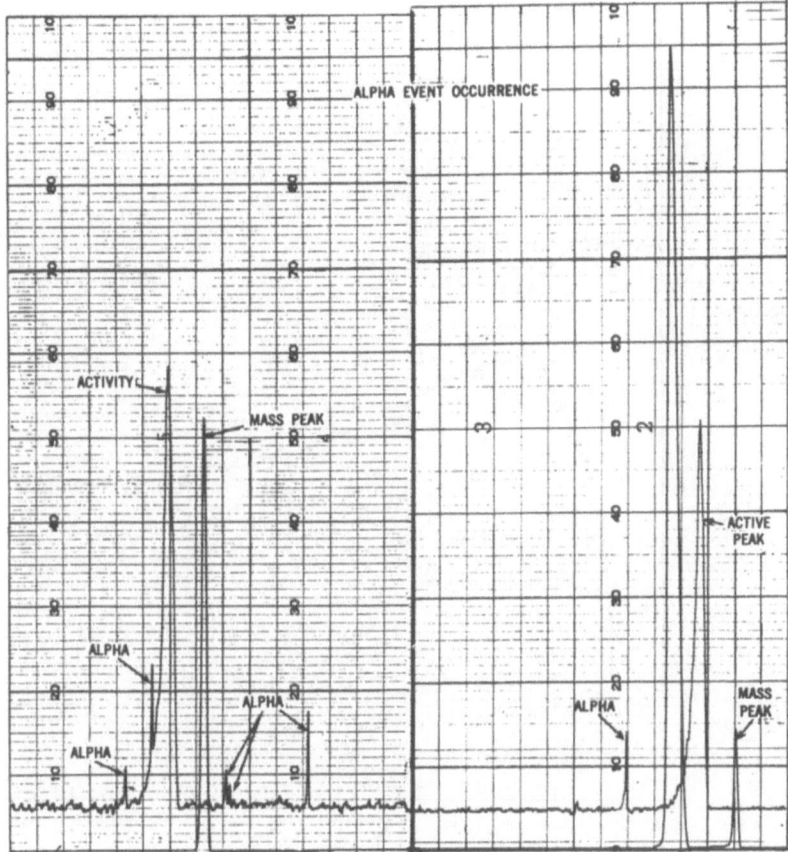

Fig. 4. Alpha-event occurrence.

and H^3 has 5700 [15]. Their short-path beta emission can be efficiently measured in a relatively small chamber. A 275-cc chamber provides a suitable compromise between efficiency and volume, so that it enters the chamber in a "plug" relative to the size of the chamber. A 100-cc chamber was tested under equivalent flow conditions and found to be 21% efficient for C^{14}.

Tritium can be measured with about 75% efficiency in a 275-cc ion chamber. However, for a given activity, the signal is about one-third of that for C^{14} because of tritium's much lower energy per disintegration.

We find that 0.3 nCi (about 13 dis/sec) will provide a signal

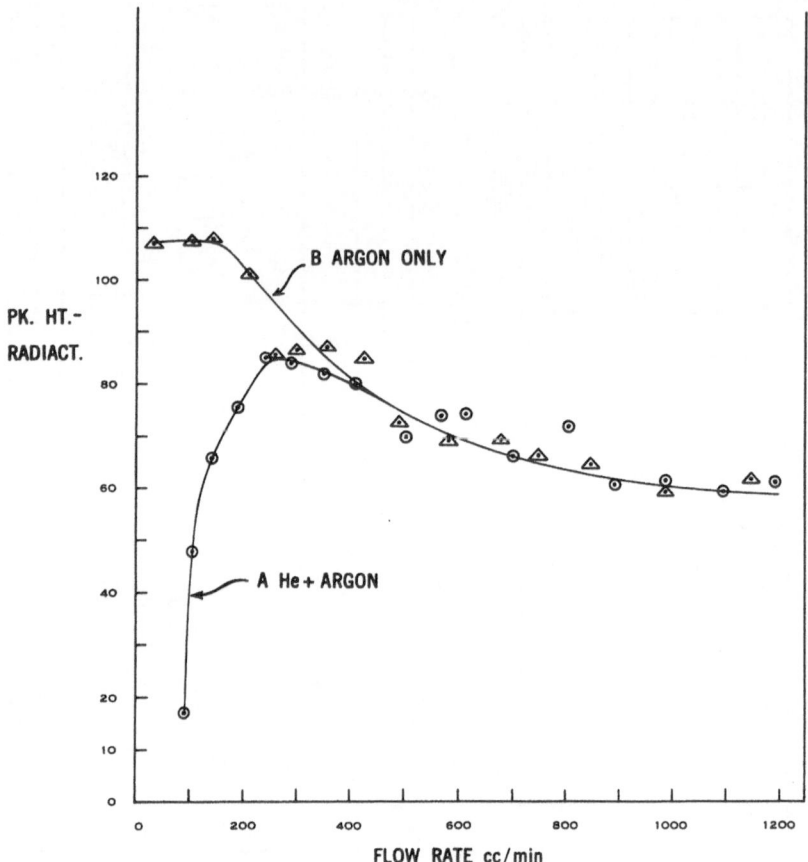

Fig. 5. Radioactivity peak height vs. flow rate.

to peak-to-peak noise ratio of 1 when the electrometer is critically damped. Peak-to-peak noise is averaged from intervals ten times the period of the amplifier. This noise level amounts to 3×10^{-16} A.

In addition to being a function of the diluent gas used, the peak height is a function of the flow rate of the diluent gas. Figure 5 is a plot of total flow rate (carrier plus diluent) vs. the peak height of the activity for 4-μliter injections of C^{14}-labeled benzene. Curve A near 100 cc/min refers to pure helium. As argon is added, the peak height increases until the flow reaches about 250 cc/min, total. At this point, the ion-pair formation is due entirely to argon—beta collision, since further increase in argon flow does not increase the peak height. Curve B was plotted from results obtained using argon only as both carrier and purge gas. Peak heights decrease from above 250 cc/min flow because the benzene peak has a finite width, and some leaves before all has entered at the faster flow rates.

CALIBRATION

So much for the instrument itself. Work with this instrument has turned up a very interesting area, which I will describe. It has to do with the calibration of the ion chamber for quantitative work.

Advantage may be taken of the fact that the absolute amount of activity measured by a suitable ion-chamber technique is independent of any characteristics of the molecule of which the unstable isotope is a part, which fact provides a calibration method independent of most of the usual variables in gas chromatographic work, e.g., peak position, temperature, and pressure. Thus, two molecules with completely different characteristics, such as CO_2 and cholesterol, will result in the same activity measurement if the absolute amount of C^{14} in the two samples is the same.

In this work the transient behavior of the ion-chamber detector was investigated theoretically, and the predictions were verified experimentally. By making the radioactivity measurement in the manner to be described, three important advantages can be achieved. First, a single standard compound of known activity can be used for chamber calibration, and the resultant stable calibration data applied to any other compound of unknown activity. Once established, these calibration data can be used indefinitely. Second, the analysis technique is independent of all the gas-chromatograph instrument variables except the flow rate through the ion chamber, which is easily controlled. Third, the accompanying mass record provides information leading to the specific activity of the fraction, without requiring the usual control sample.

THEORETICAL CONSIDERATIONS

In considering the current produced in an ion chamber, it

is best to consider first a static condition. If the chamber contains some suitable ionizable gas and some active compound, the theoretically possible current is expressed as

$$I = \frac{BNeS}{w} \tag{1a}$$

where I is current in amperes, N is the number of disintegrations per second occurring from the sample, S is the average energy per disintegration for the unstable isotope in the sample in electron volts per disintegration, w is the number of electron volts required to produce an electron pair in the ionizable gas being used, e is the number of coulombs per electron, and B is the chamber efficiency. Experimentally, e, S, w, and B are constants for a given unstable isotope, diluent gas, and chamber.

$S = 45,000$ average eV per disintegration for C^{14}, or 5700 average eV per disintegration for H^3; $w = 26.4$ eV per ion pair for argon; $e = 1.59 \times 10^{-19}$ coulombs per electron; and $B = 0.4$ for a 275-cc spherical ion chamber for C^{14}.

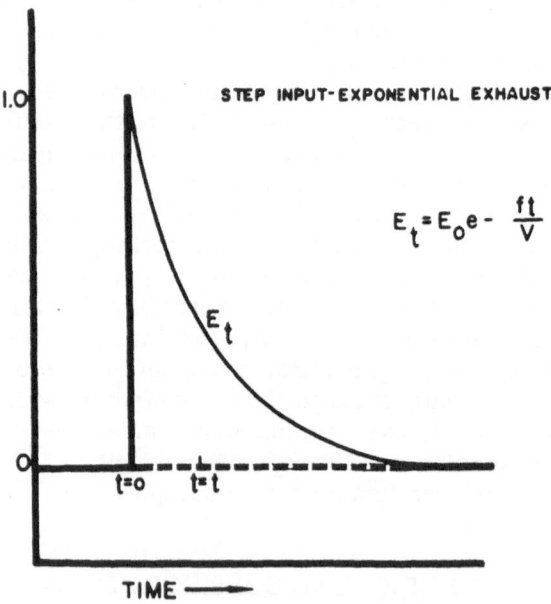

Fig. 6. Hypothetical ion-chamber response.

Equation (1a) can be rewritten as

$$N = EM \qquad (1b)$$

where E is in volts, M is a constant,

$$M = \frac{w}{BSeR} \qquad (1c)$$

and R is the value of the input resistor to the electrometer (normally $10^{12}\,\Omega$). Equation (1b) is rigorous for a static condition, that is, when all of the active molecules in a fraction are in the chamber at once. However, the active fraction entering a spherical chamber mixes completely with that already present, so that immediately after entry exhaust of the active fraction begins, and the signal E_t at any time t after $t = 0$ in Fig. 6 is given by

$$E_t = E_0 \, \exp\left(\frac{-ft}{v}\right) \qquad (2a)$$

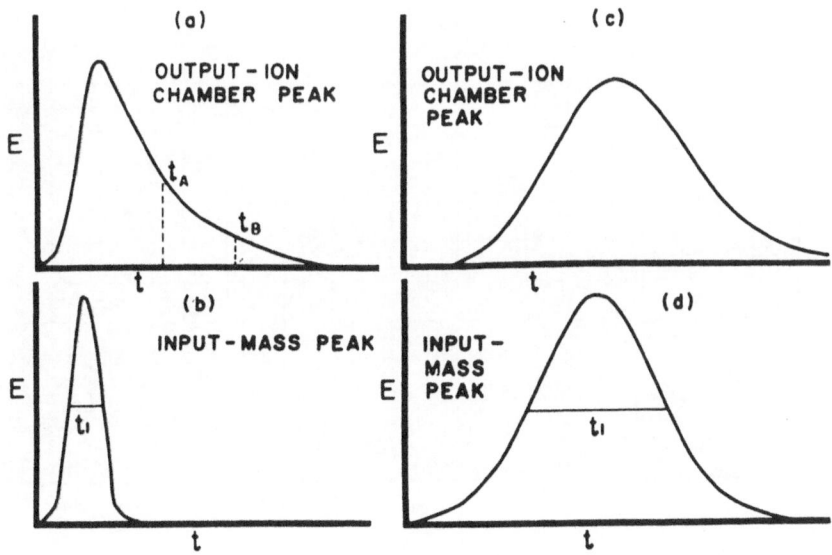

Fig. 7. Input—output profiles.

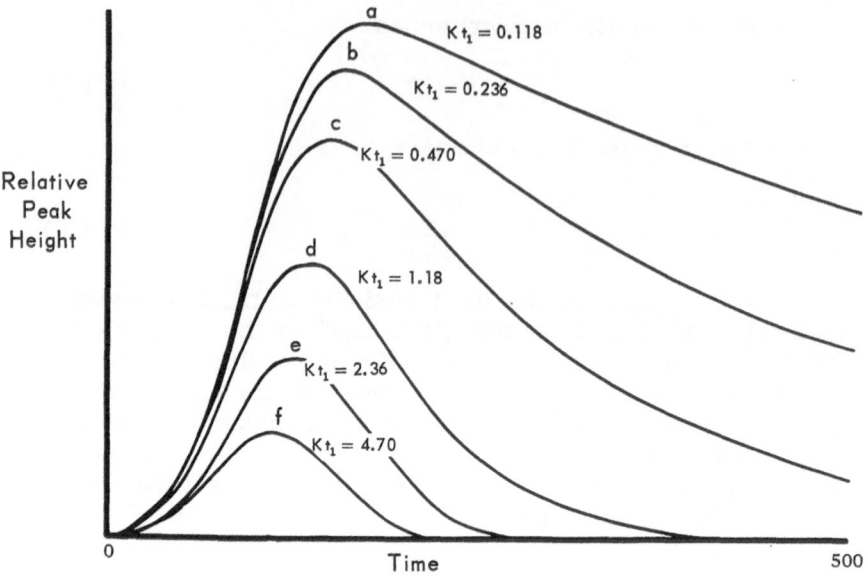

Fig. 8. Output profiles at various $K t_1$ values from computer data.

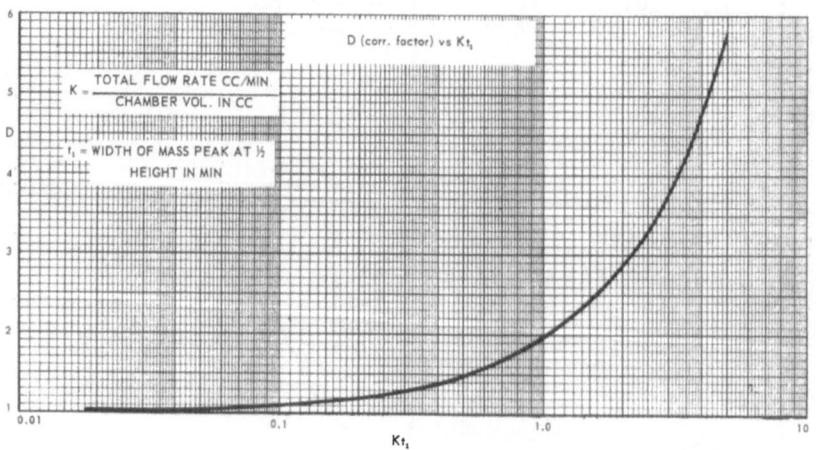

Fig. 9. Gaussian input peaks. D (correction factor) vs. $K t_1$.

where f is the flow rate in cubic centimeters per minute and v is the volume of the chamber in cubic centimeters. Thus, if the chamber volume is decreased, or the flow rate is increased, less time is required to reach E_t. This relation assumes that the electrometer response time can itself be neglected, a valid assumption under usual conditions. However, Fig. 6 assumes a "plug" sample. Of course, this is an idealization. Depending on the sample, column, etc., it may be a sharp peak providing an ion-chamber record closely approximating Fig. 6, or a peak of substantial width.

If the peak width is due mostly to diffusion spreading, we can assume a Gaussian shape, as in Fig. 7b and d. Figure 7a and c shows how the chamber output might respond to inputs like b and d. With a sharp input peak, relatively few active molecules leave the chamber before all enter, as in a. As the input peak gets broader or as the diluent flow is increased or both, the ion-chamber output curve begins to follow the shape of the input, as in c. Thus, equation (1b) cannot be applied with an input peak of finite width. However, it is not difficult in practice to correct back to the plug input by considering what effect moving a Gaussian shape through a system with an exponential decay has on the peak height observed for the exponentially decaying system. Non-Gaussian shapes will be considered also.

A transient electrical signal is modified by passing through an electrical filter. Thermal or chemical inputs are modified in analogous ways by linear processes. The principal mathematical tool for studying such processes is called convolution, and depends upon linear superposition of the effects of a series of short input pulses of varying amplitude into which the input waveform may be divided [7]. For the problem at hand, the input was considered to be a Gaussian pulse with a width at half height of t_1 minutes. The system weighting function is determined from equation (2a). The convolution was solved by the use of a RCA 301 computer. The computer program employed uses a finite series of points (200) to describe the input and produces the convolution as a finite sum. Figure 8 includes examples of ion-chamber output curves from the computer data. The curves reflect the effect of Kt_1 on the peak height of the output, where K is the ratio f/v. As Kt_1 gets larger, the peak height goes down and the shape approaches that of the input. Figure 9 shows a plot of output peak height to input integral vs. Kt_1, over a wide range of the variables. This curve may be used as the correction curve for the experimental data, the correction factor being designated D. Knowing D, we can extend equation (1b) to apply to all Gaussian-shaped inputs.

$$N = EMD \qquad (3a)$$

If we choose some readily available standard of known activity with a Gaussian-shaped chromatograph peak as the input, the following equation results:

$$N_s = E_s MD_s \qquad (3b)$$

where N_s is disintegrations per second of the standard, E_s is the observed peak height in volts, and D_s is the correction factor under the instrument conditions used. Division of (3a) by (3b) and solution for N yields

$$N = N_s \left(\frac{E D}{E_s D_s} \right) \qquad (3c)$$

This is the final form of the equation one would use for quantitative analysis of the radioactivity of an unknown.

Non-Gaussian inputs were treated in the same manner by the computer program, since the input can be arbitrary. It soon became apparent that thorough experimental verification would comprise another report. However, significant deviations from the correction curve for Gaussian inputs ($P = 1.0$) do not occur until the peaks are strongly asymmetrical ($P = 1.44$). In examining a mass peak of this shape, a working chromatographer would be very dissatisfied and undoubtedly look for a better column. With the growing improvement in columns for polar compounds, any curve other than the one shown in Fig. 9 for Gaussian inputs will seldom be required.

EXPERIMENTAL

Samples

To check the validity of equation (3c) under actual experimental conditions, four C^{14}-labeled compounds were used—

benzene, toluene, hexadecane, and octadecane. Each was tested under different Kt_1 conditions, varying both K (flow rate through chamber) and t_1 (width of mass peak).

Sample Injection

A 4-μliter sample was injected for all determinations, using the same Hamilton 5-μliter syringe each time.

Measurement of K

K is the ratio f/v, the total flow rate through the chamber divided by the chamber volume. Although this ratio can be determined by measuring the flow at the exhaust of the ion chamber, correcting it back to the temperature of the ion chamber, and dividing by the volume, these steps involve several obvious uncertainties. Calculating K from an ion-chamber record of a sharp incoming peak is more accurate. For example, if curve in Fig. 7b represents a mass peak and curve in 7a the output peak, we know that after t_A a negligible amount of sample is entering the chamber and that E is varying as $E_0 \exp(-Kt)$. Thus, by using the E values at t_A and t_B, we can calculate K. For an average, several points after t_A were taken. It is not necessary to know the activity of the sample. Any active sample will do, or even the unknown itself, as long as the mass peak is narrow enough to permit some reasonable time between t_A and t_B. Obviously, it would not be possible to measure K from curves like those of Fig. 7c or d. The K thus determined will be constant for a given flow through the chamber, and can be reproduced by restoring the flow rate at the same detector operating temperature without having to recalculate K.

Measurement of t_1

Because the measurement of t_1, the width of the mass peak at half height, is critical, measurement on a steel scale to 0.003 in. was routinely employed, since an error of 1% in t_1 results in a like error in N. The fastest peak encountered in this work was 0.09 min (5 sec), requiring a chart speed of 3 in./min for 1% accuracy.

Sample Activities

Since an independent check of the method was desired, a

TABLE I

Summary of Correction Technique Results for Gaussian Input Peaks

Compound	Number of determinations	K	t_1	D	Corrected peak height	Average deviation
Toluene	4	5.56	0.562	3.97	99.7	0.8
	9	5.82	0.326	2.79	99.7	0.7
	2	2.68	0.489	2.26	101.4	0.4
	3	7.95	0.146	2.14	100.0	1.1
	6	5.82	0.187	2.05	99.2	1.1
	3	4.55	0.146	1.65	100.3	0.1
	5	4.08	0.109	1.44	99.4	1.5
	3	2.91	0.146	1.42	100.7	0.6
	3	2.92	0.112	1.32	98.6	0.2
	3	2.68	0.090	1.35	99.6	0.8
(total)	41				99.9*	
						1.2†
					1.3‡	
Benzene	6	5.57	0.140	1.76	97.7	0.9
	6	4.18	0.143	1.59	99.0	1.2
	5	2.38	0.146	1.35	99.3	1.1
(total)	17				98.6*	
						1.2†
					2.0‡	
Hexadecane	1	6.72	1.072	8.00	132.3	—
	3	6.72	0.364	3.34	128.8	1.9
	4	5.17	0.364	2.79	131.3	1.0
	2	3.40	0.364	2.20	130.9	1.0
(total)	10				130.8*	
						1.8†
					2.2‡	
Octadecane	4	5.43	0.525	3.71	93.6	1.0
	6	5.43	0.227	2.18	94.9	1.2
	4	5.43	0.210	2.11	92.3	0.3
	7	5.43	0.115	1.61	93.1	1.1
(total)	21				93.5*	
						1.4†
					2.5‡	

*Mean.
†Standard deviation, S_k.
‡Standard deviation, S_0.

known amount of each of the compounds was counted with a liquid scintillation counter. Since each sample contained only one of the four active compounds, the scintillation counter values could be used to check the results obtained on the gas chromatograph. Three samples of each compound were counted.

Table I summarizes the results for all four compounds. Thus, K was varied from as low as 2.3 up to 8, and t_1 varied from 0.090 to 1 min, the D values ranging from 1.2 to 8. In practice, K will usually be constant at about 5, resulting in a range of Kt_1 values of from 0.3 to 3 for the usual chromatograph peaks.

The errors which limit the accuracy of this method accrue from two general sources. The first is the activity peak height measurement, which is a function of sample activity, injection reproducibility, and the linearity and reproducibility of the ion chamber and electrometer. The second is the correction itself, which is a function of how accurately K and t_1 are known and how well the assumptions are met on which the calculations are based. We are primarily interested in the latter errors and the validity of the method itself.

If the curve in Fig. 9 is plotted on linear paper, the result is nearly a straight line whose slope varies between 0.90 and 1.15. This means that a 1% error in Kt_1 results in a 1% error in D and finally N. Thus, both K and t_1 should be known to within 1%. K can be determined to this accuracy easily by averaging many observations. The time spent in making several estimations is not excessive, since once calculated, the K value can be used until operating conditions are changed. However, t_1 will differ from fraction to fraction. Since 0.003 in. is a practical limit in measuring a distance on chart paper, we can calculate the minimum required chart speed to permit the measurement of t_1 to within 1%. For a 1-sec peak width, the chart speed should be about 10 in./min. It is improbable, especially for the type of samples used in radioactivity work, that a peak faster than 6 sec will be encountered, and most will be longer. For a 6-sec peak, a chart speed of 3 in./min is required. Broader peaks can be recorded at proportionately slower chart speeds.

Standard deviations were calculated according to the methods of analysis of variance. S_k indicates the standard deviations resulting from sample injection, mass peak area estimation, and electrometer uncertainty, and refers to devia-

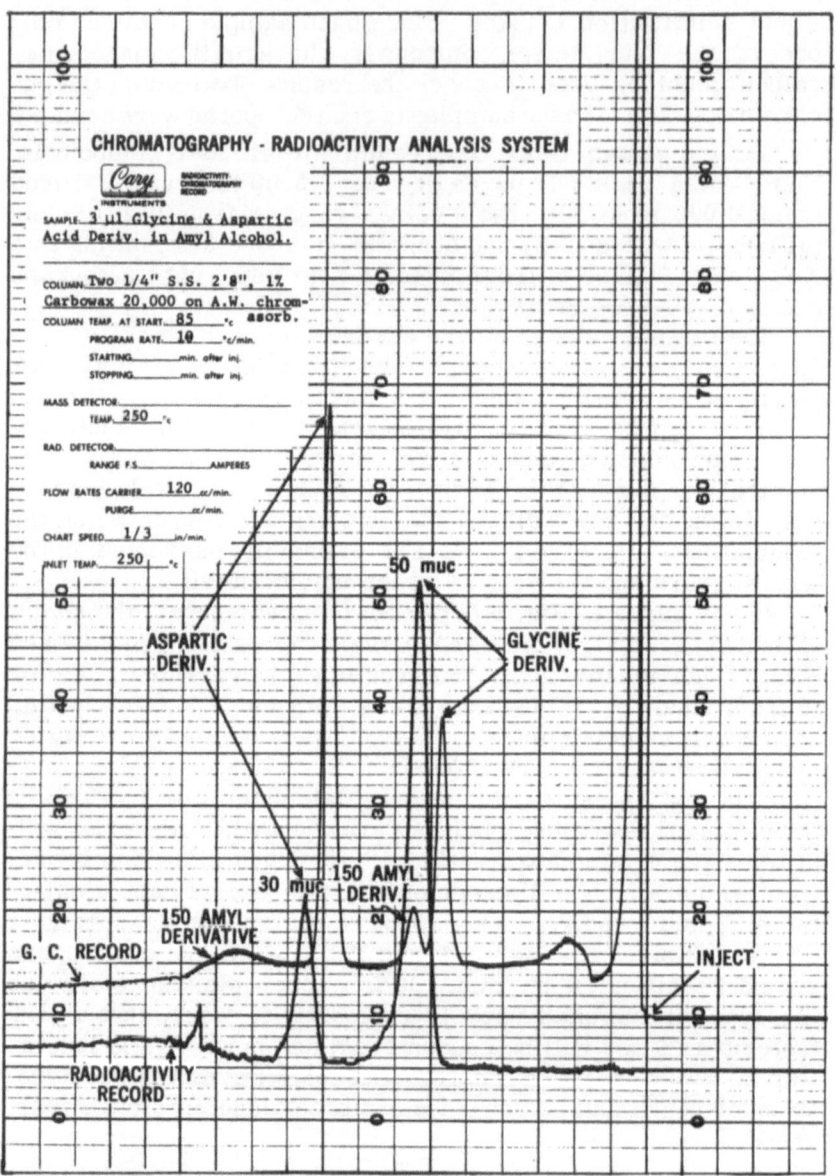

Fig. 10. Chromatography—radioactivity analysis system.

tions within a given set of K and t_1 values; thus, it does not test the correction method itself. S_0 is an indication of deviations among sets of data of different K or t_1 values or both. The square root of the variance for each compound type was calculated, where the variance is

$$S_m^2 = S_0^2 - S_k^2$$

The resulting values of S_m are: toluene 0.5; benzene 1.45; hexadecane 1.27; and octadecane 2.07. Their weighted average is 1.25%, which estimates the error inherent in the method plus other uncontrolled variables. Hence, the method is as good as the ability to sample in gas chromatography, and can doubtless be improved by study of technique and further replication of samples.

Table II shows the correlation among the four compounds from the scintillation counter data, arbitrarily using toluene as the standard. The deviations again show a relative error of around 1%.

APPLICATIONS

Since much of the interest in this type of detection is in the medical field, several examples were developed for this area.

Amino Acids

Figure 10 is a chromatogram of C^{14}-labeled derivatives of

TABLE II

Correlation among Compounds

Compound	mV	mV/μliter	nCi/μliter*	nCi/μliter†	Deviation
Toluene‡	299.7	74.9	22.7	22.7	—
Benzene	295.8	74.0	22.4	22.6	0.2
Hexadecane	391.8	98.0	29.7	29.6	0.1
Octadecane	280.5	70.1	21.2	21.4	0.2

*Calculated from ion-chamber results using toluene as standard.
†Measured with scintillation counter.
‡Standard.

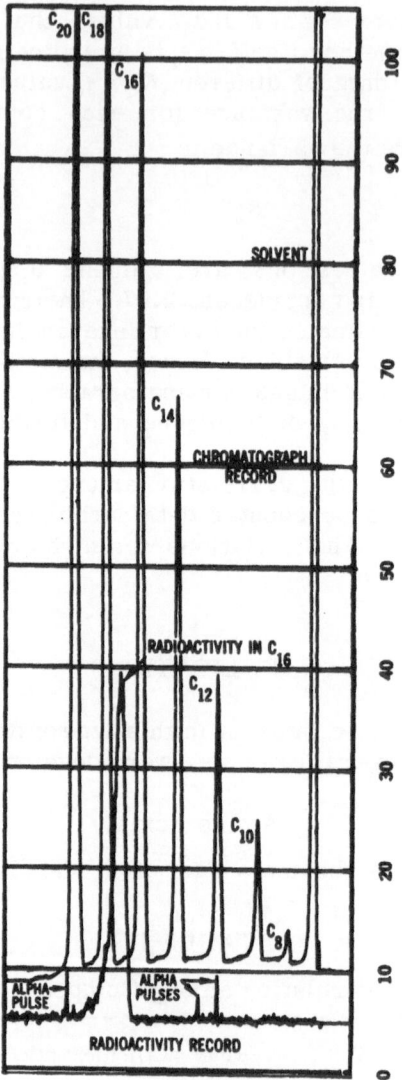

Fig. 11. Mixture of methyl esters of saturated fatty acids with C^{14} tagged palmitate having activity of 10 nCi/μ liter.

aspartic acid and glycine. These were measured on 8-ft, $^1/_4$-in.-diameter stainless steel columns filled with 1% Carbowax on acid-washed Chromosorb.

Fatty Acids

Figure 11 shows a mixture of fatty acid esters. The C^{16} component was labeled, the radioactivity peak corresponding to 10 nCi. This chromatogram was produced on 2-ft, $^1/_4$-in.- diameter stainless steel columns filled with SE 30 on hexamethyl disilazane-treated Chromosorb. The column temperature was programmed from 150 to 290°C at 12°C/min. The detector temperature was 250°C.

Steroids

Figure 12 is a chromatogram of desmosterol and cholesterol and unidentified impurities. Cholesterol was the only labeled compound. The column temperature was programmed from 250 to 350°C using 4-ft, $^1/_4$-in.-diameter stainless steel columns filled with SE 30 on hexamethyl disilazane-treated Chromosorb. The detector temperature was 295°C.

Some recent publications using the instrument described here include steroid work by Castle, Blondin, and Nes [3] and fatty acid ester work by Levis and Meade [12]. Yang and Ingalls [19] studied the effect of exposing polystyrene to tritium by the Wilzbach gas exposure method. Burr and Goodspeed [1] studied the role of benzene in the radiolysis of cyclohexane—benzene mixtures.

We have found that operating the ion-chamber system in close proximity to the beta-ray ionization mass detectors does not affect the performance.

Karmen [10] has reported observation of so-called false peaks for unlabeled compounds. We have also observed these peaks for very polar compounds. We have found that they do not occur above about 100°C for any sample charge. The chamber can be operated at temperatures up to 300°C so that the peaks are easily eliminated by always maintaining chamber temperatures in excess of 100°C.

Recently, we ran across a very interesting application, that of total radioactivity measurement. The calibration curve data developed earlier led us to the realization that, at low carrier flow rates and with very sharp chamber-input peaks, no correction at all of the peak height would be necessary. These conditions are fulfilled by using pure argon carrier at 60 cc/min and removing the column. The injected sample is then swept almost immediately into the ion chamber, and the

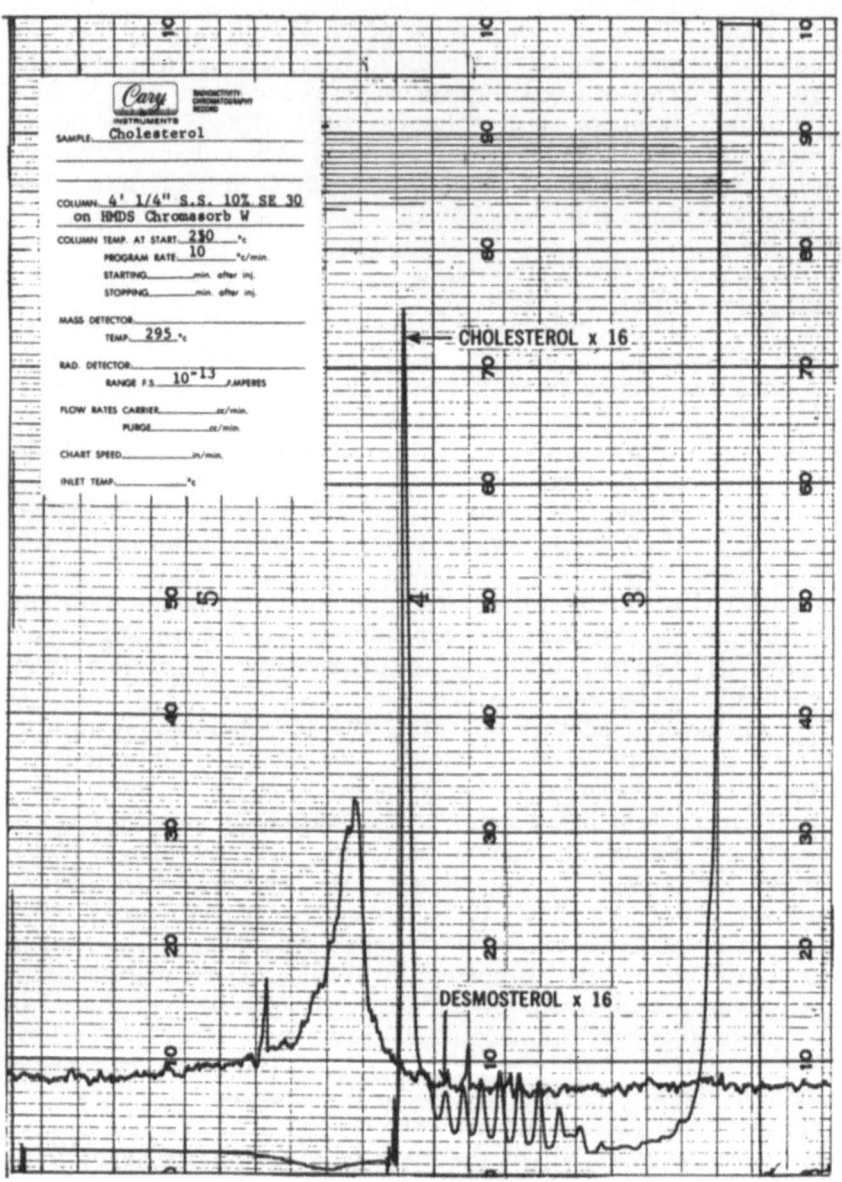

Fig. 12. Chromatogram of desmosterol and cholesterol and unidentified impurities.

peak height on the recorder indicates the total activity of the sample. The only correction necessary is for chamber efficiency, which can be easily calculated by this method with a standard. The chamber can be rapidly swept out by opening the toggle valve for the purge gas. Samples can be measured in about 30 sec this way. We have, using a very simple system, trapped the sample at the exit of the chamber so that little sample is used up in the measurement. We are going to look into this further, but we know already that it is a very convenient way of measuring total activity of a sample, accurate to 1%.

ACKNOWLEDGMENT

Scintillation counting was performed through the courtesy of J. Meade of the University of California at Los Angeles. We are indebted to R. C. Hawes, Applied Physics Corp., for suggestions during the experimental work and for reading the manuscript.

REFERENCES

1. J. G. Burr and F. C. Goodspeed, Role of benzene in the radiolysis of cyclohexane—benzene mixtures, J. Chem. Phys. 40 (5): 1433–1436 (1964).
2. F. Cacace, Nucleonics 19: 45–50 (1961).
3. M. Castle, G. Blondin, and W. R. Nes, Guidance for the origin of the ethyl group of β-sitosterol, J. Am. Chem. Soc. 85: 3306–3308 (1963).
4. H. E. Dobbs, J. Chromatog. 5: 32–7 (1961).
5. H. J. Dutton, Pittsburgh Conference on Analytical Chemistry and Applied Spectroscopy, March (1961).
6. T. B. Evans and J. E. Willard, J. Am. Chem. Soc. 78: 2908 (1956).
7. M. E. Gardner and J. L. Barnes, Transients in Linear Systems, Vol. 1, John Wiley & Sons, Inc., New York (1961), pp. 262–263.
8. A. T. James and E. A. Piper, J. Chromatog. 5: 265 (1961).
9. A. Karmen, L. Giuffrida, and R. L. Bowman, J. Lipid Res. 3: 44 (1962).
10. A. Karmen and J. Winkelman, J. Anal. Chem. 34: 1067–71 (1962).
11. R. J. Kokes, H. Tobin, Jr., and P. H. Emmett, J. Am. Chem. Soc. 77: 5860 (1955).
12. G. M. Levis and J. F. Meade, An alpha-hydroxy acid decarboxylase in brain microsomes, J. Biol. Chem. 239:1, 77–80 (1964).
13. A. Weissberger, Techniques of Organic Chemistry, Vol. 1, Part 4, Interscience, New York (1959), p. 3359.
14. A. Weissberger, ibid., p. 3363.
15. A. Weissberger, ibid., p. 3364.
16. I. M. Whittemore, Bio-Organic Chem. Quart. Rept. UCRL-9408: 49–50 (1960).
17. R. Wolfgang and C. F. Mackay, Nucleonics 16: 69–73 (1958).
18. R. Wolfgang and F. S. Rowland, Anal. Chem. 30: 903–906 (1958).
19. J. Y. Yang and R. B. Ingalls, Tritium beta–decay induced reactions in the polystyrene fluff, J. Am. Chem. Soc. 85:3920–3923 (1963).

An Inexpensive, Dual-Detector Gas Chromatograph Suitable for Temperature Programming

John A. Perry

Sinclair Research, Inc.
Harvey, Illinois

A gas chromatograph has been developed which can be built with simple shop facilities out of parts costing about $100.00. The instrument incorporates both thermistor and filament detectors, permits temperature programming, and affords a very stable base line. The instrument could also accommodate high-temperature gas chromatography, up to the softening point of silver solder—over 500°C.

INTRODUCTION

The manifold utility and unusual simplicity of gas chromatography as an analytical tool makes a gas chromatograph a desired item in the modern chemical laboratory. A continuing effort has been made at Sinclair Research to provide those gas chromatographic instruments which are commercially unavailable. One such instrument, accommodating both analytical and relatively large-scale preparative work, has been described [1].

The instrument reported in this paper meets several requirements: (1) fabrication without undue trouble from inexpensive parts and with minimum use of the machine shop, (2) accommodation of packed, coiled columns, (3) production of an unusually stable base line, (4) accommodation of temperature programming [2], and (5) choice of either thermistor or filament detectors at the time of column connection [3].

Fig. 1. Top view, complete instrument.

Fig. 2. Top view, lid on, bottom section.

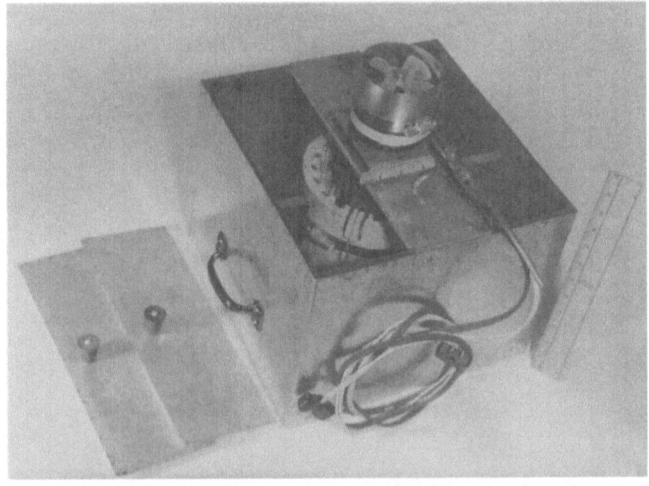

Fig. 3. Top–quarter view, lids off, top section.

DESIGN

The instrument (Fig. 1) has two sections—a high–thermal-mass, heavily insulated, double-walled bottom section (Fig. 2); and a low-thermal-mass, lightly insulated, double-walled, removable top section (Fig. 3). The bottom section is roughly a cube about 10 in. along each edge; it contains detectors, sample-injection, some flow-control facilities, and column mounting. The top section is also roughly a 10-in. cube; it carries a fan and a cylindrical heater.

All walls and lids are made of 24-gauge galvanized sheet metal. The shell of the bottom section consists of two boxes. The inner box is separated from the outer by 2 in. of solid insulation at the bottom and 1 in. at the sides. The bottom section has a removable lid which carries $\frac{1}{2}$ in. of insulation. In operation, the bottom section is filled with gravel. Rapid temperature changes within the section are thus prevented,. helping to stabilize the base line. Also, good sample volatilization and transfer to the column are aided because the sampling lines are prevented from cooling by the gravel insulation. For Fig. 4, the gravel and lid were removed to show the interior of the bottom section.

Fig. 4. Top view, lid off, racks out, bottom section.

To the floor of the inner box is bolted a broad "U" (Fig. 5) made of three sections of $\frac{1}{2}$-in. steel plate. A square Hevi-Duty 750-W heater (Fig. 6, 20) is the heat source for the bottom section. The horizontal base of the U has two 1-in. diameter holes to pass the heater connections. The U base also has six tapped holes to receive 8—32 bolts which hold the U to the floor, and which hold the heater (Fig. 6, 20) and the steel plates (Fig. 6, 18 and 19) to the U. Leads for the heater are brought out through the middle of the 2-in. insulation under the floor of the inner box. The thick parts (Fig. 6, 15—17) of the U carry heat from the heater to the detectors; the detectors are thus kept at a considerably higher temperature than the surrounding gravel.

Above the flat, square heater, and bolted against it, are two metal squares of the same area as the heater. One of these is a square (Fig. 6, 19) of 16-gauge stainless steel; this faces the heater coils. Above the stainless steel square is another square (Fig. 6, 18) of $\frac{3}{16}$-in. steel plate. The upper steel plate oxidizes and blisters under intense heat and would eventually short out the heater coils were it not for the intervening protective stainless steel. The upper $\frac{3}{16}$-in. steel plate forms part of the sample injection assembly.

Fig. 5. Iron U.

Fig. 6. Functional components of bottom section.

A smaller, vertical piece (Fig. 6, 21) of $\frac{3}{16}$-in. steel plate is welded to the horizontal $\frac{3}{16}$-in. plate, Thus, the $\frac{3}{16}$-in., two-piece, steel-plate assembly (Fig. 6, 18, 21) is in most intimate contact with the heater and at the highest temperature in the bottom section of the gas chromatograph. The means for sample injection is welded to this assembly.

The carrier-gas, heat-exchanging coils (in Fig. 6, the carrier gas enters at 2 and leaves at 4) are bolted to the horizontal plate just over the heater. The comparatively very hot carrier gas then enters through a tee (22), which is welded to a 90°-bend of $\frac{3}{8}$-in. stainless steel tube (3), which in turn is silver-soldered to the hot plate assembly (18-21). Inside the $\frac{3}{8}$-in. tube runs a $\frac{1}{8}$-in. stainless steel tube; this tube is continuous from just inside the end (28) of the $\frac{3}{8}$-in. tube to the union (11) to which the column is fastened. (The top ferrule of the Imperial Hi-Seal tee (22) is drilled out to pass the $\frac{1}{8}$-in. tube.) The $\frac{1}{8}$-in. tube is centered within the $\frac{3}{8}$-in. tube by a spiral of #16 Nichrome wire.

The sample is injected from outside the box through a septum into the $\frac{1}{8}$-in. tube. The sample-receiving end of this tube is flared both for centering the tube and for guiding the syringe needle. The septum is pressed by a nut (24) against the end (28) of the $\frac{3}{8}$-in. tube; the outside of the nut (24) is countersunk as a needle guide, and a brass plate is pressed onto the nut, dissipating heat from the septum into the air and lengthening the service life of the septum. The flare of the $\frac{1}{8}$-in. tube is recessed about $\frac{1}{4}$ in. from the sealing surface (28) of the $\frac{3}{8}$-in. tube. This provides clearance between the end of the $\frac{1}{8}$-in. tube-flare and the septum which is partially extruded into the $\frac{3}{8}$-in. tube. In this way, the septum does not prevent the carrier gas from issuing around the edge of the flare. The sample is deposited within the hot-gas-bathed, $\frac{1}{8}$-in. tube, vaporized, and carried in the same, continuous tube directly to the column.

The column is mounted to two of the three $\frac{1}{8}$-in. bulkhead unions (9-11). These unions are long enough to project through the lid of the lower section and are welded to the strap-iron support (23), which is in turn bolted to brackets fastened to the inner box. As can be seen from Fig. 2, one end of the column is to be fastened to the sample-entry port marked "Col.". The other end is to be fastened either to the port (10) leading to the filament detector ("Fil.") or to the port (9) leading to the thermistor detector ("TH"). In this way, the operator very simply selects either filament or thermistor detection.

The filament and the thermistor detectors are silver-soldered into prepared $1/4$-in. Imperial steel Hi-Seal tees, as indicated in Figs. 7 and 8. The detector-equipped tees (Fig. 6, 5—8) are then mounted and held by set screws in notches cut in the steel plates (15, 17) which make up the sides of the steel U. The filament detectors (5, 7) are held in one plate (15); the thermistor detectors (6, 8) in the other, (17). This inexpensive method of preparing and using detectors furnishes sensitive, low dead-volume detectors which permanently withstand temperatures up to the melting point of silver solder—over 500°C. The detectors are massively mounted and very well insulated thermally. They are, therefore, stable to vibration and to most environmental temperature changes. These detectors are also easily dismounted, inspected, or repaired.

The leads from these detectors are silver-soldered to ordinary plastic-coated, hook-up wire, are carried out through the oven walls inside ceramic thermocouple-lead insulators (24-27), and are fastened to externally mounted Amphenol fittings. Any suitable bridge power supply and signal attenuator is then connected to the desired Amphenol fitting.

The exit gases from the sample detectors are carried out through the oven walls for flow measurement by vent lines

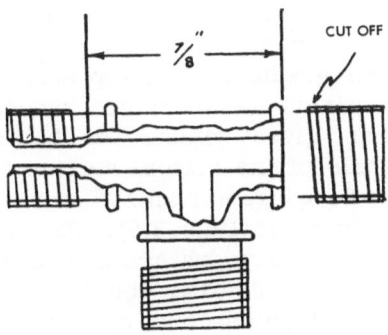

Fig. 7. Thermal-conductivity detector mount. (Thermistors: 9/32-in. hole, 7/8 in. deep and 3/8-in. hole, 1/16 in. deep. Filaments: 7/32-in. hole, 7/8 in. deep and 11/32-in. hole, 1/16 in. deep. A 1/4-in. Imperial hi-seal steel tee.)

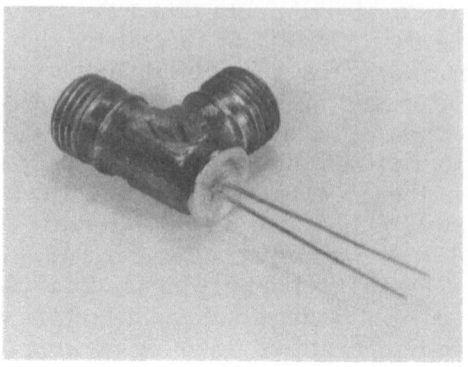

Fig. 8. Modified tee carrying silver-soldered detector.

(12, 13). These vent lines normally are $\frac{1}{4}$ in. OD to prevent bubbling of the high-boiling liquids which condense on the cooler walls of the vent tubes. (The lines in the assembly shown in Fig. 6 are $\frac{1}{8}$ in. OD, the better to prevent remixing enroute to fraction collectors.)

The reference detectors are supplied with a separate stream of carrier gas which enters at a port (1), flows in series through detectors (8, 7), and exits through a tube (14) for flow measurement. The exit reference carrier gas can be led from the exit port (14) back into the port (13), thus bathing the idle thermistor sample detector in helium, rather than air, at the high temperatures used with filament detection.

Whereas the bottom section is designed to maintain stable, steep temperature gradients, the top section is designed to accommodate very rapid temperature changes in time and to be isothermal at any given instant. The shell of the top section fits over the shell of the bottom section (Figs. 1, 9) and is made of the same 24-gauge galvanized steel sheet, formed into two walls about $\frac{1}{4}$ in. apart (Figs. 3,9). The walls are separated top and bottom by spacers, but are uninsulated. Metal strips center two Hevi-Duty 4808-104, 450-W heater sections connected in parallel, forming one cylindrical, 900-W heater which is hung from the lid of the top section and which surrounds the column.

The coil of the column can have a maximum outer diameter as indicated for the operator on the lid of the bottom section

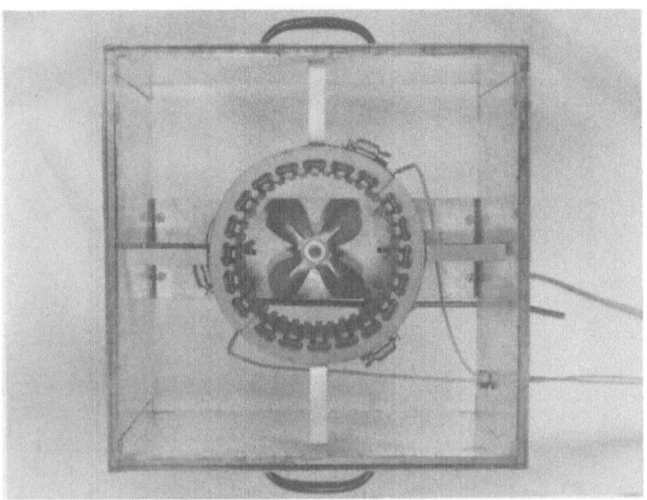

Fig. 9. Bottom view, top section.

as shown in Fig. 2. A one-thickness heater-filling helix of maximum diameter would be about 30 ft long, adequate for the great majority of applications; alternatively, a two-thickness coil can be used. During heating, air within the closed top-oven is circulated by a fan through the axis of the column coil and back around the outside of the heater. For rapid cooling, auxiliary house air can be blown in through two $^1/_4$-in. copper tubing ports (Figs. 1, 9). The heated air leaves through easily removable lids (Figs. 1, 3, 9). Heating with 115-V input yields a reproducible, and surprisingly, linear temperature rise of 20—25°F/min in going from about 100 to about 400°F. Cooling to 100°F from over 400°F takes about 10 min. The course of such a heating and cooling cycle is shown in Fig. 10.

Thermocouples are permanently installed on the detector, on the sample inlet point just before the sample enters the column, and in the top oven. A terminal board (Fig. 1) provides electrical access to these couples. The couples are protected in transit through the walls of the bottom section by $^1/_4$-in. OD copper tubes held in place with flared ends.

A working installation is shown in Fig. 11. This unit is being used with temperature programming so that special means must be taken to prevent the changing pressure drop across the column from affecting the flow through the column.

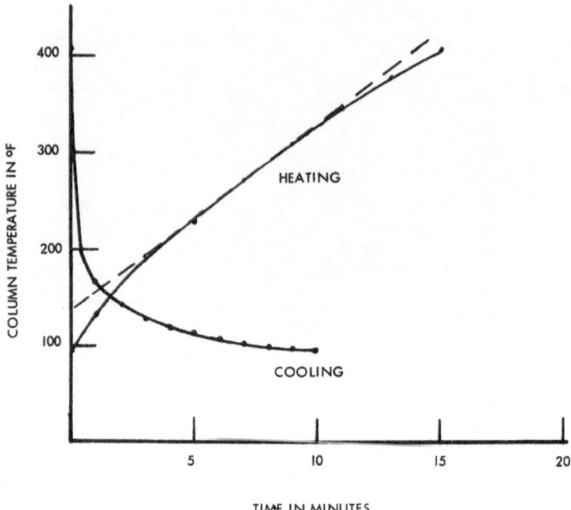

Fig. 10. Typical heating and cooling curves.

Fig. 11. Complete working instrument.

One means of achieving constant flow is to install a Moore Products constant differential type flow controller, Model 63 BV-L, upstream from the column. The reference gas is taken off upstream from the controller with flow regulated by a needle valve. Therefore, constant flows of gas cross the reference and sample detectors, helping to achieve a stable base line.

PERFORMANCE

The simplest method of temperature programming is to impose some selected voltage across the column heater at the moment when the sample is injected, or at some time definitely relately to that moment. The chromatogram is then recorded as the column temperature rises. The illustrations shown

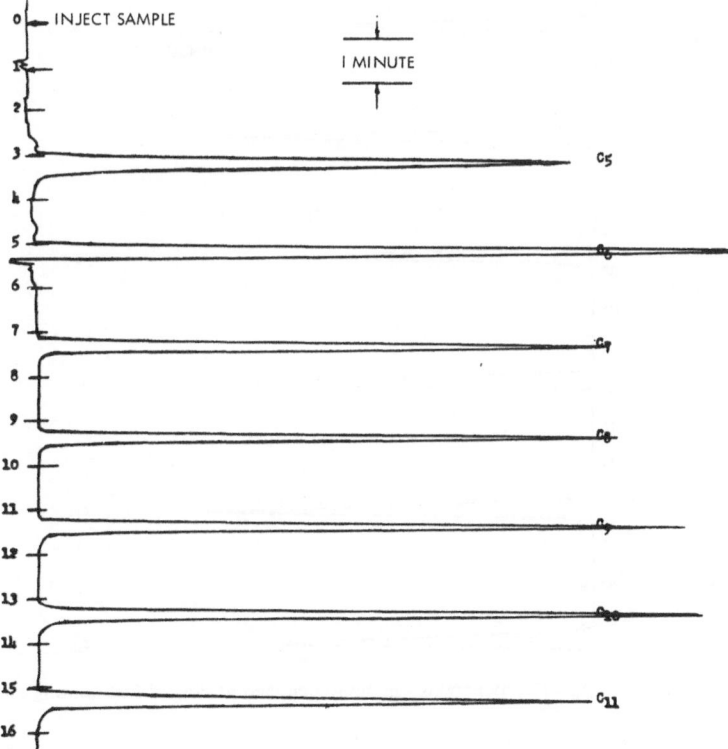

Fig. 12. C_5 —C_{11} paraffins chromatographed at full heating rate.

were made with this type of temperature programming. In
Figs. 12 and 13, the chromatograms of a mixture of five- to
eleven-carbon, normal paraffins are presented. In each, the
voltage was imposed across the heater just after the sample
was injected; in Fig. 12, the voltage was 100 V, in Fig. 13, 90.

Over ten of these instruments have been built and some have
been in use for well over a year. All have been found highly
satisfactory and easy to use.

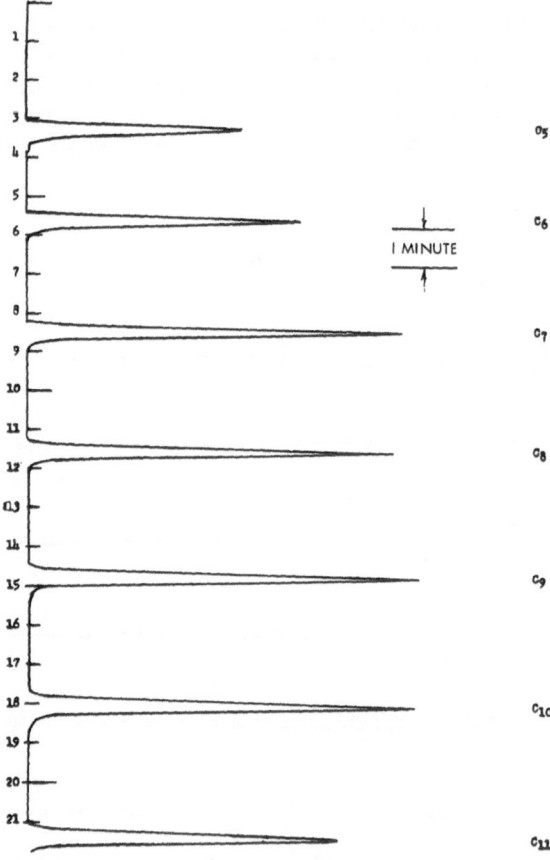

Fig. 13. C_5—C_{11} paraffins chromatographed at 80% of full heating rate.

COSTS

Fabrication of the instrument takes from two to four man-weeks. A few approximate direct costs follow:

Detectors
      Filaments. . . . . $ 5 (per pair)
      Thermistors . . . . 26 (per pair)

Heaters
      Bottom. 6
      Top 14

Fan Motor and Blades 4
Amphenol Fittings. 4
Bulkhead Unions 8

Auxiliary Flow Controller . . <u>30</u>

TOTAL. . . . . $ 97

ACKNOWLEDGMENT

The many helpful suggestions of Mr. Lester Morse concerning the details of construction are gratefully acknowledged.

REFERENCES

1. J. A. Perry, in Noebels, Wall, and Brenner (eds.): Gas Chromatography, Academic Press (1961), p. 183.
2. A. J. Martin, C. E. Bennett, and F. W. Martinez, in Noebels, Wall, and Brenner (eds.): Gas Chromatography, Academic Press (1961), p. 363.
3. C. P. Cowan and P. J. Stirling, in Coates, Noebels, and Fagerson (eds.): Gas Chromatography, Academic Press (1958), p. 165.

A New Method of Pyrolysis

Theron Johns and Robert A. Morris

Beckman Instruments, Inc.
Fullerton, California

The usefulness of pyrolysis in gas chromatography is determined by the uniqueness and repeatability of the pattern produced and the convenience of the method. In order to produce a unique repeatable pattern, it is necessary to be able to control and repeat the factors of sample size, time and temperature of pyrolysis, sample dispersion, and atmospheric conditions in the pyrolysis chamber. Careful attention must also be given to the column system used and the operating parameters of the gas chromatograph. The results of a study directed towards repeatable and convenient pyrolysis are presented.

INTRODUCTION

As pointed out by Perry [5], two basic limitations of gas chromatography are its application to materials of low volatility and its poor qualitative capability. Pyrolysis has been used extensively to overcome the first limitation, and recent work indicates that a real potential exists for qualitative analysis by pyrolysis gas chromatography.

In 1954, Davison et al. [1] showed that polymers could be identified by chromatographing the pyrolyzate. Subsequently, a large number of papers have been given in this field. Reference to this work is given in Perry's review paper. Janak [3] showed that pyrolysis can be used in combination with gas chromatography for the quantitative analysis of barbituric acid from urine by pyrolyzing a urine specimen. Keulemans and Perry [4] found that the pyrolysis pattern of hydrocarbons could be used to determine the structure of the hydrocarbons. Dhont [2] showed similar results for alcohols.

The successful application of pyrolysis for the identification of nonvolatile and volatile materials requires that the method of pyrolysis be capable of handling solid, liquid, and even vapor

361

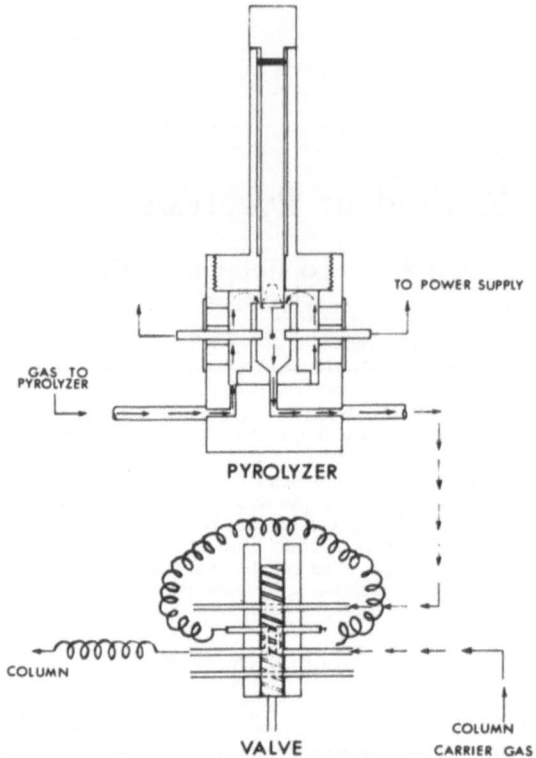

Fig. 1. Schematic of pyrolysis system.

samples. It is also necessary to be able to precisely repeat the conditions of pyrolysis, if the fragmentation pattern is to be useful. Additionally, the method used for pyrolysis should not interfere with the routine operation of the gas chromatograph.

The two chief methods used to date employ either filaments that can be heated very rapidly electrically or hot tubular reactors which are generally preheated to the desired temperature and the sample then injected. This paper describes a third method, arc pyrolysis.

EXPERIMENTAL AND RESULTS

The details of equipment needed for arc pyrolysis are shown in Fig. 1. Two electrodes are spaced a suitable distance apart, mounted in an insulating chamber which can be closed, and connected to power supply which regulates the voltage, current, and time that the arc is on. The power supply could deliver a current of $0 - 150$ mA to the electrodes. The voltage capability was 5000 V.

The sample, which was either a solid or a liquid, was positioned between the two electrodes by a sample holder which is connected to a removable piston. The length of the sample holder was designed so as to position the sample within the arc. A gas flow is maintained in the pyrolysis chamber during pyrolysis. The exhaust from the pyrolyzer is routed through a sample loop of suitable volume. This loop is part of a sampling valve which can be actuated in order to inject the pyrolysis product into a chromatographic column or columns. All parts of the system except the upper end of the sample piston are mounted in a heated compartment, which can be set at any temperature between ambient and 280°C. The flow rate of gas through the pyrolyzer and the time of pyrolysis can be varied as desired except as limited by the volume of the sample loop of the sampling valve. Since the pyrolysis is carried out under flowing conditions, these variables must be adjusted so that actuation of the valve is accomplished at the time the pyrolysis products are in the sample loop.

The following experimental conditions were maintained for the work reported in this paper:

Current . . . . .55 mA
Gas flow through pyrolyzer . . . . .10 ml/min of helium
Time of pyrolysis . . . . .20 sec
Volume of sample loop . . . . .3 ml
Temp. of pyrolysis chamber and valve . . . . .140°C
Sample size . . . . .0.1 — 0.3 mg injected. Split 10:1
Column . . . . .13 ft, $\frac{1}{8}$-in. 30% Carbowax 1540
                                          on 42-60 mesh firebrick
Column temp. . . . . . . .50 — 180°C programmed at 10°C/min
                                          single column program
Column flow rate. . . . . .25 ml/min
Detector . . . . .flame ionization detector at 240°C
Attenuation. . . . . .5 · 10^{-9} A (0—1 min)
                                          2.5 · 10^{-10} A (1—14 min)
Chart speed . . . . .0.5 in./min

Approximately equal amounts (about 2 mg) of Mylar and Tygon were pyrolyzed and compared chromatographically, as shown in Fig. 2. With the column system and conditions used, a component-type separation was not obtained with the very light components formed (C_1—C_5 hydrocarbons). Additionally, any inorganic components present were not detected with the flame-ionization detector. Even so, a point-by-point comparison of the chromatograms shows significant differences both in the elution time of the components formed and in the relative amounts of components. The arc pyrolysis of Mylar shows components which peak at 1.8, 10.7, 11.7, and possibly at 9.0 min, which are not present in Tygon. Additionally, the ratios of the peaks at 2.6 and 3.3 min are significantly different. Tygon shows peaks at 5.1, 5.6, and 11.5 min, which are not present in Mylar.

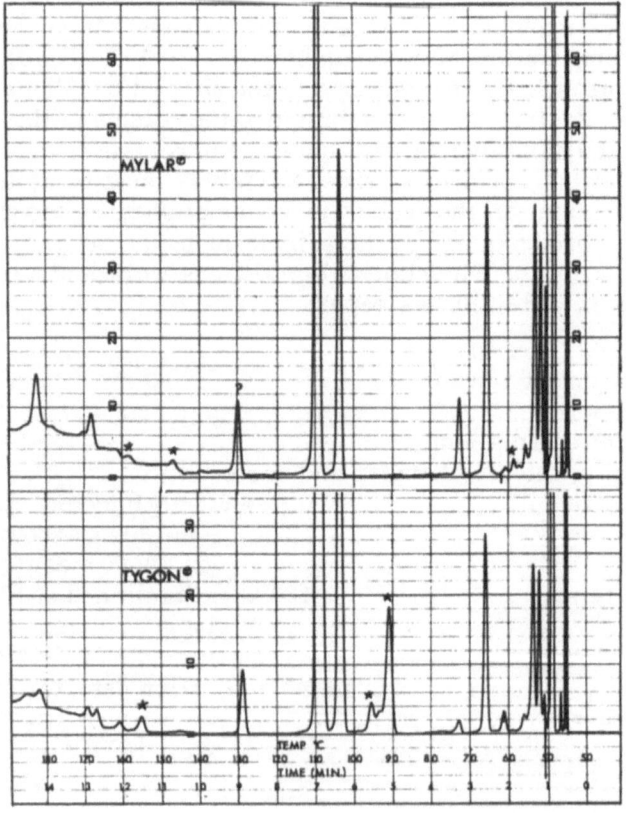

Fig. 2. Chromatogram of pyrolysis products from Mylar and Tygon.

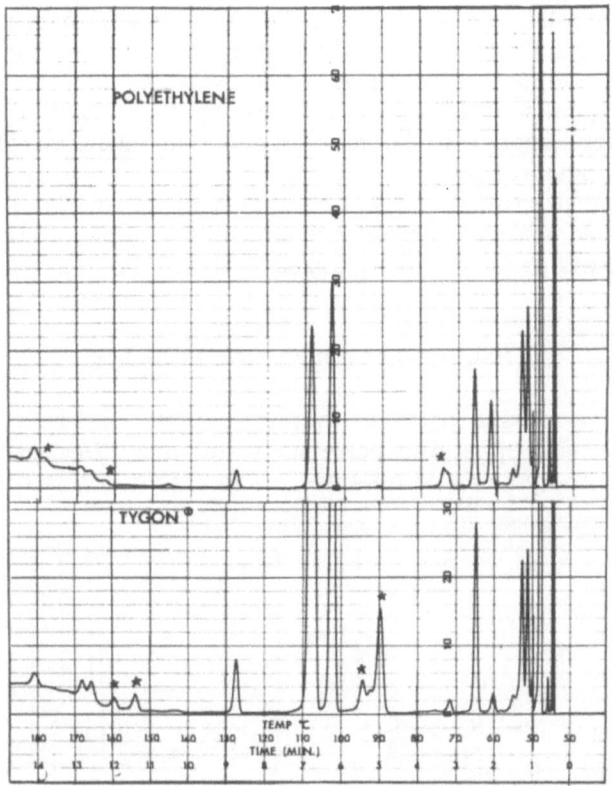

Fig. 3. Chromatogram of pyrolysis products of Tygon and polyethylene.

A comparison of Tygon and polyethylene is shown in Fig. 3. In this case, polyethylene has peaks at 3.4, 12.3, and 13.8 min, which are not present with Tygon. The ratio of the peak at 2.2 to the peak at 2.6 is also significantly different. Tygon has peaks at 5.1, 5.6, 11.5, and 12.0 min, which are not present with polyethylene.

Figure 4 is a comparison of two samples of Apiezon L at about the 1.0-mg and 2.5-mg levels with squalene. It can be seen that the pattern for Apiezon L is essentially the same in both cases except for a decrease in the magnitude of the peaks with the smaller sample. Significant differences are apparent between squalene and Apiezon L. (Note the peaks which are marked.)

A comparison of cholesterol, 7-dehydrocholesterol, and

stigmasterol is shown in Fig. 5. The chromatograms of the pyrolysis products of these three compounds are very similar; however, some subtle differences exist, as indicated by the marked peaks. It is also interesting that stigmasterol and cholesterol appear to be more alike than cholesterol and 7-dehydrocholesterol.

Chromatograms of the arc pyrolysis products of silicone rubber and natural rubber are shown in Fig. 6. In this case, the patterns produced are markedly dissimilar.

DISCUSSION AND CONCLUSIONS

The pyrolysis patterns obtained with an arc pyrolyzer indicate that repeatable patterns which are different for each material tested can be obtained. The extent of this difference

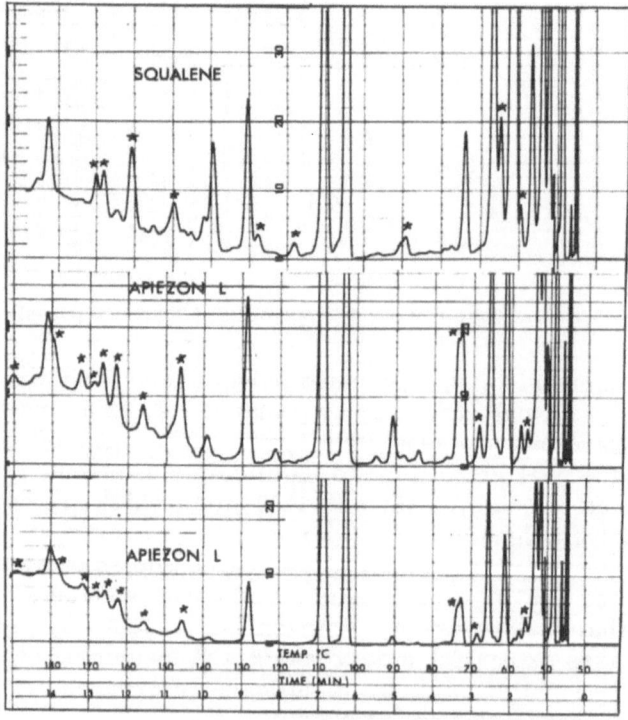

Fig. 4. Chromatogram of pyrolysis products of squalene and Apiezon L.

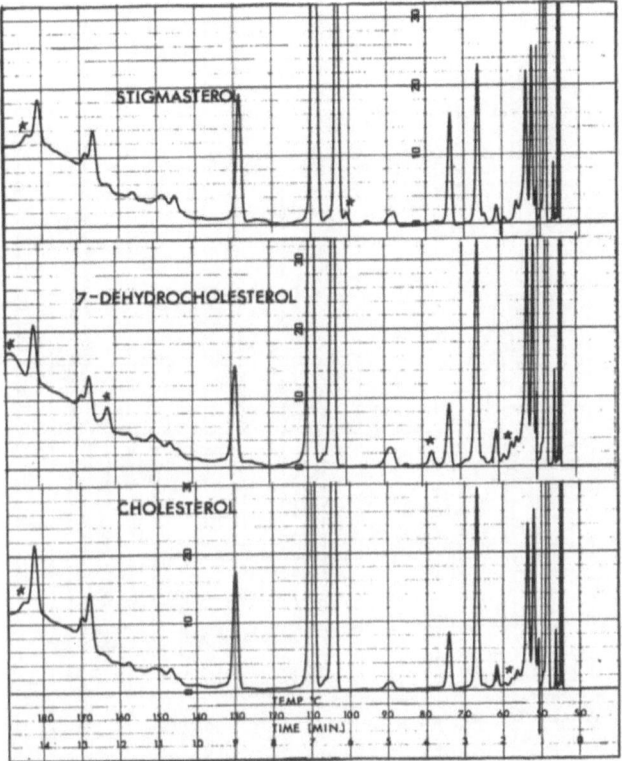

Fig. 5. Chromatogram of pyrolysis products of stigmasterol, 7-dehydrocholesterol, and cholesterol.

in the pyrolysis patterns for certain materials can be substantially increased through proper selection of columns and operating conditions. It is expected that a combination of an adsorption column and a partition column with proper valve switching and temperature programming will be the preferred approach in many cases. It is also likely that, for the difficult case of determining a minor impurity, effective use can be made of a selective column. In this way, if the minor component has a substituent group which is likely to produce a particular compound, a column can be selected which would give a resolved peak for that compound.

It has been shown that repeatable pyrolysis patterns can be obtained with an arc pyrolyzer. This technique is easily adaptable to the pyrolysis of solids, liquids, and vapors.

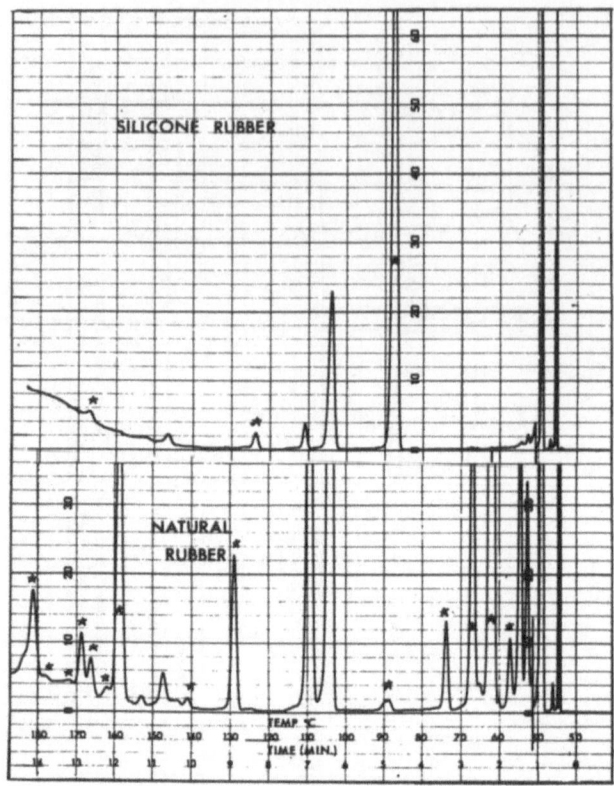

Fig. 6. Chromatogram of pyrolysis products of silicone rubber and natural rubber.

REFERENCES

1. W. H. T. Davison, S. Slaney, and A. L. Wragg, Chem. and Ind. (London): 1356 (1954).
2. J. H. Dhont, Nature 192: 747 (1961).
3. J. Janak, in R. P. W. Scott (ed.): Gas Chromatography 1960, Butterworths, London (1960), p. 387.
4. A. I. M. Keulemans and S. G. Perry, in M. Van Swaay (ed.): Gas Chromatography 1962, Butterworths, London (1963), p. 351.
5. S. G. Perry, J. Gas Chromatogr. 2: 54 (1964).

Gas Chromatographic Determination of Total Oxygen in Organic Materials

F. L. Boys and D. D. Dworak

Sinclair Research, Inc.
Harvey, Illinois

A sample is pyrolyzed in the presence of carbon to quantitatively convert oxygen to carbon monoxide. Helium elutes the pyrolysis products through a molecular sieve column for separation of the carbon monoxide and through a katharometer for quantitative measurement. Analysis time is 15 — 30 min, detectability is 30 ppm, no blank is involved, and less than 50 mg of sample is used.

INTRODUCTION

Advances in petroleum technology have necessitated the determination of oxygen in samples at concentrations considerably below 0.1%. The classical Schütze—Unterzaucher method for determining oxygen is a tedious procedure, and even with the various improvements that have been reported, is not suited for a large number of low-concentration oxygen determinations. It appears doubtful that any of the volumetric, titrimetric, or gravimetric finishes used with the Unterzaucher carbon pyrolysis technique are practical at much less than 0.1% oxygen.

Rapid and sensitive fast neutron activation methods for the nondestructive determination of total oxygen content have been reported by Gilmore [1] and Stallwood [2]. Sample size, equipment space, and expense requirements are disadvantages of this method.

The known separation and detection capabilities of gas chromatography indicated that a combination of the Unterzaucher carbon pyrolysis with gas chromatography should be applicable to the problem.

369

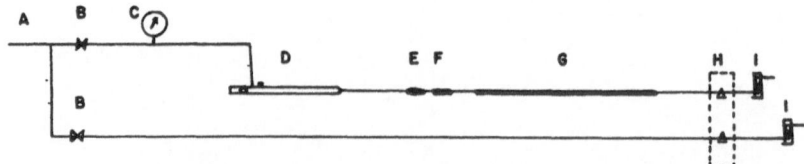

Fig. 1. Flow schematic: (A) helium at 10 psi, (B) needle valve, (C) pressure gauge, (D) pyrolysis tube, (E) Ascarite $\frac{1}{4}$ by 1 in., (F) Indicating Drierite $\frac{1}{4}$ by 1 in., (G) 5A molecular sieve $\frac{1}{4}$ by 12 in., (H) thermistor detector, and (I) flowmeter.

EXPERIMENTAL

Initial experiments utilized the conventional Unterzaucher pyrolysis train, with helium flush, connected to a refrigerated molecular sieve column used as a trap. After pyrolysis was completed, the trapping column was switched to the chromatograph and warmed to elute the pyrolysis products. In the first runs, the blanks were high and erratic. However, by eliminating porous tubing, improving flow geometry, and pretreating the helium carrier gas, a blank approximating 50 ppm was achieved with a detectability of about 50 ppm, each based on a 40-mg sample. The above work was done independently, but is quite similar in approach and success to a method detailed by Suchanec [3].

Even though the above method is a great improvement over the conventional Unterzaucher method in time and sensitivity, the presence of a blank is detrimental to analysis at low concentrations. The obvious means of eliminating the blank is to use the carbon-filled pyrolysis tube as the inlet of the chromatograph, with the eluent passing directly through the column to the

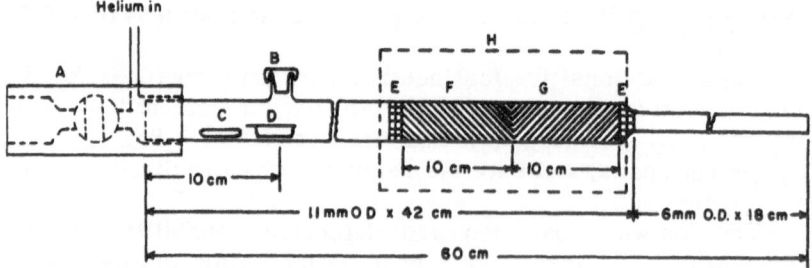

Fig. 2. Pyrolysis tube: (A) modified Hoke 30201-1 ball valve, (B) rubber septum, (C) glass-covered iron rod, (D) platinum boat, (E) quartz wool, (F) quartz chips, (G) 20-mesh Cabot Spheron 9 carbon, and (H) furnace.

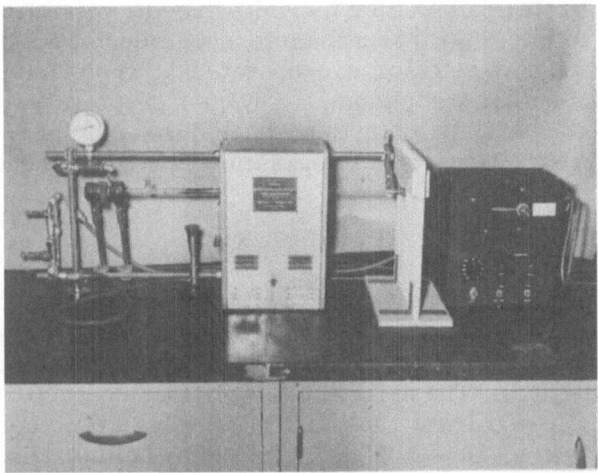

Fig. 3. Oxygen analyzer.

detector, with no trapping involved. Assuming all the hydrogen atoms are converted to methane, a 40-mg sample would be equivalent to about a 20-ml gaseous sample injection; however, with the molecular sieve column, resolution is no problem, so the carbon monoxide peak may be fairly broad. This system proved satisfactory and the final form is shown schematically in Fig. 1, with the pyrolysis tube shown in Fig. 2. Figure 3 shows that the Unterzaucher pyrolysis unit and the chromatograph have merely been added together, rather than being integrated into a single unit.

EQUIPMENT

The pyrolysis tube is constructed of quartz and packed as shown in Fig. 2. A ball valve is modified to allow helium entrance immediately after the ball, and to allow the valve to be connected to the inlet of the pyrolysis tube with an O-ring seal. An O-ring seal connects the pyrolysis tube exit to a glass tube containing 1 in. each of Ascarite and Indicating Drierite, then to a $\frac{1}{4}$ OD by 12 in. copper column containing Linde 5A molecular sieve. From the column, the resolved pyrolysis products go to the thermal-conductivity detector of the chromatograph. Helium carrier gas is controlled by a two-stage regulator followed by individual needle valves to control the flow rate to the sensing and reference sides of the detector. The column and detector are operated at room temperature in an air-conditioned room. Any gas chromatograph, with thermal-conductivity detection, should be suitable, as flow geometry, sensitivity, and temperature control are not critical.

PROCEDURE

The pyrolysis tube is packed as shown in Fig. 2, inserted into the furnace, and purged with helium. While purging, the furnace temperature is raised to 1050°C to remove water and other volatiles. The pyrolysis tube is then connected through the Ascarite—Drierite tube to the column and detector with a helium flow of 80 — 100 ml/min. With helium also flowing through the reference side of the detector, the chromatograph power supply is turned on and the base line allowed to stabilize. The system is conditioned by charging two 50-μ liter hydrocarbon samples. To calibrate the instrument, a 50-μ liter sample of air (equivalent to 0.0137 mg oxygen, assuming dry air at 25°C and 76 cm pressure) is injected through the septum from a 50 μ liter Hamilton gastight syringe. Only argon, nitrogen, and carbon monoxide should be eluted from the column, as shown in Fig. 4. The sensitivity of oxygen per unit area of the carbon monoxide peak is calculated. Liquid samples are charged, from a 10 or 50 liter Hamilton syringe, through the rubber septum into a platinum boat. The boat is maneuvered with a glass-covered iron rod and an external magnet to a position about $1\frac{1}{2}$ in. in front of the furnace. The glass-covered iron rod is moved back from the boat, and a flame is then used to vaporize the sample from the boat into the pyrolysis zone. Complete vaporization should take 10 — 20 sec; longer vaporization time will give an unduly broad carbon monoxide peak; a shorter vaporization period risks lack of sufficient contact time in the pyrolysis tube for complete conversion of all oxygen to carbon monoxide. The technique is not difficult to develop.

For solid samples, the boat is removed from the pyrolysis tube through the ball valve with a hooked wire, and the sample weighed into the boat. The filled boat is then inserted through the ball valve and the system purged (for about 3 min) to remove air. The boat is then moved forward and the sample vaporized as described for liquid samples.

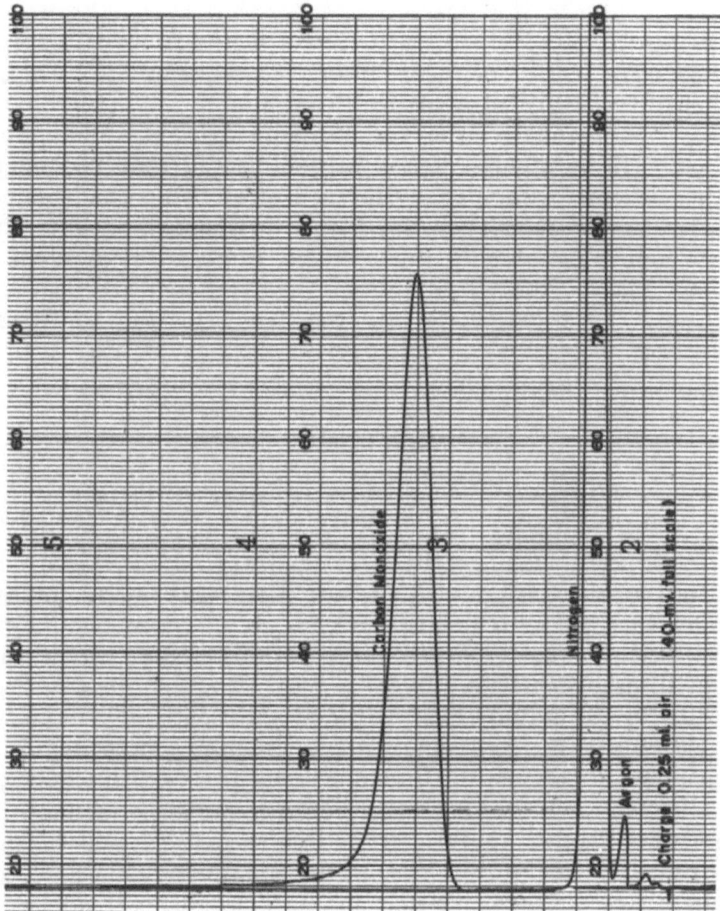

Fig. 4. Air calibration.

Oxygen concentration is calculated knowing the carbon monoxide peak area, the instrument sensitivity, and the sample weight. For liquid samples, sample volume and specific gravity are used to determine the sample weight.

Each sample pyrolyzed deposits a small amount of coke at the head of the pyrolysis unit, resulting in an increasing pressure drop through the instrument. The needle valve is adjusted between runs to re-establish the initial flow rate used for calibration. After approximately thirty sample runs, the pressure required to maintain the proper flow will approach 5 psig. When this occurs, the quartz chips are removed, the carbon plug is broken up with a rod, and new quartz chips are added. Occasionally, the vaporization area is burned free of carbon. The pyrolysis tube is purged, heated, then reconnected to the chromatograph ready for use after being conditioned by running two hydrocarbon samples. Without the quartz chips, the pressure buildup is much more rapid. For reasons that are not known, the system is only 95% as sensitive at the end of thirty runs as it is for the first run; this decrease in sensitivity is gradual as consecutive runs are made.

DISCUSSION

The short sections of Ascarite and Indicating Drierite protect the molecular sieve column and function only if a new pyrolysis tube has not been completely purged before connection to the column. The 5A molecular sieve column is packed with 18 – 40 mesh material to give a minimum pressure drop. The 12-in. column separates carbon monoxide from all other volatile pyrolysis products, as shown in Fig. 5; the life of the column has averaged six months. It is possible to activate a 5A molecular sieve column so thoroughly that carbon monoxide shows signs of being irreversibly adsorbed.

Starting with the equipment cold, a period of about 3 hr is required to reach equilibrium so that an analysis may be made. In practice, the pyrolysis unit is usually held at a stand-by temperature of 900°C, so that an analysis may be made with an hour's notice. The pyrolysis temperature of 1050°C has been found to quantitatively convert oxygen to carbon monoxide in all samples received. Twenty-mesh Cabot Spheron 9 carbon was used to give a small pressure drop in the pyrolysis tube. For convenience, air is used as the primary calibration standard. Benzoic acid, run either as a solid or in solution, verifies the air calibration.

Fluorides give high results because the fluoride attacks the quartz tube, releasing oxygen to form carbon monoxide. Phosphorus-containing compounds tend to yield low results because of the difficulty of pyrolysis. With the conventional Unterzaucher procedure, elevated pyrolysis temperatures and longer pyrolysis times are used to approach theoretical yields

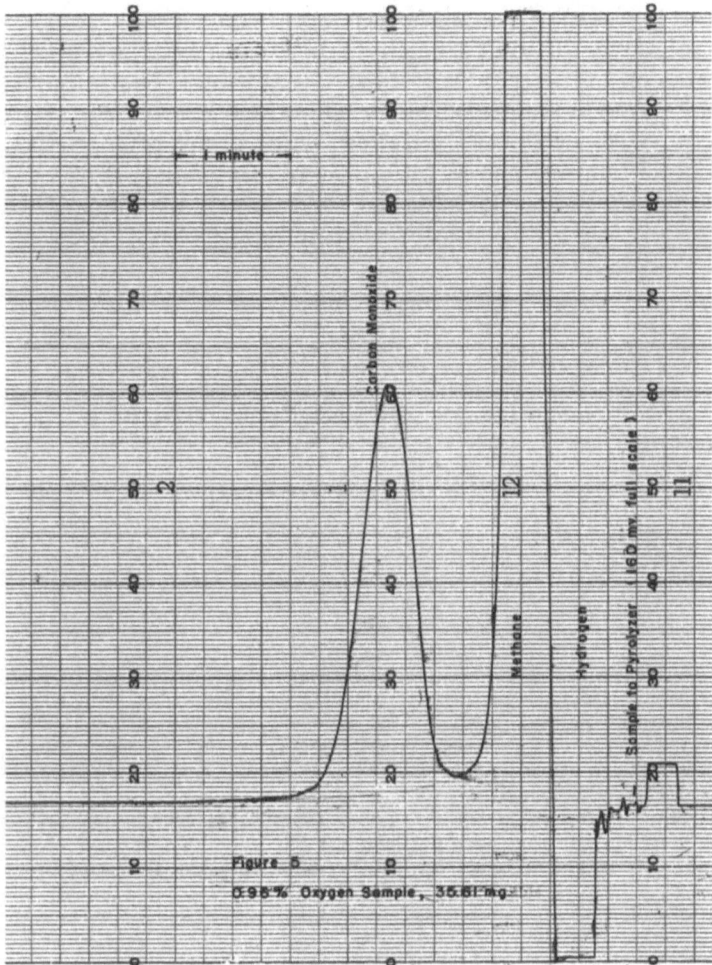

Fig. 5. 0.96% oxygen sample, 35.61 mg.

when running phosphorus compounds. No other materials are known to cause troubles.

The flow rate of 80 – 100 ml/min through the system was selected as the maximum that would give sufficient pyrolysis contact time.

The air calibrations and the liquid samples require 15 min for analysis. Solid samples, because of weighing and flushing of the system, require 30 min for analysis. Sample size is usually 10 – 40 mg. Smaller samples are difficult to measure accurately; larger samples are not required for sensitivity and are more difficult to handle because of pressure buildup as well as causing more coke formation. The majority of samples run in our laboratory contained 0.01 − 20% oxygen; however, with no helium purification or other refinements, as little as 30 ppm oxygen in a 40-mg sample is measurable. The expected difference between duplicate analyses is 5% with 1% oxygen, and 10% with samples containing 0.03% oxygen.

REFERENCES

1. J. T. Gilmore and D. E. Hull, Anal. Chem. 35: 1623 (1963).
2. R. A. Stallwood, W. E. Mott, and D. T. Finale, Anal. Chem. 35: 6 (1963).
3. R. R. Suchanec, doctoral thesis, University of Pittsburgh (1961). [Available from University Microfilms, Inc., Ann Arbor, Michigan (1962).]

Gas Chromatographic Methods for the Detailed Study of Controlled-Temperature Polymer Degradation

Robert T. Conley

Seton Hall University
South Orange, New Jersey and

Martin Company
Baltimore, Maryland

A wide variety of methods have been suggested for the analysis of polymer pyrolyzates. In general, these techniques have gained wide acceptance for qualitative identification of homopolymers and copolymers and, in some instances, the quantitative determination of specific complex-polymer compositions. Attempts to apply these methods to the study of the product compositions produced from controlled-temperature polymer degradation have not been particularly successful. Variations in the state of the sample, quantity of sample, and method of heating and controlling the temperature have been examined in some detail in an effort to delineate the changes in these variables as a function of the volatile products produced on controlled-temperature pyrolyses.

This report will discuss in detail the results of this study. Examples drawn from a wide range of polymeric substances, from homopolymers and copolymers of the vinyl types to condensation polymers, will be discussed. The use of gas chromatographic procedures in conjunction with other instrumental methods, such as infrared spectrophotometry, X-ray, and thermogravimetric analyses, will be presented in an effort to exemplify the necessary correlations which are needed in order to adequately assess the value of pyrolytic data in establishing the mechanisms of polymer degradation.

INTRODUCTION

In the investigation of the mechanisms by which polymers degrade under thermal and oxidative conditions, it is necessary to examine the changes in composition of the solid phase and as well the products found in the vapor phase due to volatilization. Such studies are of considerable importance with the

advent of space flight and ballistic missiles. Resins of highly
complex structures, such as phenol-formaldehyde polycondensates, have been used successfully in the ablative heat
shields which protect the re-entry vehicle. Since the thermal
and oxidative degradation of such polymers are a part of the
process of ablation, a knowledge of the degradation behavior
of the resin over a wide range of temperatures would be a
valuable aid in the overall understanding of the complex
reactions that occur during ablation. For some time, we have
been examining a large number of resins under thermal and
oxidative conditions in the temperature range 120 — 1000°C.

The problem is best exemplified by a specific resin system and for this report the phenol-formaldehyde polycondensates seem most appropriate on the basis of presently
available data. The approach to the problem resolves itself
into a study of the resin decomposition at elevated temperatures
as a function of the heat exposure of the resin sample. The
monitoring of chemical changes that occur in the solid phase
and composition studies of the volatilized products using gas
chromatography seemed the most promising approach to obtaining the desired degradation data. We have recently reported
[1-6] the oxidation reactions of phenolic systems at temperatures up to 220°C. The key information was gained by examining polycondensates of known phenol-formaldehyde ratios, at
varying degrees of curing, at elevated temperatures in air by
continuously monitoring the changes in the infrared spectra of
film samples. Both qualitative information and quantitative
kinetic data could be obtained concerning the introduction of
new chemical species, the oxidation of methylol groups, and
chain-scission processes. Alternately, Heron [7] has reported
data of these resin systems at elevated temperatures by pyrolytic degradation with gas chromatographic separation of the
volatile products. In addition, thermogravimetric analyses
indicated that the rate of oxidation of phenolic resins increased
with increasing oxygen concentration and with decreasing
particle size (increasing surface area) of the initial sample.
Other studies of a less definitive nature have been reported
by Ouchi and Honda [8], Madorsky and Strauss [9], and Friedman [10]. In none of these studies [7-10] was a mechanism
of degradation accurately established as a function of structure.
Indeed, the major drawback in almost all previous work has
been the lack of information concerning the nature of the resin

immediately prior to thermal exposure. Figures 1 and 2 delineate the knowledge at hand from our initial spectral examination and related resin studies [1-6].

In view, then, of this problem, it was pertinent to investigate, using a variety of techniques, the chemistry of the resin system at elevated temperatures. Infrared spectroscopy for solid-phase composition changes, gas chromatography for volatile product compositions, thermogravimetric analyses and elemental analyses for weight loss and changes in residue composition, and X-ray analyses of the carbon char produced at high temperatures were deemed most appropriate for gaining the desired information. For the present report we shall concern ourselves most particularly with the methods, procedures, and results of the gas chromatographic studies.

EXPERIMENTAL

The resin samples were prepared from reagent-grade phenol and formaldehyde by the procedure described by Conley

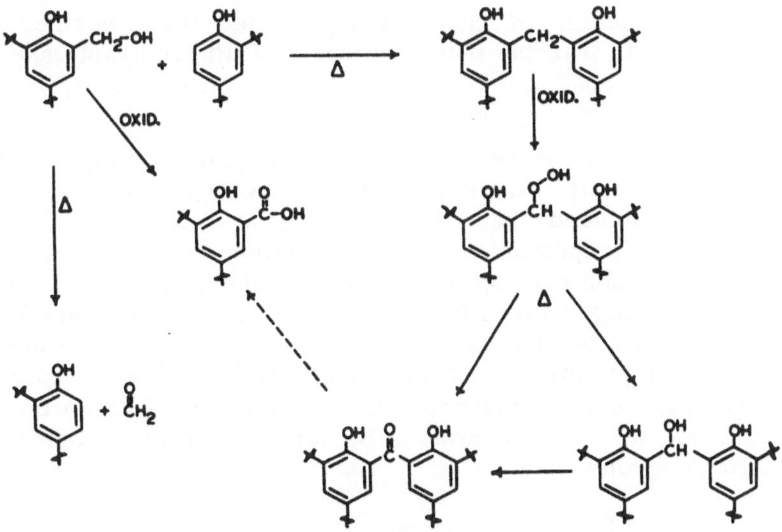

Fig. 1. The initial oxidation reactions occurring during the air oxidation of phenolic resins at elevated temperatures.

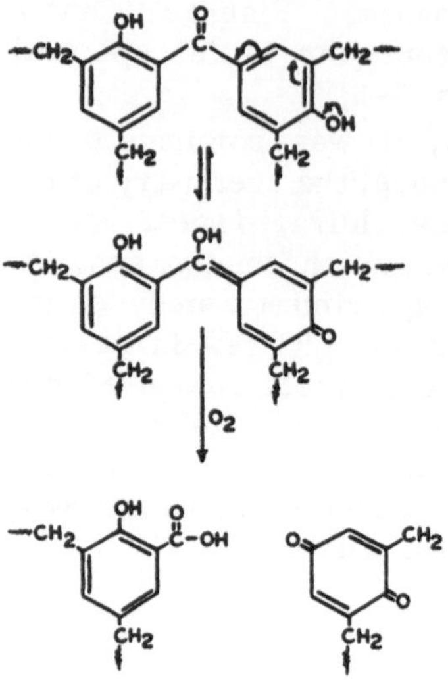

Fig. 2. Chain-scission reactions, a secondary process, of phenolic resin oxidation.

and Bieron [1] for a typical base-catalyzed phenolic resin. The cure time for the resin samples was generally 3 hr at 120°C. The curing atmosphere for the resin was: (1) high vacuum (10^{-5} torr), (2) air, and (3) inert atmosphere (argon). The exact method of resin curing varied for a number of samples and will be noted in each of the relevant cases.

Infrared Studies

Resin samples were cured directly on polished salt plates. The cured resins were examined with either a Perkin-Elmer model 21 infrared spectrophotometer, or a Beckman IR-5A infrared spectrophotometer. Both instruments were equipped with sodium chloride optics. After the plates were subjected to various temperatures for a prescribed interval of time, the solid residues on the plates were examined spectrophotometrically. In addition, residues from the pyrolysis studies, involving gas-phase chromatography studies of volatile product mixtures, were also examined by infrared spectroscopy for functional-group changes.

Gas-Phase Chromatographic Studies

The liquid resin was cured in a 0.035-in.-diameter by a 0.625-in.-long capillary tube at 120°C in air for 3 hr. The resin was then removed from the capillary, weighed, and inserted into a quartz capillary tube. This tube was inserted into the coiled nichrome ribbon and the whole assembly placed into the pyrolysis chamber (Fig. 3). The pyrolysis chamber was wrapped with heating tape and allowed to come to thermal equilibrium while being purged with helium. The desired temperature of the run and the heating time were set on the pyrolysis control unit and current was then applied to the ribbon. A F & M model 500 gas-phase chromatograph was used to separate the products of degradation. Two columns were used: a Carbowax 1500 column for the condensable products (benzene, toluene, and cresols) and a molecular sieve 5A column for the noncondensable products (CH_4, CO, CO_2).

Figure 4 exemplifies in block diagram form the complete chromatographic apparatus: the pyrolysis chamber (Fig. 3), the pyrolyzer control, and the inlet heater. The inlet heater was an adaptation of the F & M fraction delivery apparatus. This heater was added to the apparatus to prevent condensation of the reaction products in the inlet line to the chromatograph. The pyrolyzer control unit is unique for studies of this type. A block diagram and the complete schematic of the time-temperature controller are shown in Figs. 5 and 6. This unit

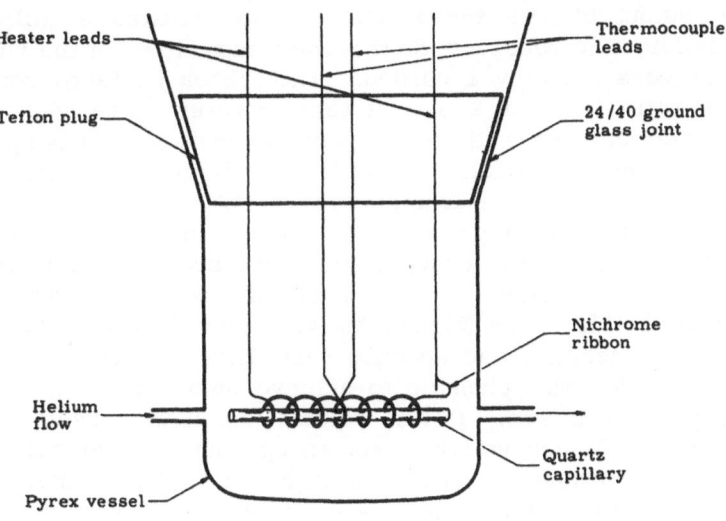

Fig. 3. Pyrolysis chamber.

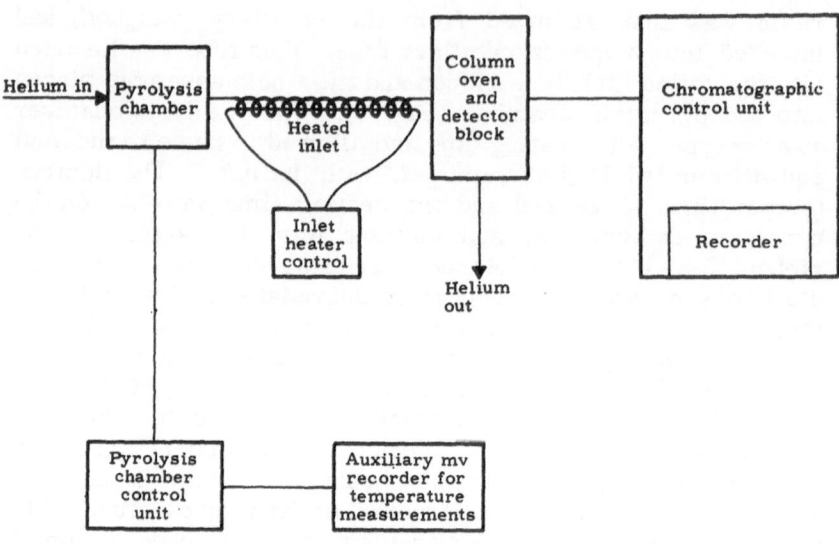

Fig. 4. Chromatographic arrangement.

controlled the temperature of the nichrome sample heater by the use of a feedback circuit. The control signal was taken directly from a thermocouple touching the nichrome ribbon. The output of this thermocouple also actuated a millivolt potentiometric recorder, so that the temperature of the ribbon could be accurately recorded. The ribbon achieved control temperature within 2 sec at temperatures up to 800°C. At lower temperatures, there was some overshoot, but a temperature plateau was reached within 3 sec. Figure 7 indicates this temperature rise as a function of time for both a low-temperature and a high-temperature run. From the linear portion of the high-temperature curve, a maximum heating rate of 310°C/sec was calculated. It can be shown, using the method described by McAdams [11] and known values of the thermal conductivity, density, and specific heat, that the maximum time required for the phenolic to achieve 99.9% of its final temperature is 2 sec. In this calculation, it was assumed that the thermal conductivity was independent of temperature, whereas, in fact, it increases with increasing temperature. This will tend to make the time even less than 2 sec.

The resin was heated and held at a constant temperature for a total time of 10 sec in the helium gas stream; at the end

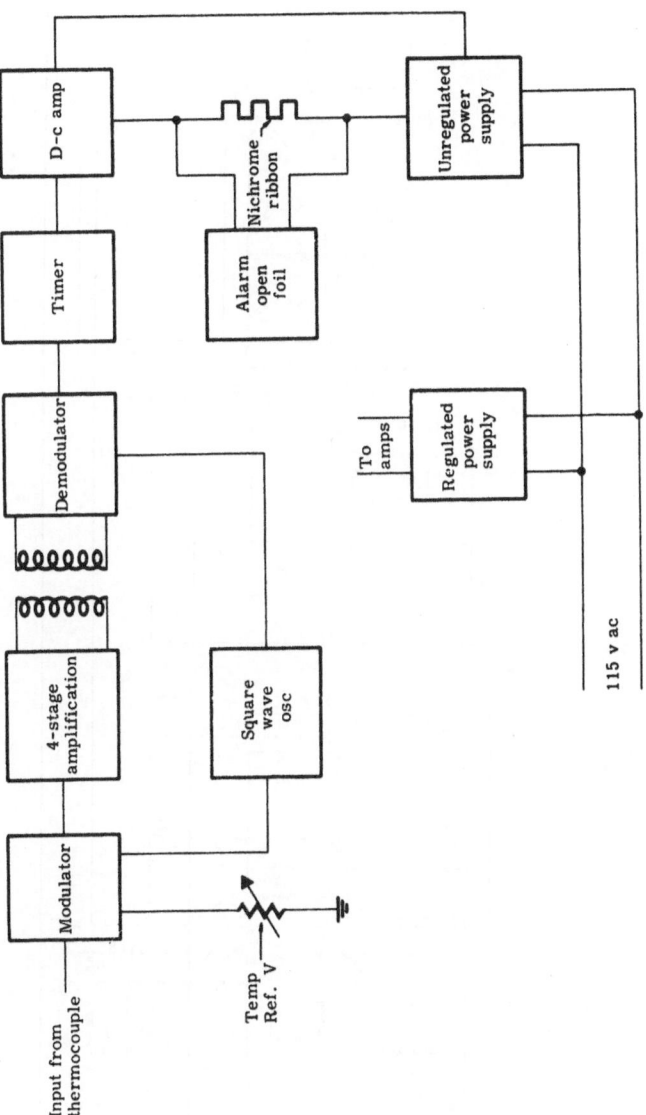

Fig. 5. Time—temperature controller.

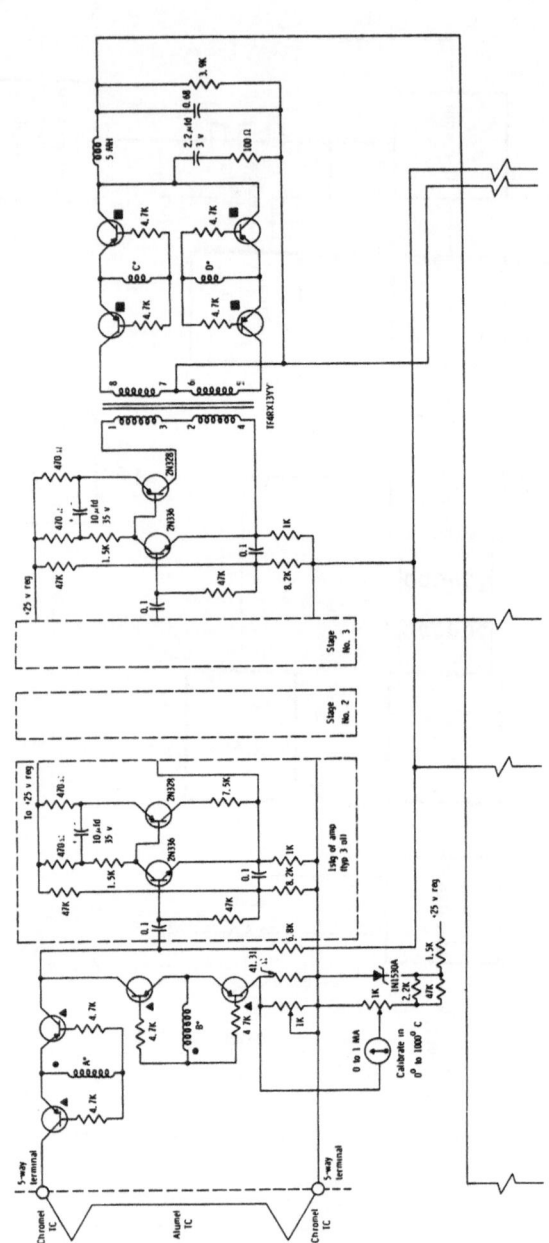

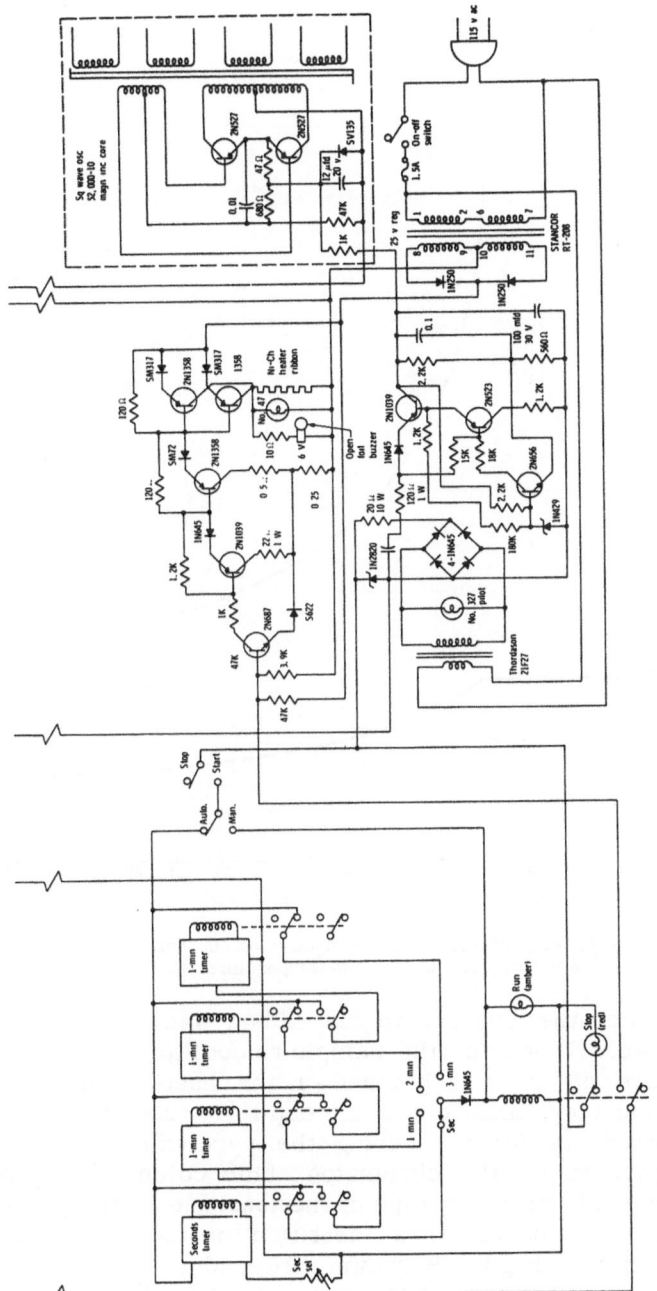

Fig. 6. Wiring diagram for time—temperature controller. All capacitors in microfarads unless otherwise noted; all resistors $\frac{1}{2}$ W, 5% unless otherwise noted. (▲), (■) 4 matched 2N527; (*) secondary windings of square—wave oscillator transformer.

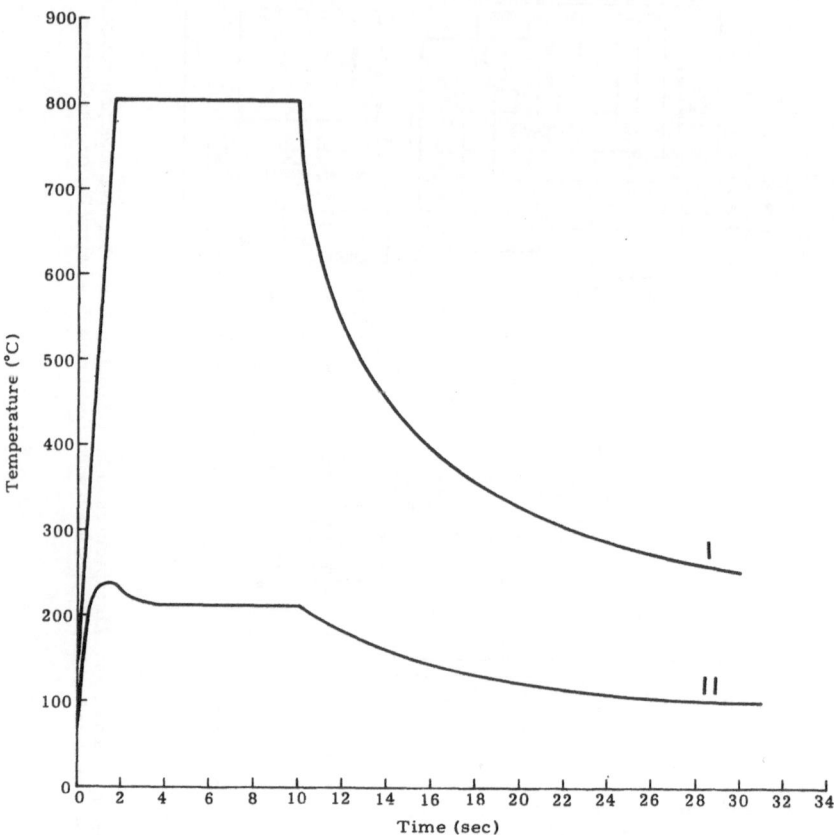

Fig. 7. Temperature of pyrolysis chamber heater versus heating time—(I) high-temperature run; (II) low-temperature run.

of this time, the current to the ribbon was cut off. It took approximately 7 sec for the sample to cool to one-half of the temperature of the flat portion of the heating curve. The residue was then taken out of the capillary tube and weighed.

The products formed during the pyrolysis of the resin were separated on the chromatographic column. Figure 8 shows a typical chromatograph of the noncondensable pyrolysis products. The peaks were identified on the basis of their retention time. Figure 9 shows a typical chromatograph of the condensable products. The first peak in Fig. 9 was assigned to the total noncondensable products, since the retention time

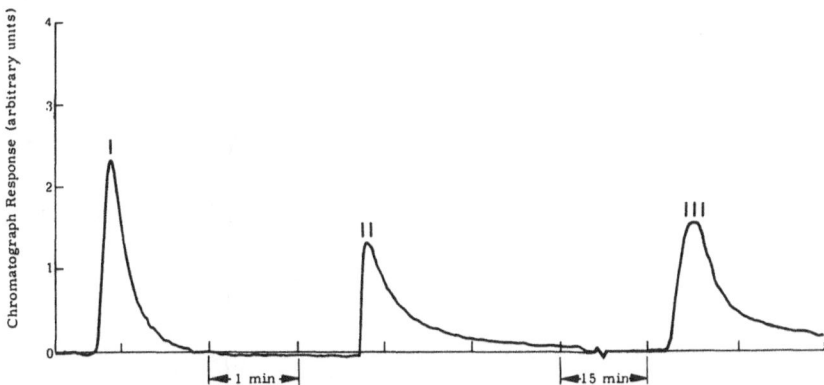

Fig. 8. Chromatograph of a pyrolysis run on a molecular sieve 5A (30/60 mesh) column. Column conditions: $\frac{1}{4}$-in.–diameter × 12-ft copper tubing; block temperature, 240°C; injection-port temperature, 255°C; initial column temperature, 50°C; chart speed, 1 in./min; helium flow rate, 100 cc/min; helium back pressure, 55 psi; bridge current, 100 mA; program rate, 7.9°C/min; (I) CH_4; (II) CO; (III) CO_2.

for this peak represents the dead time of the chromatograph. The second and third peaks were identified as benzene and toluene on the basis of their retention times. Water and paraformaldehyde were identified both by retention times and by infrared analysis of the trapped material and were associated with the broad peak that exhibits tailing. The peak preceding the orthocresol—phenol peak has not been positively identified because of the inability to trap a sufficient quantity of material for infrared analysis. It is likely that the peak is 2,4-dimethylphenol, which conclusion is based on a comparison of retention time with that of an authentic sample. The orthocresol and phenol peak was identified both by infrared analysis and by its retention time when compared with pure phenol and orthocresol samples and various mixtures of these materials. Infrared analysis of the final peak suggested that it was a mixture of paracresol, and some other cresol or xylenol-type material.

Fifty percent of the observed weight loss in a sample could be accounted for on the basis of the total amount of phenol, orthocresol, paracresol, carbon monoxide, carbon dioxide, methane, benzene, toluene, and benzaldehyde detected. The remaining 50% was due to water and paraformaldehyde and a high-molecular-weight, nonvolatile residue. The water and paraformaldehyde have been identified qualitatively, but no

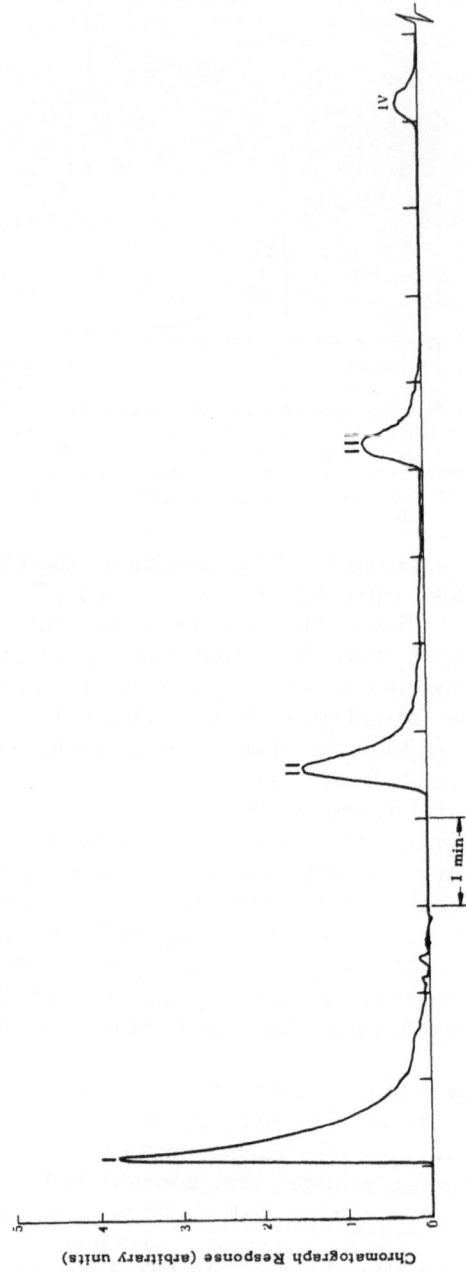

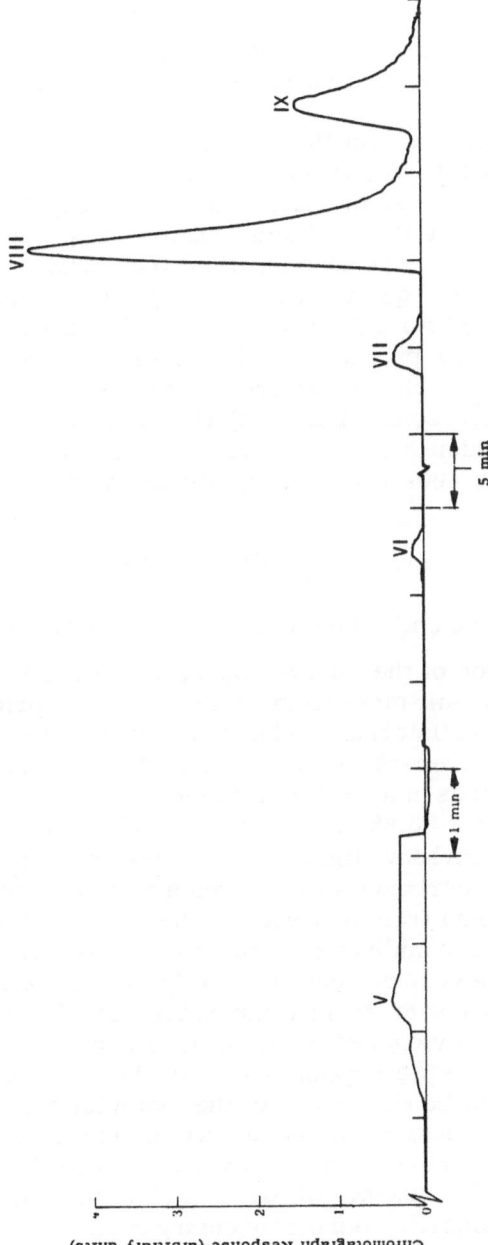

Fig. 9. Chromatograph of a pyrolysis run on a Carbowax 1500–25 wt.% chromosorb P (60/80 mesh) column. Column conditions: 1½-in.–diameter × 12-ft copper tubing; block temperature, 230°C; injection–port temperature, 222°C; initial column temperature, 50°C; chart speed, 1 in./min; helium flow rate, 100 cc/min; helium back pressure, 55 psi; bridge current, 100 mA; program procedure, hold temperature constant for 6 min after injection at 50°C, then program at 5.6°C/min to 200°C and hold temperature constant until analysis is over. (I) noncondensables; (II) benzene; (III) toluene; (IV) unidentified; (V) water and paraformaldehyde; (VI) benzaldehyde; (VII) 2,4–dimethylphenol; (VIII) phenol + o–cresol; (IX) p–cresol.

quantitative measure of their weight has been made. Visual observation of the reaction cell after degradation revealed a yellow, nonvolatile residue, confirming that part of the unaccounted weight was due to a high-molecular-weight, nonvolatile residue.

As an alternate technique, a strip of nichrome ribbon was coated with the resin, cured as described, and the pyrolysis carried out in the same manner as previously described with the exception that the nichrome ribbon strip was used as the heating element. Generally, this technique was far less reliable than the quartz-tube method, due primarily to the fact that the resin pulled away from the metal strip during the rapid heating procedure. The extent of pyrolysis varied widely. Due to these variations, the method was deemed impractical for controlled pyrolytic-degradation studies and therefore abandoned after evaluation of data from sixty experiments over the 400—1000°C temperature range.

RESULTS AND DISCUSSION

Resin Curing and High-Temperature Postcuring Studies

Examination of the infrared spectrum of the resin samples deposited on a salt plate from acetone solution prior to curing is of particular interest. Absorption bands due to oxidation during resin preparation were absent from these spectra. Absorption bands indicative of pre-oxidation are observed at 6.03, 5.94, and 5.85 μ, assigned to absorption by benzophenone-type carbonyl linkages, quinone-type carbonyl groups, and carboxyl—carbonyl species, respectively. These bands are generally formed from oxidation of the resin, either during its preparation or from exposure to an oxygen-containing atmosphere for extended periods at slightly elevated temperatures. After three days at room temperature in air, no detectable oxidation was evidenced from examination of the infrared spectrum. The absorption band at 9.45 μ, assigned to the hydroxyl-group bending mode of the methylol group ($-CH_2OH$), was taken as evidence for the degree of curing the sample had undergone. Complete disappearance of this band would be indicative of the absence of the methylol species, either from extended condensation during the curing cycle, or from oxidation of these groups to carboxyl species during the curing process.

After curing for 3 hr at 120°C, either in a stream of high-purity argon (99.95%) or under high vacuum (10^{-5} torr), oxidation was observed in almost every sample. The spectrum showed a rather strong band at 6.03 μ resulting from the oxidation of methylene groups ($-CH_2-$) to benzophenone. The initial oxidation mechanism prior to carbonyl formation (in the absence of external oxygen) is not known. However, it was established that the presence of water, produced from further condensation during the curing process and the absorption of oxygen and water from the air by predried samples, markedly affects the rate of growth of the ketonic carbonyl group.

Further heating of the resin at higher temperatures in inert atmospheres resulted in condensation (postcuring) and oxidation of the resin. During the postcuring period, the 9.45-μ band due to methylol groups was markedly reduced. In addition to the condensation step, increased oxidation at 400°C has taken place. For example, the quinone-type carbonyl band at 5.94 μ has increased in intensity, and a broadening of the 3.0—4.0 μ region, together with a weak shoulder on the 5.94 μ band (indicative of the formation of carboxylic acids), has appeared. These reactions are indicative of extended oxidation from within the resin sample, since oxygen was carefully excluded from the sample during this heating process. Oxygen contamination by absorption at room temperature seems an unjustified explanation of this phenomenon. Treatment of the resin samples at even higher temperatures resulted in a general broadening of the bands from 6 to 14 μ due to scattering of the infrared radiation during the initial stages of char formation. Comparison of the spectra indicates remarkable differences in the residual solids as the temperature is increased above 400°C. Considerable loss of hydroxylic components, causing a narrowing and sharpening of the hydroxyl frequency, is evident. In the curve obtained by scale expansion of the residual solid, loss of aliphatic methylene, as compared with aromatic-ring, carbon—hydrogen vibration, is evident, along with additional amounts of carboxylic acid.

Freshly prepared resin samples, when filmed and cured in vacuo, would give an occasional nonoxidized resin film, but, in general, these samples were difficult to prepare. In concurrent studies, low-temperature (below 150°C) curing, nonoxidized samples, although obtainable routinely in oxygen-free nitrogen atmospheres with the aid of special curing apparatus [1–6],

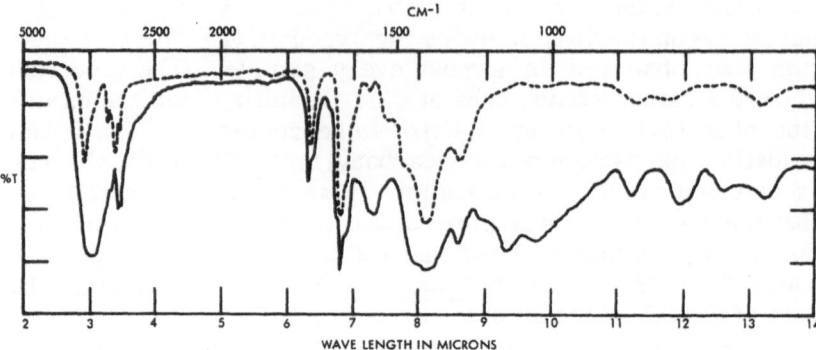

Fig. 10. Spectrum 1—resin cured for 3 hr at 120°C in nitrogen (————). Spectrum 2—resin cured for 1 hr at 400°C in vacuo (------). (Baselines are staggered for clarity.)

readily oxidized at elevated temperatures (above 200°C) in the absence of oxygen-containing atmospheres. Figure 10 shows a typical nonoxidized sample which, after postcuring at 400°C in vacuo, showed no sign of appreciable oxidation. Infrared examination of the volatile components condensed on the walls of the vacuum tube used as the curing oven revealed the presence of oxidized species of relatively high molecular weight (dimers, etc.), indicating that, while oxidation is taking place under all curing conditions, an oxidation-free residual cured resin can, in some instances, be attained. This observation would tend to support the surface oxidation phenomena reported earlier [2], together with the additional volatilization of fragments under the specific curing conditions employed.

High-Temperature Degradation Studies

In screening the techniques available for gaining information as to the extent of degradation to volatile products, it was of interest to compare the weight loss determined by thermogravimetric analysis (TGA) and that loss in weight which occurred under the high heating rates (310°C/sec) described in the experimental section for the gas-phase chromatographic analyses. As can be seen in Fig. 11, these curves vary considerably. At the low-temperature portion of these curves (0—250°C), the weight loss for the 310°C/sec heating rate is considerably lower than for the corresponding TGA curve.

This is consistent with a generalized mechanism for weight loss which can be depicted as follows:

Polymer A

C $\Big|_i$ Residue (nonvolatile) + Residue (volatile)

Polymer B

where polymer A is the starting cured resin, polymer B is the postcured material, i is the number of reaction steps required to produce the residue products and is determined by the temperature and molecular weight of the polymer, and C is extended condensation (postcuring). Weight loss at low heating rates would be expected to be greater than that observed at higher heating rates if diffusion of material of high volatility (low molecular weight) is the mechanism controlling weight loss. At higher heating rates, the volatile species would be expected to undergo thermally-induced postcuring reactions (C) since they would not have time to diffuse out of the solid matrix. At high temperatures (300—600°C) and high heating rates, more of the residue from polymer A is volatilized as it degrades thermally, hence, a larger overall weight loss. The high heating-rate, weight-loss curve was terminated at 600°C because internal stresses occurring above this temperature caused powdering of the sample. This, in turn, led to large errors in the determination of the weight losses. It should be noted here that the TGA measurements of total weight loss, when compared with the high heating-rate situations (such as encountered in re-entry problems) give residues of a higher weight.

The most significant portion of this study resulted from the study of the volatile degradation products summarized in Fig. 12 as a function of temperature and amount of pyrolysis product produced. The products, as expected for the extended oxidation of the resin, consisted chiefly of carbon dioxide and carbon monoxide. Water, paraformaldehyde, methane, and aromatic products were also formed, although all of the weight loss cannot be accounted for by the products listed. Residues which were volatilized from the sample as a result of the degradation were detectable to the chromatograph in the walls of the pyrolysis vessel and in the heated lines. Products reported by Heron [7], Ouchi and Honda [8], Madorsky and Strauss [9], and Friedman [10] included propanol, propylene,

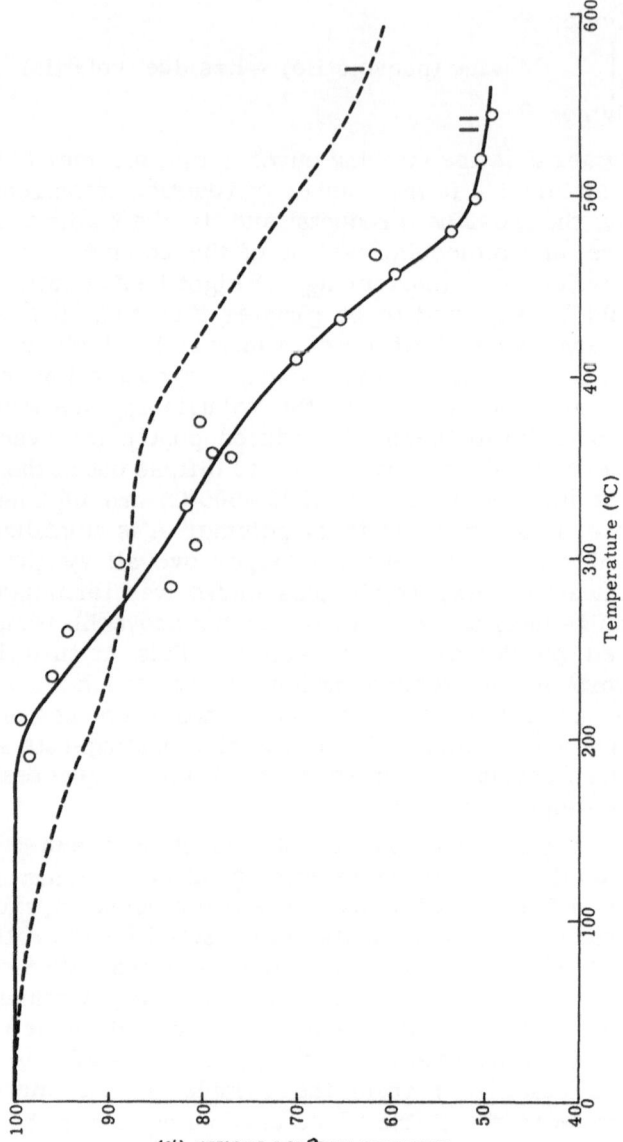

Fig. 11. Weight-loss curves for phenolic resin in inert atmosphere: (I) heating rate = 0.05°C/sec (thermogravimeric analysis); (II) heating rate ≈ 300°C/sec to indicated temperature and held at temperature for a total exposure of 10 sec; (O) determined by weighing the residue at end of run.

ethane, and ethylene. Gas-phase chromatographic attempts to detect these products, as well as the presence of hydrogen, as reported by Ouchi and Honda [8], failed even after extended work under conditions which normally detect such materials. It is likely that these investigators worked with phenolic resins from commercial sources which contain propanol as a solvent. The thermal decomposition of entrapped solvent would adequately account for the observed differences in degradation products. Since the resins prepared in this study were cured under carefully controlled conditions to remove all solvents and examined thoroughly by spectrophotometric methods for contamination, this conclusion seems reasonable.

Water and paraformaldehyde were among the major products formed on heating the phenolic resin to 400°C. Quantitative data were not obtainable, due to difficulties encountered in peak resolution and calibration of the chromatograph. It is estimated that these products were present in the same order of magnitude as the 2,4-dimethylphenol (Fig. 12). In order to correlate these results to earlier work [1], it should be pointed out that neither of these products appeared in significant quantities below 400°C. Therefore, they are products of high-temperature resin reactions and are not present as impurities in the resin sample. The infrared studies discussed in the previous section indicated they result from the loss of methylol groups in this temperature range. It is consistent to affirm the reaction sequence shown in Figs. 1 and 2 to account for the formation of these products.

This mechanism is in contrast to the mechanism proposed by Ouchi and Honda [8] for which we have been able to gain no experimental verification. The present proposal has the advantage of consistency with the observed loss of methylol groups and the lack of detectable diphenylether linkages in the infrared spectra in both this study and in the initial, lower-temperature degradation studies [1,2].

If one considers the possible modes of production of products from the data summarized in Fig. 12, two processes are seemingly in competition as the temperature is increased. In all cases, oxidation, as evidenced by the production of carbon dioxide and carbon monoxide, is the predominant reaction route. It is known from the infrared data that the initial cured resin contains methylene linkages, appreciable amounts of residual methylol groups, and dihydroxybenzophe-

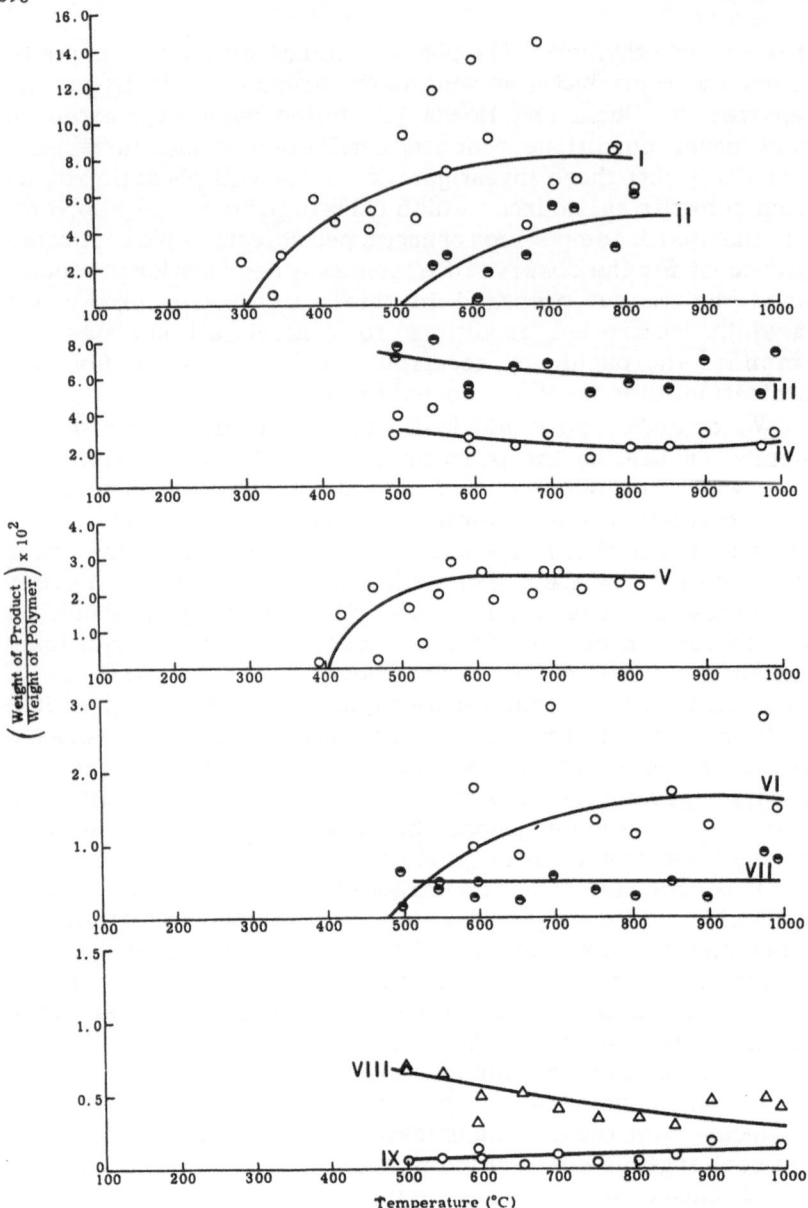

Fig. 12. Variation in pyrolysis product formation with pyrolysis temperature: (I) CO_2; (II) CO; (III) phenol + o-cresol; (IV) p-cresol; (V) CH_4; (VI) benzene; (VII) toluene; (VIII) 2,4-dimethylphenol; (IX) benzaldehyde. Abscissa, temperature in °C; ordinate, (product weight/polymer weight) × 10^2.

none linkages from oxidation during curing. The variations in the relative amounts of these species, together with the additional amounts of dihydroxybenzophenone linkages from further methylene oxidation and substituted salicylic acid moieties from methylol oxidation (Fig. 1) would tend to account for the large degree of scatter observed in the data reported for carbon dioxide and carbon monoxide formation. Since the number of reactions producing these products becomes increasingly complex as the temperature is increased and adequate experimental control of the individual concentration of each species is nearly impossible, the goal of more accurate results, at the present, seems to present an insurmountable task in analysis of this resin system.

Despite the difficulties encountered in the analysis of carbon dioxide and carbon monoxide, the formation of aromatic products, which presumably arise from a radical-bond-rupture reaction sequence, shows much less scatter in the individual-data points. Since these products are produced in smaller amounts, and the concentration of the initial reactants is more constant (since these species comprise the bulk of the resin sample), the observed deviations might be expected to be decreased considerably. A comparison of the data obtained indicates that only benzene and toluene show significant scattering in their respective product-yield determinations. The additional reactions proposed to account for the formation of the observed products are indicated in Fig. 13.

From the initial oxidation reactions (Figs. 1 and 2), supported by the infrared data and previous work [1,2], it is possible to extend the degradation-reaction sequence to account for all of the observable products. Route I of Fig. 14 summarizes the oxidative degradation processes in a generalized form. Particular substitutions of methylene and hydroxyl are not indicated. This is done for reasons of simplicity and because the choice between the various substituted species, insofar as degradation differences are concerned, has not been possible. Route II (Fig. 14) indicates the formation of phenol, cresol, and other methyl-substituted species, together with methane. It should be noted that the formation of methane occurs in increasing amounts above 400°C, whereas the other species of a cresol and xylenol type are generally found in relatively constant amounts or even in slightly decreasing amounts, as the temperature increases. Benzene, toluene, and benzalde-

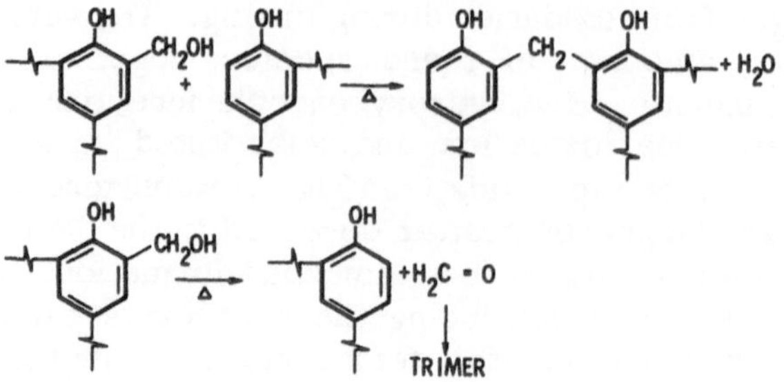

Fig. 13. Postcuring reactions—the formation of water and paraformaldehyde.

hyde are of particular interest in this respect, since the loss of hydroxyl from the aromatic ring is a necessary requirement for the formation of these materials (Route III, Fig. 14). The hydroxyl radical, in turn, would represent a source of oxygen for further oxidation reactions and, as well, a source of water through hydrogen extraction from any available methylene linkage. In none of these reactions does the hydrogen atom become significantly available for combination to form the hydrogen molecule. Therefore, it seems unlikely that the hydrogen observed in previous work [8] was produced from the phenolic resin system. Further substantiation of this seems to be borne out in the observation of methane, which is apparently formed via a hydrogen abstraction process by methyl radicals. The radical concentration is never high enough to recombine to form ethane. The oxidation reactions shown in Fig. 14 are consistent with the formation of the initial reactants proposed in Figs. 1 and 2. The decarboxylation and decarbonylation to form carbon dioxide and carbon monoxide are consistent with the respective temperatures at which each product begins to appear. The remaining reactions are generalized to show the formation of phenol, cresols, and higher phenolic species. Since these materials are produced at the lower temperatures in appreciable amounts, it is doubtful that they arise from the nonoxidized, postcured resin. Most likely these products are formed from dihydroxydiphenylmethane and slightly higher homologs (low polymers which have been terminated) entrapped in the cured resin system.

Char Formation

The char-forming reaction (Fig. 15) is best explained

through the initial formation of a quinone-type linkage. The production of this linkage during degradations of phenolic systems is known from infrared studies [1]. At temperatures above 450°C, decomposition to produce char and carbon monoxide, via ring scission, is rapid. This is supported by the observation that the carbon monoxide expected from this degradation does not appear at lower temperatures. Further, in this temperature range, carbon char is found in the residual products of the degradation. The char formation is first detected visually and substantiated for materials heated to higher temperatures (600°C) by the presence of a graphite-like line in the X-ray pattern of the residues. This reaction also accounts for the observation of hydroxyl and carbonyl groups at 700°C in the charred material.

An interesting aspect of the present work is that, at no point up to temperatures of at least 500°C, is it necessary to invoke thermal nonoxidative degradation of the completely cured, nonoxidized, cross-linked polymer (postcured polymer). This is consistent with the observed fact that the postcured polymer shows no change in its infrared spectrum after prolonged heating to 450°C in vacuo. The postcured polymer has been shown to have a higher oxidative stability than the nonpostcured, partially oxidized resin. For example, it has been shown previously [2] that the rate constants, as determined by infrared spectrophotometric monitoring of the ingrowth of benzophenone-carbonyl groups, are quite large. In contrast, rate constants obtained on nonoxidized samples from this study (using the same method), cured at 450°C in high vacuum, are considerably smaller (a factor of roughly 33). This comparison is summarized in Table I.

For the nonoxidized, 450°C, cured samples of this study, a plot of $-\log K$ versus the reciprocal of the absolute temperature T gives a straight line whose slope, when multiplied by $2.303 R$, gives an activation energy of approximately 16.7 kcal. This value corresponds favorably with the value for the same oxidation reaction obtained earlier [2] on the resins cured at 120°C. It is also interesting to mention here that the rate constants obtained on the postcured resin (450°C) and the activation energy determined are in the range of those obtained in a study of the oxidative degradation of polybenzyl (14.5 kcal) [5].

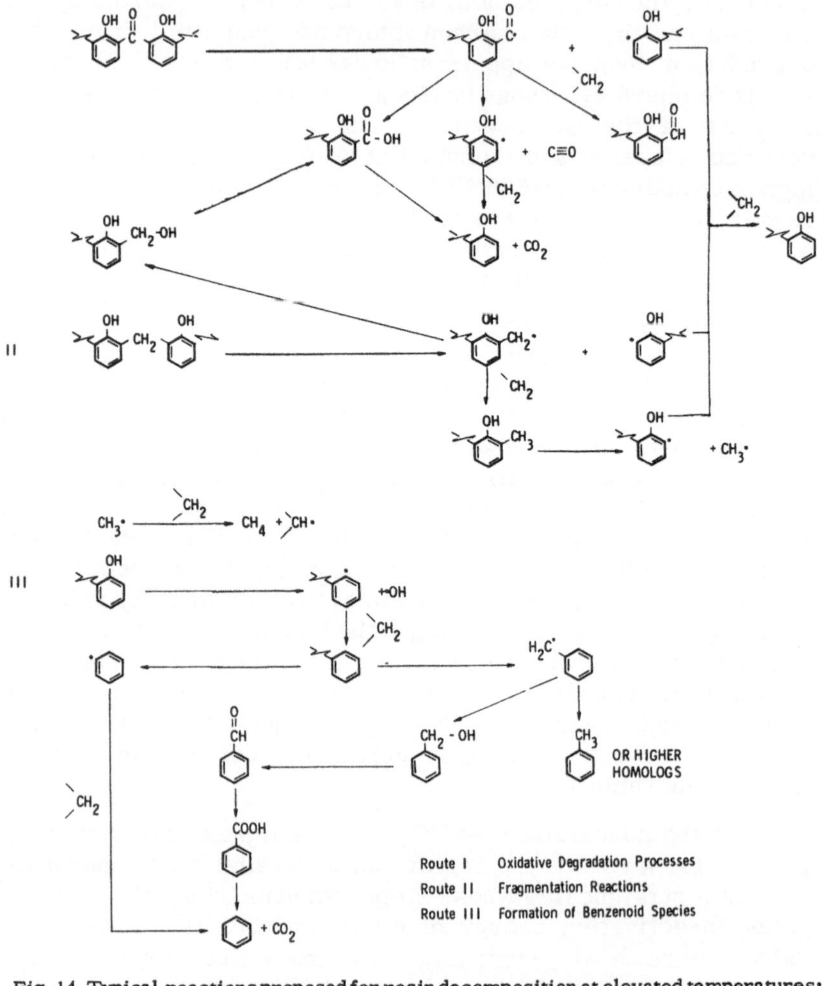

Fig. 14. Typical reactions proposed for resin decomposition at elevated temperatures: Route I—oxidative degradation processes; Route II—fragmentation reactions; Route III—formation of benzenoid species.

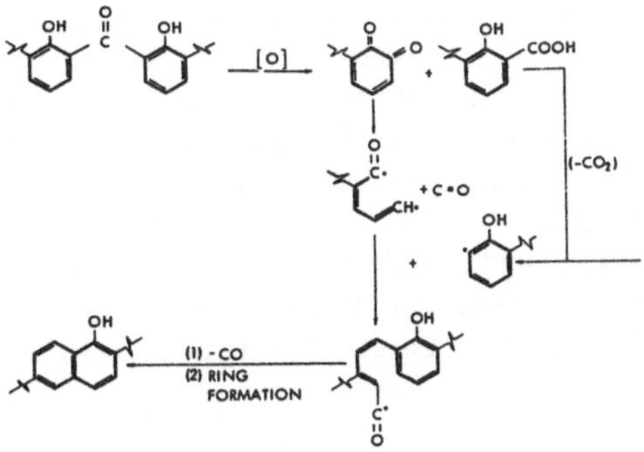

Fig. 15. Postulated char-forming reactions based on carbon monoxide formation.

CONCLUSIONS

It should be noted from this study that the evaluation of data for the thermal and oxidative degradation of resins of this type requires the investigator to invoke a large number of experimental techniques and methods to obtain enough data for a reasonable working postulate. From the gas chromatographic procedures outlined, we believe that it is clearly demonstrated that when the sample becomes the limiting factor in the analyses, rather than the oven, careful control becomes more and more important. Kinetic information remains, at present, outside the limits of analysis by gas chromatography

TABLE I

Comparison of Rate Constants Obtained for the Initial Oxidation Reaction of Phenolic Resins as a Function of Curing Temperature

		Rate constant $(\text{min}^{-1} \times 10^{-2})$	
Temperature (°C)	$1/T$ $(°K^{-1} \times 10^{-3})$	resin-curing temperature	
		120 °C	450 °C
180	2.21	6.88	0.22
		7.67	0.28
200	2.12	17.5	0.61
		17.8	0.56
220	2.03	37.6	1.12
			1.44

due to the critical nature of sample size and temperature control. Pyrolysis-unit design and control presently give excellent qualitative information, but studies of solid—gas reaction rates remain outside of the scope of the common pyrolysis techniques. The advantages of the technique, at present, should not be overlooked. The analyses of this type of volatile has heretofore rested with such techniques as mass spectrometry (identification is made in many cases by mass alone) and frequently the data have been misinterpreted [10].

ACKNOWLEDGMENTS

The author is indebted to Dr. William Jackson and his co-workers for their efforts in this and related investigations. The author wishes to thank Mr. Eric Straus of the Martin Company for his interest, helpful discussion, and many suggestions concerning this and related problems.

REFERENCES

1. R. T. Conley and J. F. Bieron, J. Appl. Polymer Sci. 7, 103 (1963).
2. R. T. Conley and J. F. Bieron, J. Appl. Polymer Sci. 7, 171 (1963).
3. R. T. Conley, J. Appl. Polymer Sci. (in press).
4. W. M. Jackson and R. T. Conley, J. Appl. Polymer Sci. (in press).
5. R. T. Conley, J. Appl. Polymer Sci. (in press).
6. R. T. Conley, Conference on Stability of Plastics, Soc. Plastics Engrs., Washington, D. C., June (1964).
7. G. F. Heron, Conference on High-Temperature Resistance and Thermal Degradation of Polymers, Soc. Chem. Ind. (London) Monograph 13, 475 (1961).
8. K. Ouchi and H. Honda, Fuel 38, 429 (1959).
9. S. Madorsky and S. Strauss, WADC Tech. Rept., 59-64, June (1959).
10. H. L. Friedman, Technical Information Series, No. R60sD380, General Electric Space Science Laboratory, May (1960).
11. W. McAdams, Heat Transmission, McGraw-Hill, New York (1954), p.31.

NMR Spectroscopy

NMR—Fun Chemistry

C. L. McGehee and C. H. Summers

Columbian Carbon Company
Lake Charles, Louisiana

This paper describes how NMR is used in this laboratory. The main emphasis is on the determination of the structure of new organic compounds by this method in conjunction with other disciplines. One example of quantitative analysis and one example of a percent hydrogen determination of samples are discussed in detail.

INTRODUCTION

One of the things NMR can be used for is the determination of the structure of organic compounds. Other techniques, besides NMR, include infrared spectrophotometry, mass spectrometry, gas chromatography, and elemental analysis. We feel safe in formulating a structure only when all of these methods are in agreement. Gas chromatography is most valuable in predicting the purity of the sample. Mass spectrometry is used to determine the molecular formula of the compound to be studied. This is done by determining the mass number of the compound and the ratio of the peak height at the mass number of the molecular ion to that at the molecular ion plus one. For instance, the mass number of both $C_{10}H_{18}$ and $C_9H_{14}O$ is 138, but the ratios of the peaks at mass number 138 to those at 139 differ.

EXPERIMENTAL RESULTS

All NMR chemical shifts presented in this talk are in ppm downfield from tetramethylsilane.

NMR is useful in locating double bonds in organic molecules. Figure 1 shows the spectrum of 1,5-cyclooctadiene in CCl_4. Note the rather narrow band at 2.38, which is due to the protons

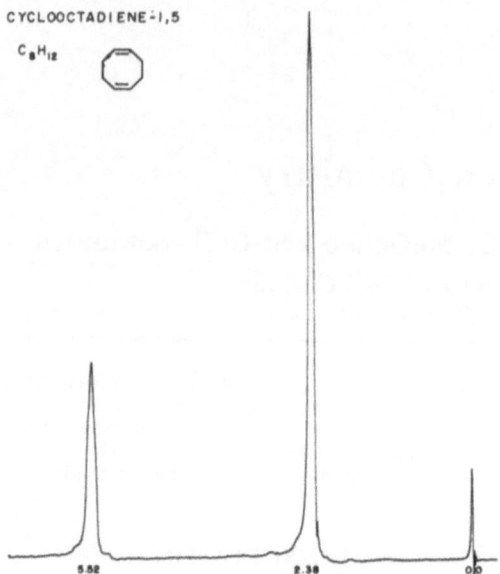

Fig. 1. NMR spectrum of 1,5-cyclooctadiene in CCl_4.

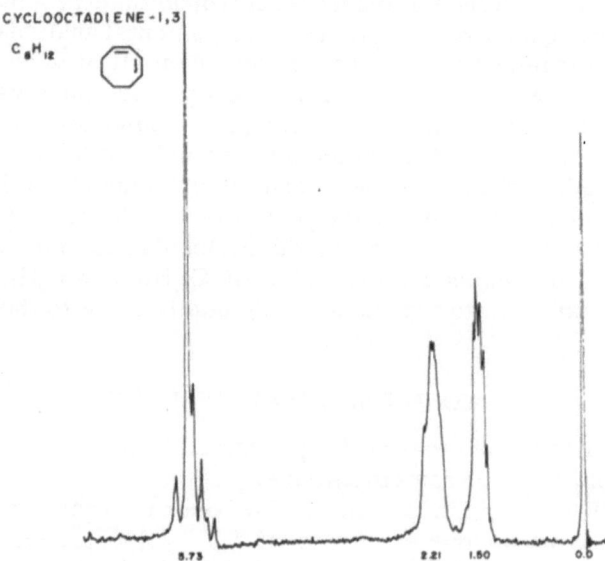

Fig. 2. NMR spectrum of 1,3-cyclooctadiene in CCl_4.

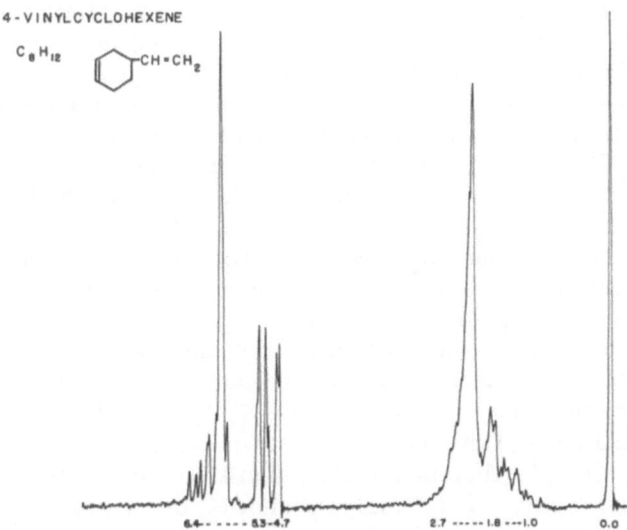

Fig. 3. NMR spectrum of 4-vinylcyclohexene in CCl_4.

on methylene groups alpha to double bonds, and the broader band at 5.52, due to olefinic protons. These bands integrate in the ratio 8:4.

Figure 2 shows the spectrum of 1,3-cyclooctadiene (an isomer of 1,5-cyclooctadiene) in CCl_4. The spectra of the two compounds are quite different. In the spectrum of the 1,3-isomer, the band at 1.50 is due to protons on methylene groups beta to double bonds, the one at 2.21 is due to those on methylene groups alpha to double bonds, and the one at 5.73 is due to the olefinic protons. These protons integrate in the ratio of 4:4:4.

A third isomer of C_8H_{12} is 4-vinylcyclohexene, the spectrum of which in CCl_4 is shown in Fig. 3. The band in the 1.0–1.8 region is due to protons on methylene groups beta to double bonds, that in the 1.8–2.7 region is due to protons on methylene and methine protons alpha to double bonds, that in the 4.7–5.3 region is due to the terminal olefin protons, and that in the 5.4–6.4 region is due to nonterminal olefinic protons. These peaks integrated in the ratio 1.6:5.2:2.0:3.2, or 2:5:2:3.

Another eight-carbon, olefinic hydrocarbon is 1,7-octadiene. Figure 4 shows the spectrum of this compound in CCl_4. Spectra

such as these are esthetically pleasing to the NMR spectroscopist.   The band at 1.38 is due to the protons on methylene
groups beta to double bonds, that at 2.04 is due to protons on
methylene groups alpha to double bonds, those at 4.76 and 5.00
are due to the olefinic protons on the ends of the molecule, and
that at 5.71 is due to the other olefinic protons.

Figure 5 shows the spectrum of the aromatic hydrocarbon
triethylphenylmethane in CCl_4.   The band at 0.64 is due to
methyl protons, that at 1.69 is due to methylene protons, and
that at 7.20 is due to aromatic protons.   They integrated in the
ratio 9.1:6.0:4.9, or 9:6:5.   Mass spectrometry showed the
compound to have a parent peak at mass number 176, and infrared spectrophotometry showed a compound with a singly-substi-
tuted benzene ring.

An example of the use of NMR in quantitative analysis con-
cerns a sample that had been submitted for analysis by gas
chromatography.   This analysis showed the sample to contain
62% benzene along with the desired products.   The organic
chemists knew this was impossible, so they redid the experiment and submitted a second sample.   Gas chromatographic

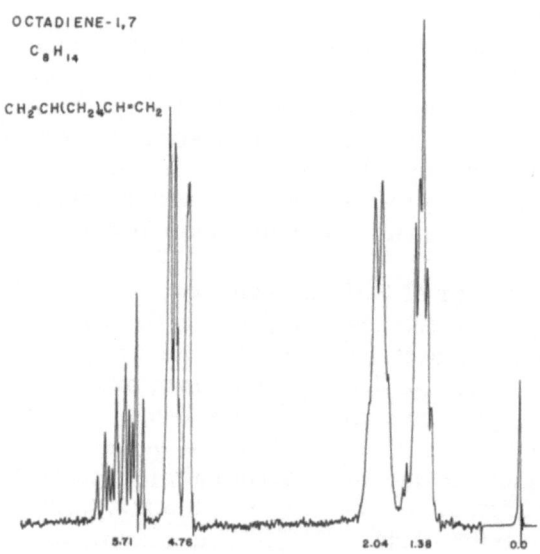

Fig. 4. NMR spectrum of 1,7-octadiene in CCl_4.

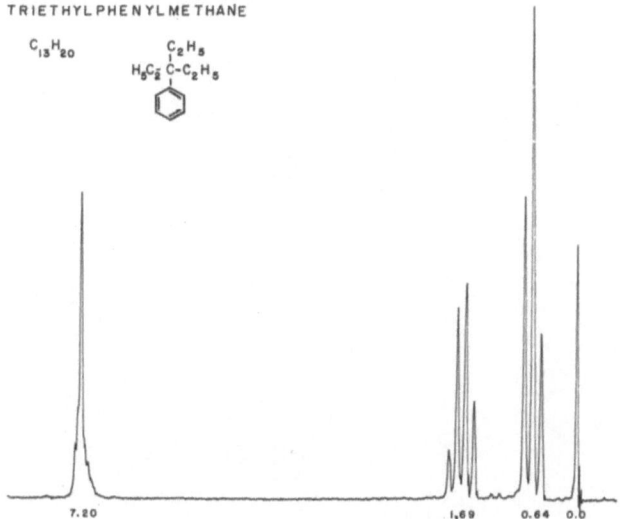

Fig. 5. NMR spectrum of triethylphenylmethane in CCl$_4$.

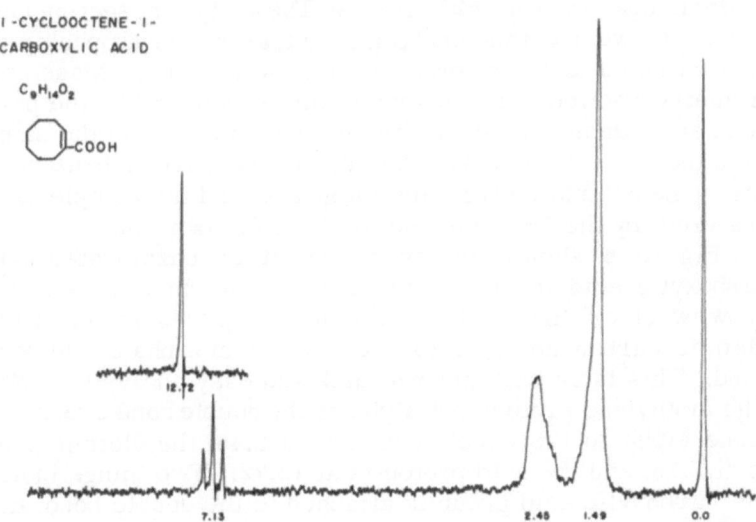

Fig. 6. NMR spectrum of 1-cyclooctene-1-carboxylic acid in CCl$_4$.

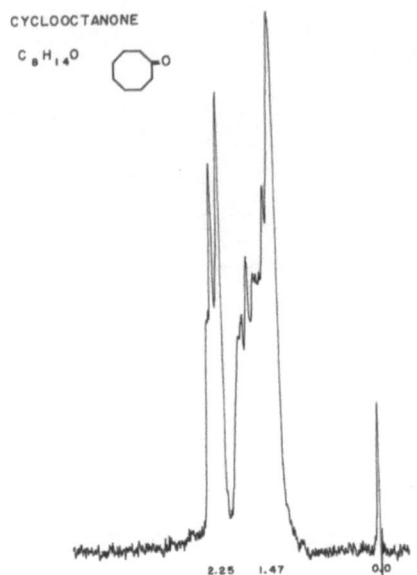

Fig. 7. NMR spectrum of cyclooctanone in CCl₄.

analysis again showed 62% benzene. The analytical section joined forces to verify this analysis. Infrared spectrophotometry showed benzene to be present in large amounts. Mass spectrometry showed a large peak at mass number 78—the parent peak of benzene. As all of the major compounds in the sample have the same molecular formula, the mole percent benzene can easily be calculated from the integrals, and the sample can be analyzed, by the NMR method, to 64 mol.% benzene.

Figure 6 shows the spectrum of an unsaturated cyclic carboxylic acid in CCl₄. Here the organic chemists wanted to know whether the carboxylic acid group was attached to an olefinic carbon atom or to a carbon atom alpha to the double bond. This is an example of a quick and easy analysis by NMR. The methylene protons not alpha to the double bond are at 1.49, those alpha to the double bond are at 2.45, the olefinic proton is at 7.13, and the acid proton is at 12.72. Two things indicate the carboxylic acid group is attached to the double bond; there is only one olefinic proton when the sample is integrated, and the olefinic band is a triplet.

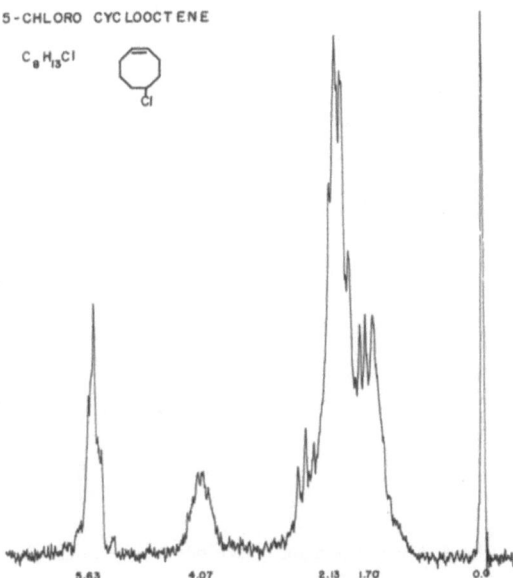

5-CHLORO CYCLOOCTENE

$C_8H_{13}Cl$

5.63   4.07   2.13 1.70   0.0

Fig. 8. NMR spectrum of 5-chlorocyclooctene.

Figure 7 shows the spectrum of cyclooctanone in CCl_4. Here, the band at 2.25 is due to the methylene protons alpha to the carbonyl group, and the band at 1.47 is due to the other methylene protons. They integrated in the ratio 4.2:9.8, or 4:10.

Figure 8 shows the spectrum of another cyclic, eight-carbon compound, 5-chlorocyclooctene. The band at 5.63 is due to the olefinic protons, that at 4.07 to the proton on the carbon atom attached to the chlorine, that at 2.13 to those protons on methylene groups alpha to the double bond or the carbon atom attached to the chlorine, and that at 1.70 to the other methylene protons. They integrate in the ratio 2:1:8:2.

Figure 9 shows the spectrum of cyclododecene in CCl_4. The multiplet at 5.35 is due to olefinic protons, that at 2.06 to the protons on methylene protons alpha to the double bond, and that at 1.36 to the remaining methylene protons. They integrated in the ratio 1.6:4.1:16.3, rather than 2:4:16, because analysis of the sample by gas chromatography revealed it to be only 91% pure.

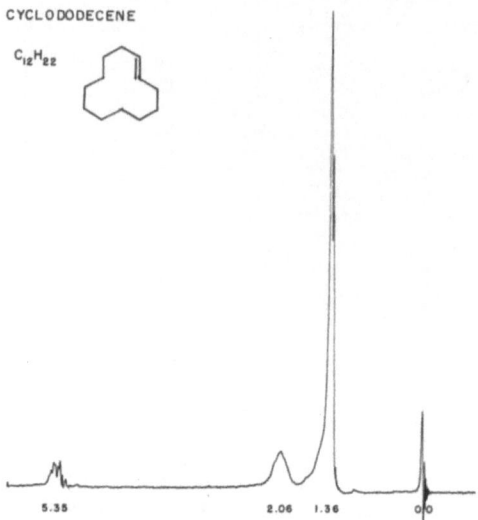

Fig. 9. NMR spectrum of cyclododecene in CCl$_4$.

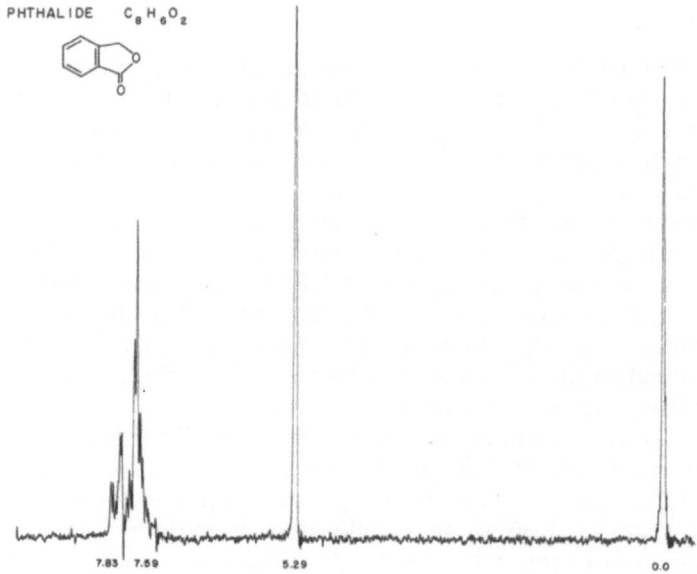

Fig. 10. NMR spectrum of phthalide in CCl$_4$.

We shall now discuss the spectra of some aromatic compounds. Figure 10 shows the spectrum of phthalide in CCl_4. The band at 7.83 is due to the aromatic proton closest to the carbonyl group, that at 7.59 is due to the remaining aromatic protons, and that at 5.29 is due to the methylene protons. They integrate in the ratio 1:3:2. Note how far downfield the methylene protons are shifted compared to those discussed previously.

Figure 11 shows the spectrum of ortho-toluic acid in CCl_4. The acid proton appears at 12.62 (note that the sweep was offset 300 cps to fit this on the chart), the aromatic proton closest to the carboxylic acid group appears at 8.15, the other aromatic protons appear at 7.36, and the methyl protons appear at 2.71. They integrate in the ratio 1:1:3:3.

Figure 12 shows the spectrum of another aromatic compound in CCl_4. The aromatic protons are at 6.86, the methylene protons are at 2.73, and the methyl protons are at 2.26. They integrate in the ratio 6:4:12. The pattern of the aromatic proton region is characteristic of 1,2,4-trisubstitution. A downfield doublet of doublets appears in all of the 1,2,3-trisubstituted benzenes we have examined, and a 1,3,5-compound would have been most unlikely as a product.

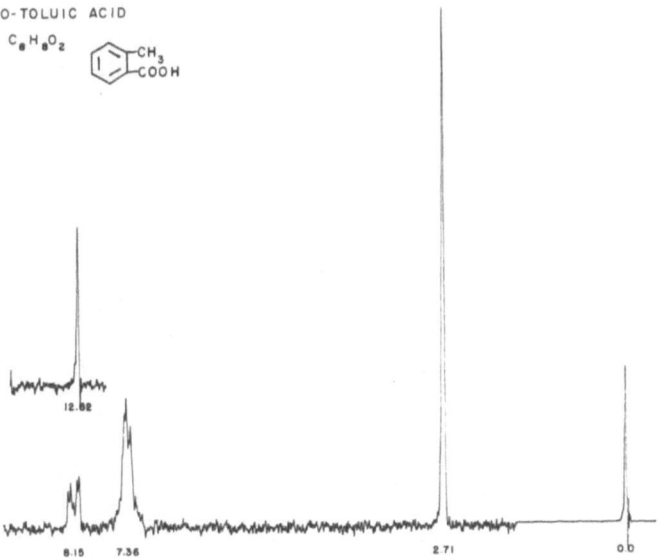

Fig. 11. NMR spectrum of ortho-toluic acid in CCl_4.

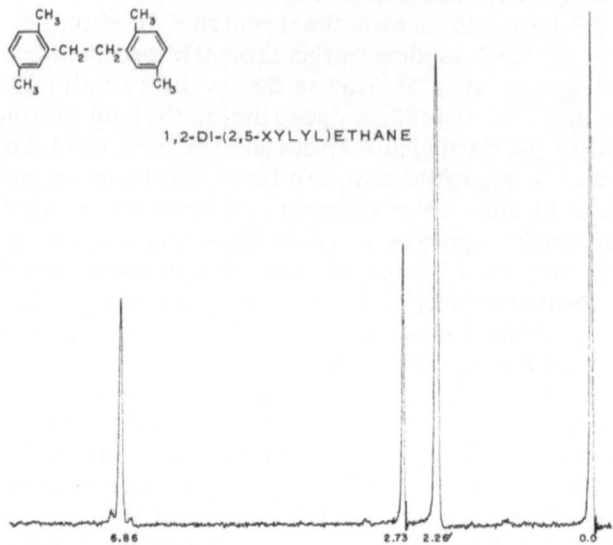

Fig. 12. NMR spectrum of 1,2-di-(2, 5-xylyl) ethane in CCl$_4$.

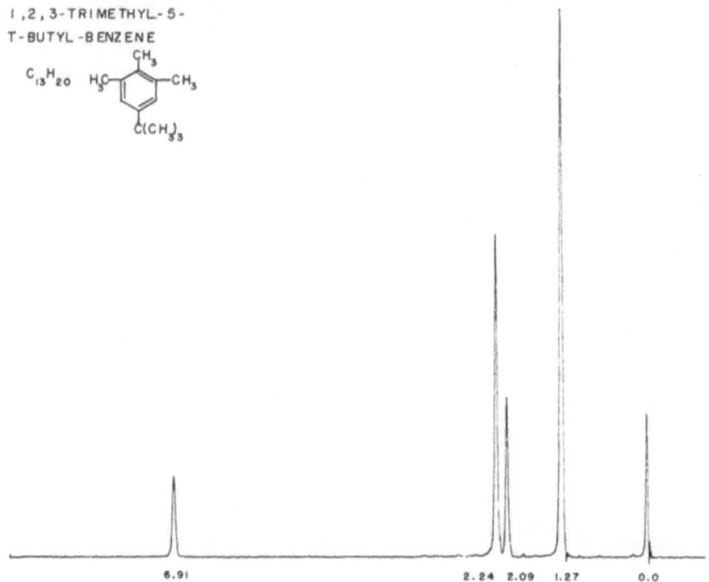

Fig. 13. NMR spectrum of 1,2,3-trimethyl-5-tert.-butylbenzene.

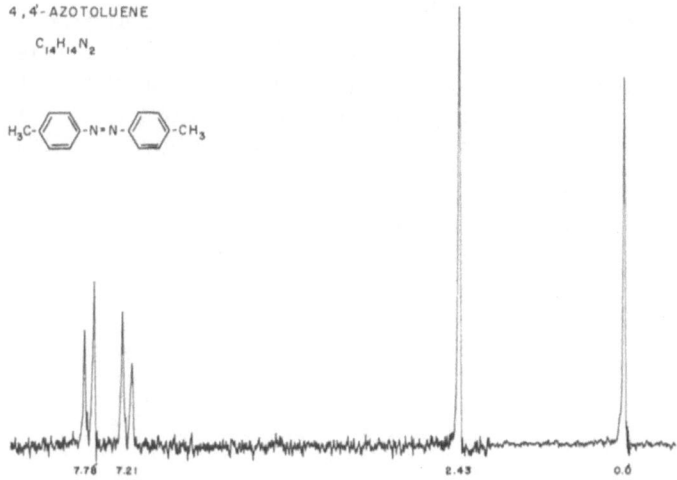

Fig. 14. NMR spectrum of 4,4'-azotoluene in CCl$_4$.

Figure 13 shows the spectrum of 1,2,3-trimethyl-5-tert.-butylbenzene. Here, the aromatic protons are at 6.91, the meta-methyl protons are at 2.24, the para-methyl protons are at 2.09, and the tert.-butyl protons are at 1.27. They integrate in the ratio 2:6:3:9.

Figure 14 shows the spectrum of 4,4'-azotoluene in CCl$_4$. The peak at 2.43 is due to the methyl protons, the doublet at 7.21 is due to the aromatic protons closest to the methyl group, and the doublet at 7.78 is due to the aromatic protons closest to the azo group. This brings up an important point. CCl$_4$ is the best solvent to use in NMR work. If a more polar solvent is needed, CDCl$_3$ or some deuterated oxygen-containing compound is used. The organic chemist sometimes synthesizes compounds that are not very soluble in even these compounds. Examples of these are para-substituted azo compounds containing chlorine, bromine, or iodine. The spectra we have obtained for these compounds are not very good, because there is a low signal-to-noise ratio for even saturated solutions. The fluorine-containing compound is soluble.

NMR can also be used to calculate percent hydrogen in a sample. The easiest way to do this is to use an internal standard. Here it is best if the peaks due to the internal standard and those due to the sample do not overlap. For example, a hydrocarbon sample showed NMR bands from 1.0–5.5 ppm. Benzophenone (used as the internal standard) shows NMR bands in the 6–7 ppm region. The relative integrals were 228 and 184 for 0.8023 g of benzophenone and 0.3055 g of sample. As benzophenone contains 5.53% hydrogen, the sample contains 11.71% hydrogen as shown in the following calculation.

$$\frac{0.8023 \text{ g}}{0.3055 \text{ g}} \cdot \frac{184}{228} (5.53\% \text{ H}) = 11.71\% \text{ H}$$

Emission, Flame, and Atomic Absorption Spectroscopy

Extraction and Flame Spectrophotometric Determination of Palladium and Rhodium

Howard C. Eshelman, John Dyer,
and James Armentor*

University of Southwestern Louisiana
Lafayette, Louisiana

Solvent extraction methods were used to prepare the solutions of the palladium and rhodium chelates that were investigated. Toluene, chloroform, 4-methyl-2-pentanone, and pentyl acetate were used as extraction solvents. The effectiveness of several different chelating agents for palladium and for rhodium was investigated. The optimum pH was determined for each of the extraction systems that was used. The extraction of the salicylaldoxime chelate of palladium into 4-methyl-2-pentanone was found to be quantitative at pH 3. The chelate of rhodium produced by sodium diethyldithiocarbamate was completely extracted into 4-methyl-2-pentanone at pH 8.

Studies were carried out to determine the optimum combustion conditions for each of the organic extracts. The position of maximum emission intensity in the flame mantle was determined for each of the solutions of the chelates of palladium and of rhodium. Emission readings were taken at 363.5 mμ for palladium, and 369.2 mμ for rhodium. When a 4-methyl-2-pentanone, rather than an aqueous, solution of palladium is aspirated into the flame, emission intensity is increased 21-fold. The emission sensitivity for rhodium using the ketone aerosol is 27 times as great as the sensitivity obtained when an aqueous aerosol is used. The results of these studies made it possible to formulate a method for the separation and determination of palladium and rhodium when they are present in the same sample.

INTRODUCTION

There is very little information available in the literature which pertains to the application of flame photometry to the analysis of the platinum metals. Dr. Paul Gilbert has stated, "That no one has used flame photometry for determining the platinum metals—at least ruthenium, rhodium, and palladium—

---

*Present address: Shell Oil Co., Pasadena, Texas.

is the more surprising in view of the analytical difficulties that beset these elements." [4].

In the present investigation, a flame spectrophotometric method for palladium and for rhodium has been developed. Palladium is isolated from most of the other elements by a single extraction with a 4-methyl-2-pentanone solution of salicylaldoxime, after which the organic phase is aspirated directly into the flame. Rhodium is then extracted from the aqueous phase into 4-methyl-2-pentanone. The rhodium is extracted as the diethyldithiocarbamate chelate.

The method of Gilchrist and Wichers [5] has been used for many years in the analysis of the platinum metals. However, the flame spectrophotometric method for palladium and rhodium has the advantage of speed and can be used for samples containing them in the microgram range. The colorimetric method of Forsythe and his co-workers [3] gives excellent results for palladium, but the procedure is time-consuming.

The effects of the following experimental variables were studied: flows of oxygen and acetylene, the optimum ratio of flows, the emission intensity from different regions of the flame mantle, and several cations and anions.

EXPERIMENTAL

Apparatus

The flame spectrophotometer and associated equipment that was used have been described previously [2].

Reagents

The standard solutions of palladium and rhodium were made using spectrographically analyzed palladium (II) chloride and rhodium (III) chloride obtained from J. Bishop and Co. The weight of the chloride needed to furnish 500 mg of the metal was calculated from the assay value. This weight was dissolved in 0.025M hydrochloric acid and diluted to 500 ml with the same solvent.

Salicylaldoxime, 1.5% was prepared by dissolving 1.5 g of salicylaldoxime in 4-methyl-2-pentanone and diluting to 100 ml with additional solvent.

Sodium diethyldithiocarbamate, 8% was prepared by dissolving 8 g of the compound in demineralized water and diluting to 100 ml.

Procedure

Use 20 ml of a solution containing palladium (II) chloride and rhodium (III) chloride. The concentration of the palladium and the rhodium should be 5–100 $\mu g/ml$. Add a known volume (usually 20.0 ml) of a 1.5% solution of salicylaldoxime in 4-methyl-2-pentanone and agitate briefly. Finally, adjust the pH of the aqueous phase to 3.2 ± 0.2 with 0.5M aqueous ammonia. The base should be added in small portions and with thorough mixing. Let stand for one hour with occasional shaking. Quantitatively separate the two phases. Reserve the aqueous phase for the determination of rhodium. Aspirate the organic phase into the flame and record the palladium line emission at 363.5 $m\mu$ and the flame background at 362.0 $m\mu$. Bracket the sample with suitable aliquots of the standard palladium solution which have been carried through the same extraction procedure.

Add 20.00 ml of 4-methyl-2-pentanone and 15 ml of 8% diethyldithiocarbamate to the aqueous phase that was reserved for the determination of rhodium. Shake for 1 min, then add 2M ammonium acetate buffer, pH 8, in three 5-ml portions and shake for 5 min. Let stand for 2 hr with occasional shaking. Decant the organic phase into the sample cup. Aspirate the organic phase into the flame and record the rhodium line emission at 369.2 $m\mu$ and the flame background at 367.0 $m\mu$. Bracket the sample with suitable aliquots of the standard rhodium solution which have been carried through the diethyl-dithiocarbamate extraction procedure.

RESULTS AND DISCUSSION

Slit Width

With the line-rich spectrum of palladium and of rhodium, the minimum slit width that is compatible with the sensitivity requirements should be used. With a photomultiplier tube, sufficient emission intensity was obtained with a slit width of 0.030 mm, which corresponded to a band width of 0.42 $m\mu$ at one-half maximum emission intensity. Little additional resolution is gained by the use of a narrower slit width [1].

Oxygen and Acetylene Flows

The effect of increasing acetylene flow rate, at constant oxygen flow rate, upon the emission intensity of palladium is shown in Fig. 1. A slight maximum in emission intensity occurs from 2.6–3.0 ft^3/hr for acetylene, while the oxygen flow rate is maintained at 6.0 ft^3/hr. The optimum ratio of oxygen to acetylene flows is from 2.3:1 down to 2.0:1. A similar effect is noted for other oxygen flow rates.

The emission intensity of the palladium 363.5-mμ line as a function of oxygen flow rate at constant acetylene flow rate is shown in Fig. 2. The emission readings were erratic for oxygen flows less than 5 ft^3/hr. The emission readings increased pronouncedly as the oxygen flow rate was increased from 5 to 8 ft^3/hr; additional increase in oxygen flow produced no significant change.

Figure 3 shows the effect upon the palladium emission intensity produced by changing the oxygen and acetylene flow rates so that the flow ratio is kept constant at 2:1. The emission intensity of the palladium line increases about 30% as the oxygen flow rate is increased from 5 to 10 ft^3/hr. This increase in oxygen flow rate approximately triples the aspiration rate of the organic extract.

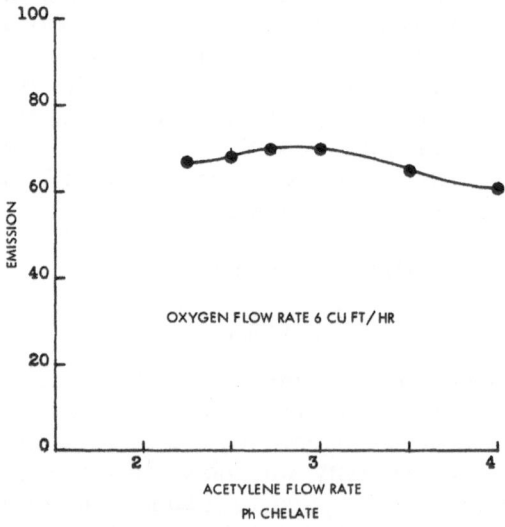

Fig. 1. Intensity of palladium emission as a function of acetylene flow rate at constant oxygen flow. Palladium, 20 μg/ml 4-methyl-2-pentanone.

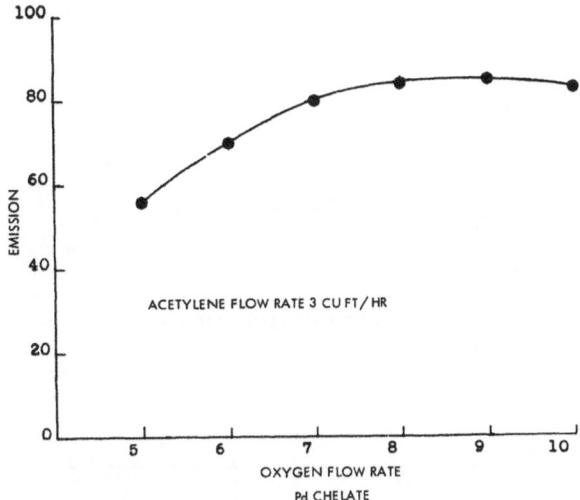

Fig. 2. Intensity of palladium emission as a function of oxygen flow rate at constant acetylene flow. Palladium, 20 μg/ml 4-methyl-2-pentanone.

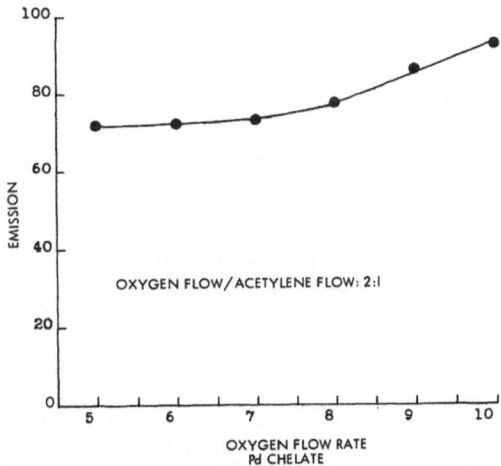

Fig. 3. Intensity of palladium emission as a function of oxygen and acetylene flows at the optimum combustion ratio. Palladium, 20 μg/ml 4-methyl-2-pentanone.

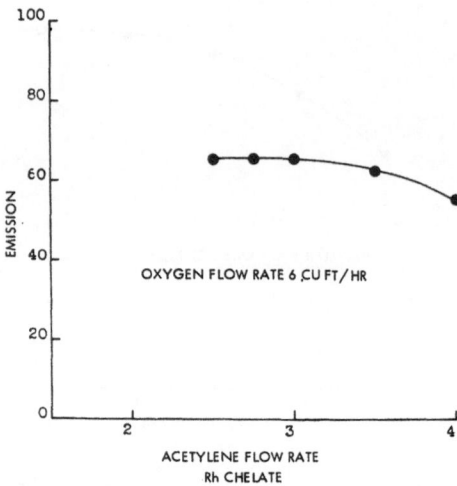

Fig. 4. Intensity of rhodium emission as a function of acetylene flow rate at constant oxygen flow. Rhodium, 10 μg/ml 4-methyl-2-pentanone.

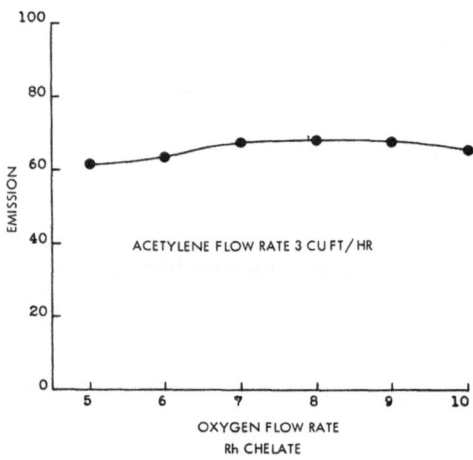

Fig. 5. Intensity of rhodium emission as a function of oxygen flow rate at constant acetylene flow. Rhodium, 10 μg/ml 4-methyl-2-oentanone.

Emission intensity of the rhodium 369.2-mμ line as a function of acetylene flow rate at constant oxygen flow is shown in Fig. 4. With the oxygen flow rate constant at 6 ft^3/hr, the maximum emission intensity is obtained when the acetylene flow rate ranges from 2.5–3.0 ft^3/hr. The last value corresponds to a flow ratio, oxygen to acetylene, of 2:1.

Increasing the oxygen flow rate at constant acetylene flow rate has only a slight effect upon the emission intensity of rhodium, as is shown by Fig. 5.

The effect upon the rhodium 369.2-mμ line produced by changing the oxygen and acetylene flow rates in such a manner that the flow ratio is kept constant at 2:1 is shown in Fig. 6. An increase in the oxygen flow rate from 5 to 10 ft^3/hr increases the emission intensity of the rhodium a little more than 30% and increases the aspiration rate of the sample almost 3-fold.

The Emission Intensity in the Different Regions of the Flame Mantle

The use of an adjustable burner mount made it possible to observe the emission intensities of palladium and rhodium at different heights in the flame. The positions of the burner ranged from 10 mm above normal to 10 mm below normal. The permanent position of the burner in its housing, as furnished

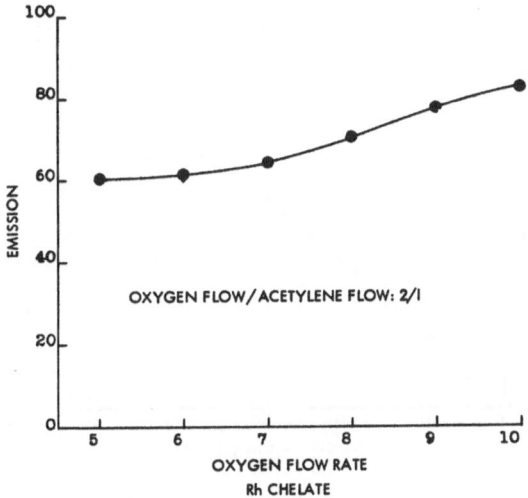

Fig. 6. Intensity of rhodium emission as a function of oxygen and acetylene flows at the optimum combustion ratio. Rhodium, 10 μg/ml 4-methyl-2-pentanone.

by Beckman Instruments, is considered the normal position. The emission intensity of palladium and rhodium increased as the burner was raised from 10 mm below normal to 5 mm above normal, then decreased as the upward movement of the burner was continued. This position of maximum emission intensity applied to toluene, 4-methyl-2-pentanone, and pentyl acetate aerosols. The burner position that gave maximum emission intensity with a chloroform aerosol was 8 mm above normal.

Working Curves

The working curve for palladium shown in Fig. 7 and the curve for rhodium shown in Fig. 8 were obtained using 4-methyl-2-pentanone solutions of their chelates. Slight self-absorption of the palladium 363.5-mμ line begins at 20 μg/ml and continues up to 200 μg/ml, becoming more intense above that concentration. Self-absorption is absent for rhodium solutions only when their concentration is less than 10 μg/ml.

Interferences

In these studies, solutions containing known amounts of palladium or rhodium and the various test ions were carried

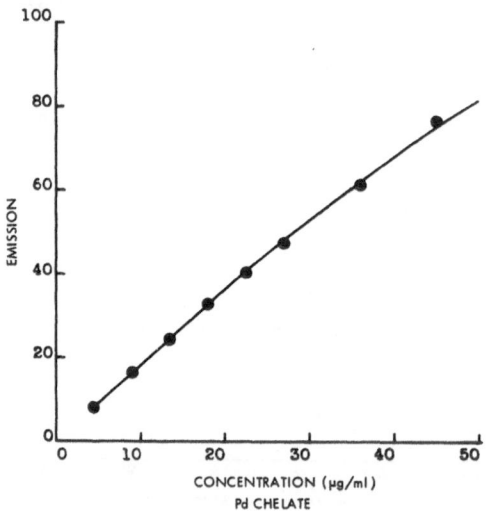

Fig. 7. Emission intensity of palladium line at 363.5 mμ from a 4-methyl-2-penta-none solution, shown as a function of palladium concentration.

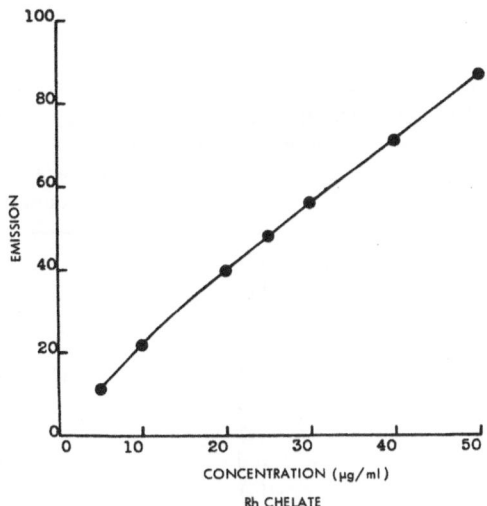

Fig. 8. Emission intensity of rhodium line at 369.2 mμ from a 4-methyl-2-pentanone solution, shown as a function of rhodium concentration.

singly through the extraction procedure. No attempt was made to differentiate between the effect of the interferent upon the extraction and upon the emission intensity of the metal. The influence of various acids upon the palladium emission intensity is shown in Table I. No serious interference was experienced from any of the acids at the concentrations that were tested. The effect of higher acid concentrations was not investigated because, in most cases, a mixture of concentrated acids is used to dissolve the palladium or rhodium sample and then the solution evaporates to near dryness. The data contained in Table II show that the deleterious effect of the acids upon the rhodium emission intensity is much more severe than that observed for palladium. Hydrochloric acid is the only one that is completely innocuous.

Table III shows the effect of some diverse ions upon the emission intensity of palladium. It is probable that the apparent discrepancies for the two concentrations of iron and ruthenium is due to the fact that, with the more concentrated solution, a solid separates at the interface of the two phases during extraction, and this inhibits to some extent the extraction of palladium.

The influence of some diverse ions upon the rhodium emis-

TABLE I

Influence of Various Acids Upon Palladium Line at 363.5 mμ. Palladium Present, 10μg/ml

Acid tested	Concentration, M	Change in emission intensity, %
Acetic	0.1	0
	0.5	0
Phosphoric	0.1	0
	0.5	0
Sulfuric	0.1	+ 1
	0.5	0
Nitric	0.1	−1
	0.5	+ 1
Perchloric	0.1	+ 3
	0.5	+ 2
Hydrochloric	0.1	0
	0.5	0

TABLE II

Influence of Various Acids Upon Rhodium Line at 369.2 mμ. Rhodium Present, 10μg/ml

Acid tested	Concentration, M	Change in emission intensity, %
Phosphoric	0.1	− 4.0
	0.5	−24.0
Sulfuric	0.1	+ 1.0
	0.5	+ 1.0
Nitric	0.1	+ 1.0
	0.5	− 3.0
Perchloric	0.1	− 2.0
	0.5	− 9.0
Hydrochloric	0.1	0
	0.5	0

TABLE III

Influence of Diverse Ions Upon the Palladium Line at 363.5 mμ. Palladium Present, 10 μg/ml

Interferent	Concentration, M	Change in emission intensity, %
Cobalt	2,000	− 1.5
Gold	1,000	+ 2.0
Iridium	500	− 6.0
Iron	100	+ 12.0
Iron	10,000	− 1.0
Nickel	1,000	− 2.5
Osmium	1,000	0
Platinum	1,000	+ 3.0
Platinum	5,000	+ 2.0
Ruthenium	100	+ 2.5
Ruthenium	1,000	− 4.0

TABLE IV

Influence of Diverse Ions Upon the Rhodium Line at 369.2 mμ. Rhodium Present, 10 μg/ml

Interferent	Concentration, M	Change in emission intensity, %
Cobalt	2,000	−11.0
Gold	1,000	− 1.0
Iridium	500	− 6.0
Iron	100	− 6.0
Iron	10,000	−20.0
Nickel	1,000	− 6.0
Osmium	1,000	+ 2.0
Platinum	1,000	+ 5.0
Platinum	5,000	+ 2.0
Ruthenium	100	+ 5.0
Ruthenium	1,000	+ 1.0

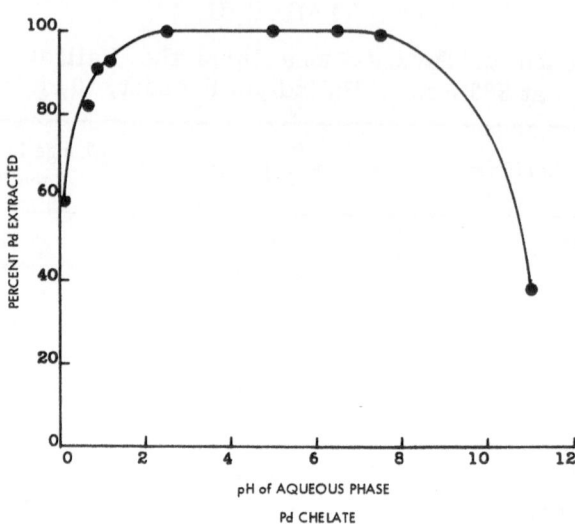

Fig. 9. Effect of pH on extraction of palladium with 1.5% salicylaldoxime in 4-methyl-2-pentanone.

TABLE V

Determination of Palladium and Rhodium
Volume of Organic Phase, 30.00 ml

Palladium taken, μg	Rhodium taken, μg	Palladium recovered, %	Rhodium recovered, %
300	720	97.0	99.3
300	820	98.5	99.3
300	900	99.0	98.2
600	660	98.3	100.0
600	1,320	98.0	99.4
1,200	660	100.0	100.0
5,000	150	98.2	93.3
5,000	300	98.4	92.4
5,000	450	99.5	97.8

sion intensity is shown in Table IV. A large change in the emission intensity is produced by some of the cations. However, it would be a simple matter to remove the interferent in most cases, e.g., the removal of the iron can be accomplished by prior extraction with TTA at low pH.

Extraction of Palladium

To circumvent the interference from other elements, palladium was extracted with a 1.5% solution of salicylaldoxime in 4-methyl-2-pentanone from an aqueous ammonia solution which was adjusted to pH 3.2 ± 0.2. 4-Methyl-2-pentanone was found to be superior to toluene and chloroform in combustion characteristics and superior to pentyl acetate in extraction characteristics.

The effect of pH of the aqueous phase upon the extraction of palladium salicylaldoxime is shown in Fig. 9. The extraction is more than 99.5% complete over the pH range 2.8–6.4. The aqueous to organic volume ratio should range from 1 to 5.

Extraction of Rhodium

The extraction of rhodium diethyldithiocarbamate into 4-methyl-2-pentanone is satisfactory in all respects only when the extraction is made with the aqueous phase strongly buffered, at pH 8, with ammonium acetate solution.

SUMMARY

The validity of the proposed method of analysis for palladium—rhodium mixtures is shown in Table V. For the nine samples analyzed, there was no error greater than 3% on the palladium analysis and only two of the rhodium analyses showed an error greater than 3%. The average error for the nine palladium determinations was 1.5%, and for rhodium the average error was 2.3%. The study of the emission characteristics of palladium and of rhodium from a 4-methyl-2-pentanone solution should be useful to anyone contemplating the use of a flame spectrophotometric method for these elements.

ACKNOWLEDGMENT

This work is part of a research program which has been aided by grants from the Petroleum Research Fund of the American Chemical Society. The authors are grateful for this assistance.

REFERENCES

1. J. A. Dean, Flame Photometry, McGraw-Hill, New York (1960), p. 124.
2. Howard C. Eshelman and James Armentor, in J. E. Forrette and E. Lanterman (eds.): Developments in Applied Spectroscopy, Vol. 3, Plenum Press, New York (1964), p. 190.
3. J. H. W. Forsythe, R. J. Magee, and C. L. Wilson, Talanta 3:324, 330 (1960).
4. Paul T. Gilbert, Jr., Analytical Flame Photometry: New Developments, Beckman Instruments, Inc., Fullerton, California (1960), p. 43.
5. R. Gilchrist and E. Wichers, J. Am. Chem. Soc. 57:2565 (1935).

The Determination of Copper, Nickel, Cobalt, Manganese, and Magnesium in Irons and Steels by Atomic Absorption Spectrophotometry

Sabina Sprague and Walter Slavin

Perkin-Elmer Corporation
Norwalk, Connecticut

The determinations of copper, nickel, cobalt, manganese, and magnesium in irons and steels are reported here. The analyses obtained for standard samples are compared with certificate values. Analytical precision is discussed, as are analytical time and operational complexity.

INTRODUCTION

Atomic absorption has been used to determine many metals in irons and steels. Elwell and Gidley [1] have discussed the analysis of steels for lead, zinc, and manganese. Belcher and Bray [2] have reported a method for the determination of magnesium in cast iron, which is now the Australian Standard Method K1, Part 20, 1964. McPherson, Price, and Scaife [3] have determined cobalt in steels. Kinson, Hodges, and Belcher [4] have developed a method for chromium in steels, and Kinson and Belcher have done extensive work on nickel in steels [5] and recently on manganese in steels [6].

When Prince, Coglianese, and Coless [7] determined magnesium, chromium, copper, nickel, and cobalt in irons and steels by flame photometry, they found it necessary in all cases to make extensive background corrections and to remove the iron by extraction. In contrast, atomic absorption is especially suited to the analysis of steels because the only sample preparation required is sample dissolution.

In this paper, we report the analyses of twelve National Bureau of Standards standard samples by atomic absorption for copper, manganese, nickel, cobalt, and magnesium. Some previous work by atomic absorption has been reported for all of these metals, except copper. Only five metals are discussed here, since time did not allow further work. However, aluminum [8], molybdenum [9], and many of the other constituents of steels have also been determined by atomic absorption.

EXPERIMENTAL

Kinson, Hodges, and Belcher [4] dissolved their steel samples in a solution of 15% orthophosphoric acid and 15% sulfuric acid. The samples were then oxidized by dropwise addition of nitric acid. This procedure, which had the advantage that the tungsten was dissolved, was also used by McPherson, Price, and Scaife [3]. However, in our laboratory, we found that this method did not dissolve all of the copper in the samples, and therefore we abandoned it. The method in which steel samples are dissolved in hydrochloric acid, followed by oxidation with nitric acid, also did not dissolve all of the material. We used a mixed acid solution containing 20% nitric acid and 80% hydrochloric acid to dissolve the samples.

To produce standards, a solution of iron was prepared by dissolving high-purity iron in the minimum quantity of hydrochloric acid. A solution of strontium chloride containing 3% strontium was prepared by dissolving magnesium-free strontium carbonate in dilute hydrochloric acid. This solution was used as a diluent for the samples containing magnesium and for the magnesium standards to suppress any interference from the anions present [2].

Standards were prepared for each metal. Copper standards were prepared from high-purity copper dissolved in nitric acid. Manganese dioxide was dissolved in hydrochloric acid and diluted to the desired level. Nickel oxide was dissolved in hydrochloric acid. Cobalt standards were prepared by dilution of a 1% solution of cobalt, as the chloride,* to the desired level. A 1% solution of magnesium, as the chloride,* was used for magnesium standards.

Two sets of standards were prepared for cobalt, nickel,

*Hartman-Leddon Company, Philadelphia, Pennsylvania.

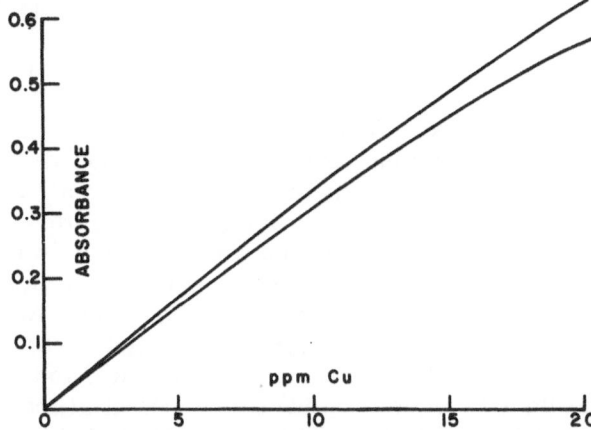

Fig. 1. Working curves for copper. Upper curve represents simple acid standards; lower curve represents copper—iron—acid standards.

and copper. One set consisted of simple acid solutions containing 0 – 20 ppm for copper and 0 – 40 ppm for cobalt and nickel. The other set of standards, at the same levels as the first set, contained, in addition, 1% iron and 15% mixed acid. The second set was used for those samples which had a very low metal content, and thus required the use of concentrated solutions of the samples.

Figure 1 shows the working curves prepared for the two sets of copper standards. The lower curve represents copper—

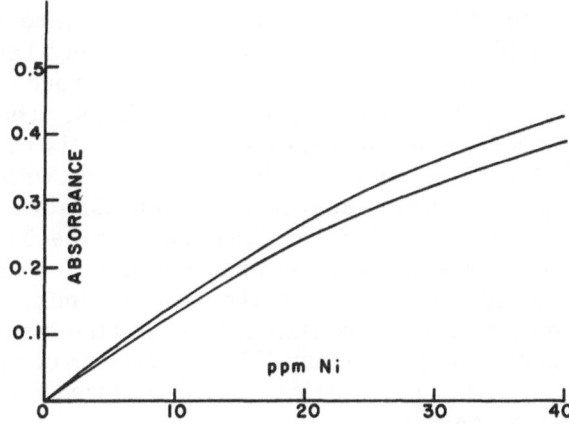

Fig. 2. Working curves for nickel. Upper curve represents simple acid standards; lower curve represents nickel—iron—acid standards.

iron—acid standards and has a slightly smaller slope than the curve obtained with copper alone. This is a result of the effect of solution density on the performance of the atomizer. Under such conditions, it is necessary to match approximately the base metal and acid content of the standards. Figure 2, showing the working curves obtained for the nickel standards, indicates that this effect is approximately constant for all elements.

Simple acid standards for manganese in the range of 0 – 20 ppm were prepared. Manganese standards in iron were not prepared, since the manganese concentration in all of the samples was so high that dilution was required. Magnesium standards in the range 0 – 2 ppm magnesium were prepared containing 1500 ppm strontium as the chloride, and 0.1% iron. The strontium was added to mask any potential interference from aluminum or silicon. The iron was added because the Australian standard method specifies it. It is probably not necessary, however, to add iron when magnesium is determined at the concentrations reported here.

The Perkin-Elmer Model 303 atomic absorption spectrophotometer was used for all the analyses. The standard conditions described in the Perkin-Elmer manual [10] were used for each metal.

PROCEDURE

Approximately 1 g of each of the twelve samples used was weighed into 100-ml beakers and dissolved in 15 ml of the mixed acid solution. Two samples of #101e were used as a check on sample preparation. The samples were heated until bubbling had stopped, and were then filtered through Whatman #541 filter paper into 100-ml volumetric flasks. The beakers and filter paper were washed thoroughly with distilled water. When cool, the samples were diluted to volume with distilled water. Aliquots of these samples were then taken for analysis.

After all of the samples were analyzed for the five metals, fresh samples were weighed out and diluted as before. These samples were then determined in the same manner. However, for these determinations, the standards were all mixed together, with the exception of magnesium. This greatly reduced the time taken in standards preparation, and introduced no apparent errors into the analysis.

ANALYSIS

The samples which contained copper in the range of 0 – 0.2% were determined against the iron-containing copper standards. Samples with copper at a level greater than 0.2% were diluted five times with distilled water, and determined against simple water standards of copper.

No sample contained manganese at a level less than about 0.2%. Those samples which contained manganese in the range 0.2 – 0.9% were diluted five times with distilled water for the analysis. Manganese in the range 0.9 – 2% was determined with twenty-times dilutions of the samples.

The range of nickel levels in the samples was wider than for any other element. Nickel in the range 0.01% – 0.2% was determined against the standards containing iron and acid. Nickel at levels from 0.2 to 2% was determined on five-times dilutions of the samples. Dilutions of 100 times were made for the samples containing nickel from 2 to 40%.

Five samples contained cobalt in the range 0.01 – 8.5%. Cobalt in the range 0 – 0.4% was determined against standards containing 1% iron. The high-level sample was diluted twenty-five times and determined against aqueous standards.

The two samples of cast iron contained magnesium at the 0.05% level. The samples were diluted ten times in distilled water and sufficient strontium chloride to provide a final concentration of 1500 ppm strontium.

RESULTS

The results for the determinations of copper, nickel, manganese, and cobalt are reported in Table I in percent metal in the solid materials. Duplicate analyses by atomic absorption on two separate sample solutions are given. In all cases, the second value was obtained using the mixed standards. The first two results for sample 101e were obtained against separate sets of standards.

The concentrations of magnesium in cast iron are reported in Table II. The values are reported in percent metal in the solid.

TABLE I

Analyses of Steel and Iron for Copper, Nickel, Manganese, and Cobalt

Sample	Type	% Copper		% Nickel		% Manganese		% Cobalt	
		NBS	AA	NBS	AA	NBS	AA	NBS	AA
159	Cr, Mo, Ag	0.181	0.170	0.137	0.147	0.807	0.83	—	—
			0.185		0.146		0.80		
130a	Pb Steel	0.027	0.026	0.010	0.010	0.753	0.76	—	—
			0.029		0.009		0.75		
50c	W, Cr, V	0.079	0.077	0.069	0.070	0.342	0.29	—	—
			0.083		0.069		0.30		
30e	Cr, V	0.094	0.0925	0.027	0.029	0.786	0.80	—	—
			0.095		0.028		0.76		
32e	Ni, Cr	0.127	0.135	1.19	1.32	0.798	0.88	—	—
			0.125		1.15		0.76		
153a	Co, Mo, W	0.094	0.089	0.168	0.179	0.192	0.17	8.47	9.25
			0.093		0.178		0.17		7.91
101e	Cr, Ni	0.359	0.36	9.48	9.8	1.77	1.85	0.18	0.18
			0.35		9.4		1.87		0.18
			0.36		9.6		1.74		0.19
462	Low alloy	0.20	0.20	0.70	0.69	0.94	0.93	0.11	0.11
			0.19		0.68		0.88		0.12
463	Low alloy	0.47	0.48	0.39	0.38	1.15	1.20	0.013	0.015
			0.47		0.39		1.13		0.014
464	Low alloy	0.094	0.094	0.135	0.14	1.32	1.33	0.028	0.031
			0.095		0.14		1.28		0.033
341	Ductile iron	0.152	0.15	20.32	18.4	0.92	0.89	—	—
			0.15		18.2		0.80		
342	Nodular iron	0.14	0.14	0.023	0.027	0.369	0.36	—	—
			0.11		0.024		0.35		

TABLE II

Magnesium in Cast Iron

Sample	Type	NBS	Atomic absorption	
341	Ductile iron	0.068	0.066	0.066
342	Nodular iron	0.053	0.050	0.050

PRECISION AND ACCURACY

Atomic absorption routinely yields a precision of 1% or better when the spectrophotometer is operated within its optimum range of 20 – 80% absorption. The agreement between the samples shown here must be considered as the precision of the whole analysis, including the sample preparation. Samples of this type, which always have a residue, can display analytical differences that are due to incomplete sample dissolution. The agreement between the values obtained by atomic absorption spectrophotometry and the certificate values supplied by the National Bureau of Standards is an adequate test of accuracy.

DISCUSSION

During the investigation of the manganese analysis, a significant interference with manganese absorption appeared

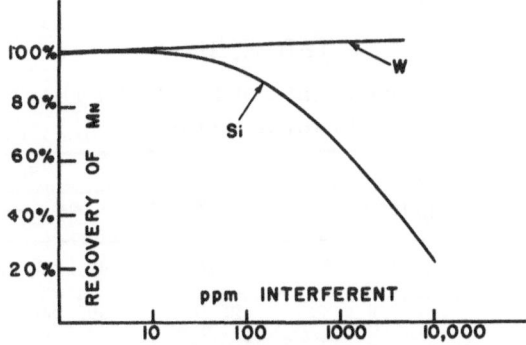

Fig. 3. Effects of various concentrations of tungsten and silicon on the recovery of manganese.

from molybdenum, silicon, and tungsten. The acetylene tank, which was nearly empty, was replaced and the interference from molybdenum and tungsten disappeared. A significant interference from silicon remained, however, as shown in Fig. 3, but was not a problem at the levels at which we were working. It must be taken into account, however, if the silicon content of the steels is greater than three times the manganese content.

In a recent publication, Belcher and Kinson [6] found no interference with manganese by silicon. In preliminary experiments, they found that chromium interfered with the absorption of manganese. By modifying their equipment so that a smaller portion of the flame is seen, they were able to remove the effect of chromium. With the improved system, no interference was found with 10 ppm manganese by 3% silicon.

The determination of the metallic constituents of steel by atomic absorption is rapid, accurate, and precise. Simple water standards can be used, except when the characteristics of the sample matrix may change the operation of the instrument. Mixed standards of the elements of interest can be used because atomic absorption is free from most interferences. In those cases where the atomizer operation is affected by the high solids content of the samples, it is necessary to match the solids content of the samples in the standards. For determinations at very low levels, where it is necessary to run the analysis at an even higher solids concentration, it may be necessary to use the method of additions [10].

In most cases, the technique is so straightforward that recipes, such as is shown in Table III for nickel, may be used.

TABLE III

Recommended Analytical Conditions for Determining
Nickel in Steels

Ni concentration range	Sample size, g/100 ml	Wavelength, A	Standards	
			Ni, ppm	% Fe
0.01 − 0.4	1	2320	40, 30, 20, 10	1
0.2 − 4.0	0.1	2320	40, 30, 20, 10	—
2.0 − 20.0	0.1	3415	200, 150, 100, 50	—

After the sample is dissolved, it is quite possible to perform analyses at the rate of 2 min per metal per analysis. This includes running the samples, preparing working curves, and interpreting the data.

REFERENCES

1. W. T. Elwell and J. A. F. Gidley, Atomic Absorption Spectrophotometry, Pergamon Press, New York (1961).
2. C. B. Belcher and H. M. Bray, Anal. Chim. Acta 26:322 (1962).
3. G. L. McPherson, J. W. Price, P. H. Scaife, Nature 199: 371 (1963).
4. K. Kinson, R. J. Hodges, and C. B. Belcher, Anal. Chim. Acta 29:134 (1963).
5. K. Kinson and C. B. Belcher, Anal. Chim. Acta 30:64 (1964).
6. C. B. Belcher and K. Kinson, Anal. Chim. Acta 30: 483 (1964).
7. L. A. Prince, F. V. Coglianese, and T. L. Coless, Welding Rev., supplement of Welding J. (N. Y.) 42 (8):347s (1963).
8. D. C. Manning, Atomic Absorption Newsletter No. 24, Sept. (1964).
9. D. J. David, Analyst 86: 730 (1961).
10. Analytical Methods for Atomic Absorption Spectrophotometry, The Perkin-Elmer Corporation, Norwalk, Connecticut (1964).

Developments in Flame Emission and Flame Absorption Photometry

John A. Dean

University of Tennessee
Knoxville, Tennessee

This work is confined largely to the significant developments in the past year in the related fields of flame emission and flame absorption photometry. Considerable attention is devoted to the subject of flames and factors affecting total atom population, and this work is concluded with a survey of recent methods for a number of the elements.

FACTORS AFFECTING TOTAL ATOM POPULATION

The exact processes occurring within the flame gases following introduction of an aerosol are still incompletely understood. If a metal M is added in small amount to the flame gases, it may interact with the gases to form various compounds, in particular, the hydroxide MOH, the oxide MO, and the hydride MH. The proportion of electronically excited forms in each of these states is a few orders of magnitude below that of the ground-state forms. In atomic absorption and flame emission photometry, the problem centers around the sufficient production of neutral atoms in the former case, and excited atoms in the latter case.

The metal hydroxides MOH are formed predominantly by a rapid, reversible, bimolecular reaction

$$M + H_2O \rightleftharpoons MOH + H \qquad (1)$$

which is balanced in the burnt gases [1]. The magnitude of this effect may be assessed roughly as follows. The hydroxides of lithium and cesium, with dissociation (to give an atom and a hydroxyl radical) energies of 102 and 92 kcal/mole (4.4 and 4.0 eV, respectively), have ratios of [Li]/[LiOH] of approximately 1:10 and [Cs]/[CsOH] of approximately 1:1 in the burnt gases. For copper, manganese, calcium, strontium, and barium, indications are that the ratio will not exceed 1:0.01.

For the oxides, the reactions

$$M + H_2O \rightleftharpoons MO + H_2 \tag{2}$$

$$M + OH \rightleftharpoons MO + H \tag{3}$$

are fast enough to set up the [M]/[MO] balance, which will be the same for both reactions because the reaction

$$H + H_2O \rightleftharpoons H_2 + OH \tag{4}$$

is balanced. Observations by Reid and Sugden [1] on the alkaline earths show that the oxides are so stable that [M]/[MO] is approximately 1:100. For copper, the ratio is about 1:0.1, while, for manganese, it is about 1:1. The foregoing observations for metal hydroxides and metal oxides were made in hydrogen–oxygen–nitrogen flames.

Persistence of neutral atoms in the plume of a flame has been studied by Zelyukova and Poluektov [2], who used an air–propane flame. Cadmium and mercury persist up to greater than 70 cm in the plume of the flame without marked attenuation, gold up to 50 cm. Silver decreases linearly to zero at a distance of 65 cm, copper drops to zero at 30 cm, and the other elements investigated (lead, magnesium, thallium, nickel, indium, and sodium) peak around 0–6 cm from the burner tip and decline rapidly to zero at about 15–20 cm distance upward in the flame gases. Studies of this type are important in deciding upon the merits of long path-length flames in atomic absorption, as contrasted with multiple traversals, and also in deciding on the position within the flame gases where interferences might be minimized. A reducing agent added to solutions of mercury salts increases the absorption signal due to mercury atoms [3].

Fuwa and Vallee [4] reported an increase in absorption sensitivity with path lengths up to 90 cm. The flame from an integral-atomizer burner was directed into and down the length of vycor tubing whose diameter should not exceed 1 cm and which should possess a reflective interior. Although at first glance this data would appear to conflict with the assertions of Zelyukova and Poluektov, the absence of significant entrained air with the subsequent maintenance of highly reducing conditions over much of the flame path undoubtedly explains the 10- to 100-fold improvement in absorption sensitivity. Circular reflections from the inner surface of the vycor tubing also enhance the absorption.

The weakness of lines of some elements may due to saturation, as pointed out by Gilbert [5]. The vapor pressure of the free metal in the flame may not permit a sufficiently large concentration of metal atoms to accumulate; thus, low emission and absorption sensitivity will result, e.g., rhenium. At 2700°K, the vapor pressure is only 3×10^{-7} atm.

FLAMES

Optimum flame conditions require a study of fuel/air or fuel/oxygen ratios. Adkins and Dean [6] have reported on correlations between fuel/oxygen ratios, height of observation in the flame for maximum emission, and excitation energy (plus dissociation energy of any metal compounds) of a number of elements atomized into acetylene−oxygen−MIBK flames. In many respects, the distribution of emission paralleled the results reported by Buell [7] for hydrogen−naphtha flames.

Flame profiles at different fuel/air ratios are important in emission and absorption. A complete study is necessary, otherwise crossovers may be overlooked. Interferences are also dependent on the gas mixture; for example, sodium interference upon magnesium exhibits a crossover at low propane flows [8]. Similar effects are found with other elements. The optimum position for absorption of aluminum is sharply localized [9]. In this instance, the detection limit by absorption is only 8 μg/ml as contrasted with 0.25 μg/ml obtained by emission methods [10]. Essentially, the same acetylene/oxygen ratio was recommended in both studies in which an acetylene-oxygen−MIBK flame was used.

The type of fuel may be critical. Willis [11] found that the strontium addition was not effective when coal gas was used as fuel, nor was it effective in an oxygen−acetylene or oxygen−hydrogen flame, whereas an air−acetylene flame was ideal for the calcium determination. Slavin, Sprague, and Manning [12] also studied calcium by absorption. While calcium may be determined at about equal sensitivity in an air–acetylene or an air–hydrogen flame, the effect of anionic interferences is much less in an acetylene flame, due probably to the higher tempera-ture and resultant greater dissociation of metal oxide and metal hydroxide (vide supra). In a fuel-rich, air–acetylene flame, the peak absorption is higher in the flame, and the effect of the phosphate suppression is somewhat reduced.

Very often a fuel-rich, hydrocarbon flame with its con-comitant reducing atmosphere will enhance the signal of an atomic line in either absorption or emission. These flames will be useful when the excitation energy of the spectral lines is 4.0 eV or lower, but the dissociation energy of the metal oxide is relatively large (greater than 5.0 or 5.5 eV).

Another method involves complexing the metal ion with a ligand which is more stable in aqueous solution than ligands leading to MOH or MO. Of general applicability is the use of organic chelating agents and solvent extraction [13-15]. In this manner, the metal atom is effectively screened from deleterious ions, yet the organic chelate readily releases the metal atom in the flame. During the chelate decomposition, the immediate environment about the metal atom is undoubtedly highly reducing. Protective chelation of magnesium with 8-hydroxyquinoline was used by Wallace [16] to overcome the interference of aluminum, silicon, sulfate, and nitrate in the atomic absorption of magnesium. This work is similar to the study by Debras-Guedon and Voinovitch [17] on the emission of aluminum and calcium in the presence of fluoride and phos-phate.

Chemiluminescence enters when the excitation energy of the metal atom exceeds approximately 4.0 eV, although the dissociation energy of the metal oxide may be quite low. In this connection, the ionization in flame gases themselves, in which the H_3O^+ radical constitutes about 90% of the ionized species, has lead to the following chain mechanism, as proposed by Green and Sugden [18], when small amounts (about 1%) of acetylene are added to hydrogen–oxygen–nitrogen flames:

$$CH + O \rightarrow CHO^+ + e^-$$ (5)

$$\Delta H^\circ \cong 0$$

$$CHO^+ + H_2O \rightarrow CO + H_3O^+ \tag{6}$$

$$\Delta H^\circ = -34 \text{ kcal/mole (1.5 eV)}$$

$$H_3O^+ + e^- \rightarrow H_2O + H \tag{7}$$

$$\Delta H^\circ = -201 \text{ kcal/mole (8.7 eV)}$$

A broad parallism exists between maximum flame ionization and emission from CH^* in the reaction zone when hydrocarbons, as well as methanol, are added to a hydrogen flame. The H_3O^+ carries indirectly the energy of formation of CO from CH. However, the excessive energy of the H_3O^+ radical cannot be liberated without the help of a third body, upon which Gilbert's [19] postulation rests, viz.

$$H_3O^+ + MO \rightarrow M^* + H_2O + OH \tag{8}$$

where

$$\Delta H^\circ = -13.1 \text{ eV} + D(MO) + E(M)$$

The postulation seems to explain most, if not all, of the abnormal excitations observed in flame emission.

Deviations of the alkali ionization from the Saha equilibrium have been ascribed by Hollander, Kalff, and Alkemade [20] to a relaxation of the establishment of the equilibrium and to chemi-ionization involving excess concentrations of flame radicals in the region directly above the reaction zone. These authors found that collisions with nitrogen molecules in carbon monoxide—nitrogen flames are the major source for the alkali ionization. They concluded that the transfer of internal energy during collisions between flame gas molecules and alkali atoms is the most likely ionization mechanism in the flames investigated and at some distance from the reaction zone. They postulated inelastic, three-body collisions with excess oxygen and carbon monoxide according to

$$CO + O + M \rightarrow CO_2 + M^+ + e^- \tag{9}$$

for ionization of the alkali metals in the region below the temperature maximum. Energy released in the formation of one CO_2 molecule (5.5 eV) generally suffices to ionize the alkali metal.

BURNERS AND ATOMIZATION

A report [21] on the premixed acetylene–oxygen atomizer–burner has now been published. The authors claim a gain of at least an order of magnitude improvement in detection limits by emission and absorption methods with this burner.

A report on the use of the unsheathed, reversed acetylene–oxygen flame has appeared [22], following a brief mention of this work early in 1962. Reversed acetylene-oxygen burns on either the hydrogen or acetylene Beckman integral-atomizer burner—the acetylene serving as the aspirating gas, rather than oxygen. The flame is blue and bushy, with a light blue, ovoid, diffuse core, riding 3–10 mm above the burner tip. This type of flame gives exceptionally low background intensity, while retaining normal output of calcium light. Since the atomizer remains cool, the incrustation due to evaporation at a warm burner tip is greatly slowed. However, this flame is more sensitive to sample flow rate than the normal flame. Operation must be at relatively low acetylene pressure and regulation of oxygen flow is important. Solution consumption is less than 1 ml/min, with the result that higher flame temperatures (about 3200°K) than usual can be achieved for aqueous solutions.

A sheathed oxygen–acetylene or oxygen–hydrogen burner is now available commercially,* and Gilbert [23] has commented on its advantages.

West and Hume [24] have used a radiofrequency plasma torch supplied with an aerosol generated in an ultrasonic atomizer system. The nitrogen plasma consisted of a brilliant pink ribbon of light, about 3 mm in diameter and 8 cm high, surrounded by a much less brillant area, perhaps 2 cm in diameter and 12 cm in height. Lines of metals requiring an excitation energy of 8.5 eV (10,000°K) were observed with an intensity high enough to be useful; the general increase in sensitivity appears to be roughly 100-fold. Solution consumption is only 0.1 ml/min. Dunken and coworkers [25] also replaced the pneumatic atomizer by an ultrasonic aerosol apparatus and reported 6- to 10-fold increase in sensitivity.

A graphite cell, heated to a high temperature, continues to be used by Russian workers for obtaining and localizing the atomic vapors of elements. It is free from some of the limita-

*Aztec Instruments, Inc., Westport, Connecticut.

tions of flames. In particular, the reducing atmosphere of the graphite cell promotes complete dissociation of the compounds of the majority of the elements. In this manner, Nikolaev and Aleskovskii [26] determined aluminum in metals and alloys. Cell temperatures range from 2200 to 2410°C.

Ground plant material suspended in a 20 vol.% solution of glycerol-2-propanol (2:1) mixture was aspirated directly into the flame by Mason [45] for the determination of potassium. A spark source continues to be used to atomize solid metallic samples in a report by Herrmann and Lang [27].

The atomization efficiency of total consumption atomizer-burners, coupled with a two-line method for measuring flame temperature, is contained in two papers by Winefordner and coworkers [28,29]. These authors found that the efficiency of sample introduction decreases with increasing solution flow rate, increases with flame height for most oxygen–hydrogen flame (although the opposite effect was observed in several oxygen–acetylene flames), and increases as the concentration of methanol increases in a methanol–water mixture. From the data of Pungor and Mahr [30], the time for atomization of a fixed quantity of liquid shows a minimum at low alcohol concentrations and low gas pressures. Apart from the viscosity of the solution, the surface tension also has an effect on the atomizing time when concentric atomizers are used. These authors found that a low atomizing pressure is desirable, as it minimizes the deformation of the filament of liquid issuing from the capillary.

SOLVENT

The use of organic solvents to increase emission readings continues. Brandenberger and Bader [31] used diethyl ether to extract and convey gallium, indium, and thallium into an oxygen-hydrogen flame. Enhancements of 15-, 10-, and 13-fold are similar to those reported earlier by Bode and Fabian [32]. Eshelman and coworkers [33] have reported on the emission of beryllium, manganese, and iron from organic solvents following their extraction as metal chelates. Pilgrim and Ford [34] used a 20:20:60 mixture of water–isopropanol–acetone to improve the detection limit for lithium (down to 0.0001%) in aluminum while using 1-g samples. Overall enhancement was roughly 5-fold. Inorganic salts remained in solution with this mixture. Manganese was determined in copper alloys and steel following an extraction with 2-thenoyl-trifluoracetone in the presence of a selective masking agent [35]. In high purity phosphorus, arsenic, and antimony, traces of lithium, sodium, potassium, and calcium were determined on the residue following sample treatment with chlorine and distillation of matrix chlorides in vacuo. The residue was dissolved in HCl and diluted with 90 vol.% acetone [36].

Organic solvents are also widely used in atomic absorption photometry. Enhancements of 3-fold are commonly achieved and are ascribed to increased efficiency of atomization.

INTERFERENCES

The amount of information on interferences is growing rapidly. Many of the interferences observed in atomic absorption photometry are identical with those which are well understood in emission flame photometry. One can only hope that excessive repetition of work will not be permitted by reviewers and editors.

Careful matching of matrix and acid concentration in standards with those prevailing in samples offers the best opportunity for achieving results essentially free from interferences in both absorption and emission flame photometry. Often, this method of evaluation entails relatively little effort, as has been demonstrated in the determination of numerous elements in the "Atomic Absorption Newsletter" distributed by Perkin-Elmer Corporation and which contains brief descriptions of work performed by Slavin, Sprague, and Manning.

High salt concentrations, 0.1M and above, often impair atomizer efficiency and, through the increased concentration of solid particles in the flame gases, may actually absorb a portion of the transmitted radiation in atomic absorption.

When interferences are too severe, their circumvention usually entails a prior removal step. Calcium, sodium, and potassium were determined in nickel—aluminum catalysts following removal of matrix metals as their oxinates [37]. However, Rains and coworkers [38] found that 10 vol.% of glycerol in 0.1M perchloric acid medium effectively eliminated many of the interferents in the determination of calcium.

Cesium in silicates has been separated on zirconium phosphate for enrichment, then the latter was dissolved in HF. Interferences were removed on strong, anion-exchange resins [39].

Herrmann and Lang [40] developed an expression for the determination of relative error in flame emission photometry,

$$\frac{C_a}{C_s} = 1 - \frac{E_a}{E_s} \exp \frac{n-1}{n} \tag{10}$$

where

$$E = k\,C^n \tag{11}$$

in which E_a and E_s are the emission readings of the analysis and of the standard solutions whose concentrations are C_a and C_s, and n is the slope of the double logarithmic plot of emission versus concentration. The relative error is larger the greater the departure of the term (E_a/E_s) from unity.

ELEMENTS

For a number of elements, the preferred lines for use in atomic absorption lie below 2100 A. For example, the selenium line at 1960 A gives a detection limit of 0.5 ppm with a hydrogen flame. For arsenic, good lines lie at 1937 and 1972 A. To utilize these lines, instruments equipped with fused silica optics and capable of purging with nitrogen or helium are mandatory.

The analytical possibilities of the line spectra of the rare earth elements and scandium in fuel-rich, acetylene–oxygen flames were evaluated by D'Silva and coworkers [41]. Nine members of this group (samarium, europium, dysprosium, lutetium, holmium, erbium, thulium, ytterbium, and scandium) possess strong lines with little or no serious interference from other members. However, in the case of lanthanum, the LaO band systems are emitted with greater intensity in stoichiometric flames, occur in regions relatively free of interference, and prove more useful.

A comparative study of the flame-photometric determination of serum calcium using hydrogen–oxygen flames has been reported by Rick and Herrmann [42]. The effects of aliphatic acids and their salts on the flame emission of calcium has been extensively studied by West [44].

Traces of cesium were concentrated from large volumes of water by batch adsorption on ammonium 12-molybdophosphate, followed by dissolution of the solids in sodium hydroxide. The cesium is then extracted with a hexone–cyclohexane solution of sodium tetraphenyl boron and the extract aspirated into the flame. The limit of detection, reported by Feldman and Rains [43], was $0.005 \mu g/ml$ of cesium of the solution burned.

REFERENCES

1. R. W. Reid and T. M. Sugden, Trans. Faraday Soc. 58: 213 (1962).
2. Yu. V. Zelyukova and N. S. Poluektov, Zhur. Anal. Khim. 18: 435 (1963).
3. N. S. Poluektov and R. A. Vitkun, Zhur. Anal. Khim. 18: 37 (1963).
4. K. Fuwa and B. L. Vallee, Anal. Chem. 35: 942 (1963).
5. P. T. Gilbert, Jr., Proc. Tenth Colloquium Spectroscopium Internationale. Spartan Books, Washington, D. C. (1963), p. 206.
6. J. A. Dean and J. E. Adkins, Jr., unpublished studies in: J. E. Adkins, Jr., Ph.D. dissertation, University of Tennessee, December (1963).
7. B. E. Buell, Anal. Chem. 34: 635 (1962).
8. R. Lockyer, in International Feigl Anniversary Symposium on Analytical Chemistry, Elsevier, Amsterdam (1963), p. 297.
9. C. I. Chakrabarti, G. R. Lyles; and F. B. Dowling, Anal. Chim. Acta. 29: 489 (1963).
10. H. C. Eshelman, J. A. Dean, O. Menis, and T. C. Rains, Anal. Chem. 31: 183 (1959).
11. J. B. Willis, Spectrochim. Acta. 16: 259 (1960).
12. W. Slavin, S. Sprague, and D. C. Manning, Perkin-Elmer Atomic Absorption Newsletter, No. 15, September (1963).
13. J. A. Dean, Flame Photometry, McGraw-Hill, New York (1960).
14. J. E. Allan, Spectrochim. Acta. 17: 467 (1961).
15. J. Ramirez-Muñoz, Rev. Inst. Hierro y del Acero. 17: No. 17, (1964).
16. F. J. Wallace, Analyst 88: 259 (1963).
17. J. Débras-Guedon and I. A. Voinovitch, Chem. Anal. (Warsaw) 5: 193 (1960).
18. J. A. Green and T. M. Sugden, Ninth Symposium on Combustion, Academic Press, New York (1963), p. 607.
19. P. T. Gilbert, Jr., Proc. Tenth Colloquium Spectroscopium Internationale, Spartan Books, Washington, D.C. (1963), pp. 203-211.
20. T. Hollander, P. J. Kalff, and C. T. J. Alkemade, J. Chem. Phys. 39: 2558 (1963).
21. R. N. Kniseley, A. P. D'Silva, and V. A. Fassel, Anal. Chem. 35: 910 (1963).
22. H. F. Loken, J. S. Teal, and E. Eisenberg, Anal. Chem. 35: 875 (1963).
23. P. T. Gilbert, Jr., Analyzer (Beckman Instruments, Inc.) 2: No. 4, 3 (October 1961).
24. C. D. West and D. N. Hume, Anal. Chem. 36.: 412 (1964).
25. H. Dunken, G. Pforr, W. Mikkeleit, and K. Geller, Z. Chem. 3: 196 (1963).
26. G. I. Nikilaev and V. B. Aleskovskii, Zhur. Anal. Khim. 18: 816 (1963).
27. R. Herrmann, and W. Lang, Arch. Eisenhüttenw. 33: 654 (1962).
28. J. D. Winefordner, C. T. Mansfield, and T. J. Vickers, Anal. Chem. 35: 1607 (1963).
29. J. D. Winefordner, C. T. Mansfield, and T. J. Vickers, Anal. Chem. 35: 1611 (1963).
30. E. Pungor, and M. Mahr, Talanta 10: 537 (1963).
31. H. von Brandenberger and H. Bader, Helv. Chim. Acta 47:353 (1964).
32. H. Bode and H. Fabian, Z. anal. Chem. 170: 387 (1959).

33. H. C. Eshelman and J. Armentor, in J. E. Forrette and E. Lanterman (eds.): Developments in Applied Spectroscopy, Vol. 3, Plenum Press, New York (1964) pp. 190–195.
34. W. E. Pilgrim and Ford, W. R., Anal. Chem. 35: 1735 (1963).
35. D. A. Johnson and P. F. Lott, Anal. Chem. 35: 1705 (1963).
36. K. H. Neeb, Z. anal. Chem. 200:278 (1963).
37. E. E. H. Pitt, Analyst 88: 399 (1963).
38. T. C. Rains, H. E. Zittel, and M. Ferguson, Talanta 10: 367 (1963).
39. O. Osterried, Z. anal. Chem. 199:260 (1964).
40. W. Lang and R. Herrmann, Z. anal. Chem. 199:161 (1964).
41. A. P. D'Silva, R. N. Kniseley, and V. A. Fassel, Anal. Chem. 36:532 (1964). 64).
42. W. Rick and R. Herrmann, Z. ges. exptl. Med. 136:307 (1963).
43. C. Feldman and T. C. Rains, Anal. Chem. 36: 405 (1964).
44. A. C. West, Anal. Chem. 36: 310 (1964).
45. J. L. Mason, Anal. Chem. 35: 874 (1963).

The Future of Atomic Absorption Spectroscopy

J. W. Robinson

Louisiana State University
Baton Rouge, Louisiana

Recent work in atomic absorption has been concerned with the consolidation of the process as a routine analytical tool. However, the sensitivities being reported are far from those within the potential of the procedure. The major weakness seems to be the atomization step. Recent advances in this and other areas are discussed.

THE PRESENT

In order to anticipate future developments in any field, it is imperative to know the state of the art at the present time. This is outlined below for the field of atomic absorption spectroscopy.

Advantages

The inherent advantages of this analytical procedure compared to other comparative procedures are as follows.

Sensitivity

The procedure is capable of high sensitivity. Many elements can be determined in the ppm and ppb range. Numerous metals can be determined at sensitivity levels considerably better than with other processes. It must be stated, however, that other metals are not very sensitive to detection by this method. However, on a broad basis, atomic absorption is capable of sensitivities at least as good as other similar analytical processes, such as X-ray fluorescence, emission spectrography, and activation analysis.

455

Interferences

The procedure is remarkably free from interferences from other components in a sample. This is inherent in the narrow spectral waveband absorbed by various atoms. Absorption by other elements at the same wavelength is virtually impossible. Indeed, the absorption wavelength is so narrow that even isotopes of the same element can be differentiated from each other.

The major source of interferences is from the cations present in the sample. Different cations form compounds with the metal being analyzed which require different amounts of energy for reduction. This influences the number of metal atoms produced in an atomizer and, therefore, the sensitivity. Consequently, different compounds of the same metal may show different sensitivities when measured in the same analytical system. This problem, of course, is well-known in flame photometry and emission spectrography. It is not so much of a problem in activation analysis and X-ray fluorescence.

Simplicity and Low Cost

The equipment and sample-handling procedures are still comparatively simple. This, in turn, reflects in the low cost of equipment ($6,000 — $8,000 complete). Compared to emission spectrography, X-ray fluorescence, and activation analysis, this is indeed cheap. Of course, flame photometry is somewhat cheaper, but generally is not as versatile an analytical process as atomic absorption.

Disadvantages

All elements in the periodic table cannot be determined by this process. The elements offering the most resistance are: (1) the nonmetals (these have absorption lines in the vacuum UV), and (2) metals which form stable compounds, e.g., oxides, in the atomizer. The latter group are particularly bad in flame atomizers. However, great strides have been made to eliminate this problem by using highly reducing flame conditions, e.g., fuel-rich oxy—acetylene flames.

Only liquid samples have been analyzed to date. When solid or gaseous samples are to be analyzed, conversion to a liquid sometimes necessitates considerable sample preparation.

Simultaneous analysis of several elements has not yet been achieved, and the interference from different cations (and some anions) can be troublesome, but perhaps the most serious obstruction to progress has been the persistent use of flames as atomizers. Although these are cheap, other forms of atomizers give better sensitivity and fewer interferences which may justify the extra expense.

Equipment

Routine-type commercial equipment is available from several different companies. All use hollow cathode sources and conventional detectors. The better instruments use modulated sources and tuned detectors. This eliminates interference from emitted radiation at the sample resonance frequency, which is at the identical frequency at which absorption takes place. Background emission can be greatly decreased by decreasing the spectral slit width visible to the detector, i.e., cutting-down the slits, but this is ineffective against emitted radiation from the element being determined.

Atomization is achieved with a flame on all commercial equipment. Although this is inexpensive, it is a barrier to progress in achieving better sensitivity and has interference from other components.

Other atomizers which have been demonstrated include flash heating, capacitor discharge lamps, L'vov furnaces, and arc discharges. Also, using flame atomizers, vycor tubes have been used to contain the flame products and atoms in the light path for extended periods of time. Although these modified atomizers have been demonstrated, they have not been made available commercially.

THE FUTURE

Predictions of the future can be based on the elimination of present limitations, and possible tailoring of equipment for emphasis in a particular use. We can expect therefore that equipment will be available which is less complicated than at present, e.g., equipment for the analysis of sodium, potassium, magnesium, and calcium in the important fields of agriculture and medicine. Simple instrumentation employing spectral filters and press-button source operation can be envisaged.

On the other hand, studies of the physical properties of atoms, such as population states and excitation mechanism, will require highly reproducible and stable equipment. This will no doubt be more complicated than that used today.

One of the most fruitful areas of improvement is that of atomization. It is true that flames are cheap and reliable, but, if we calculate the number of atoms of copper required to give 1% absorption, it turns out to be about 10^8 atoms. This calculation can be done using the formula

$$\int K_\nu d_\nu = 1\% \text{ absorption} = 0.01$$

$$\int K_\nu d_\nu = \left(\frac{\pi e^2}{mc}\right) Nf$$

where K is the adsorption coefficient at wavenumber ν, e is the charge on the electron, m is the mass of the electron, c is the speed of light, N is the number of atoms in the light path, and f is the oscillator strength. The best sensitivity limits reported for 1% absorption by copper is 10^{15} atoms. There is, therefore, a theoretical gain in sensitivity of 10^7. At least 10^4 should be attainable using suitable equipment. The key to this increased sensitivity is in better atomization and more stable sources and detectors.

Other limitations which are capable of improvement include simultaneous multielement determination—a very important aspect in most routine laboratories—and the analysis of gases and solid samples. This may be achieved by using suitable atomizers, such as electrical discharge systems or lasers.

CONCLUSIONS AND SUMMARY

The improvements listed are highly desirable from many aspects. However, a plea should be made to the manufacturers to avoid undue refinement and to keep the equipment adequate for the job in mind, but not, as has happened in other forms of instrumentation, to increase the capability for all users no matter what their needs. This can be achieved by making various models available to the public.

It is within our capability to make equipment which can (1) analyze several elements simultaneously, (2) handle solid and gaseous samples, (3) attain higher sensitivities than now available, and to make (4) simple, push-button type equipment for highly routine analysis, and (5) stable multielement hollow cathodes for the simultaneous analysis of several metals.

A Plasma-Arc Technique for the Spectrochemical Determination of Titanium and Zirconium in Molybdenum

James H. Muntz

Air Force Materials Laboratory
Wright-Patterson Air Force Base, Ohio

A plasma-arc technique has been developed for the spectrochemical determination of titanium and zirconium in molybdenum alloys. Titanium is determined in the 0.25—1.5% range and zirconium in the 0.05—0.4% range with an overall precision of about ±5% and an accuracy which approaches the precision. Sensitivity for the zirconium determination is achieved by using high gas flow and high amperage. Molybdenum is used as the internal standard and a background correction is necessary to produce a straight—line analytical curve for zirconium over the concentration range of interest.

INTRODUCTION

Research alloys of molybdenum containing titanium and zirconium have been prepared. Ingots of these materials have been sampled to check the homogeneity of the melt. Since a study of this kind could result in a large number of samples, a rapid and simple method for the analysis of titanium and zirconium was desired. Standard wet chemical methods require a separation step prior to the determination of titanium by the H_2O_2 colorimetric technique and zirconium by the phosphate gravimetric technique. Current spectrographic techniques require conversion to the oxide, addition of a buffer, and arc excitation, which is sensitive but not precise. A relatively new· spectrographic technique is the plasma arc, developed in this country by Margoshes and Scribner [2] and later modified by Owen [4]. This technique permits the use of solutions with the advantage of ease of sample and standard

461

preparation and the elimination of metallurgical background. The plasma is reported to have greater reproducibility and sensitivity than other solution techniques. Some of the uses of the plasma for spectrochemical analysis have been described by Roza and Stone [5], Landon [1], Serin and Ashton [6], and Mitteldorf and Landon [3]. The technique described here illustrates the use of the plasma arc for the determination of titanium and zirconium in molybdenum alloys and indicates the accuracy and precision that can be expected.

APPARATUS

Apparatus includes a Bausch and Lomb dual-grating spectrograph with illuminator external optics, National Spectrographic Laboratories Model "D" Plasma Arc (similar equipment is manufactured by Spex Industries), National Spectrographic Laboratories comparator—densitometer, Jarrell-Ash Co. Varisource DC arc portion, and Applied Research Laboratories photoprocessing equipment.

PROCEDURE

Sample Preparation

Dissolve 100 mg molybdenum alloy chips in 5 ml of 30% H_2O_2. Add approximately 10 ml distilled water and heat to expel the excess H_2O_2. Cool, and add 1 ml H_2SO_4. Transfer to a 50-ml volumetric flask and dilute to volume.

Preparation of Standards

Dissolve 100-mg portions of unalloyed molybdenum chips in 5 ml of 30% H_2O_2. Add approximately 10 ml distilled water and heat to expel excess H_2O_2. Add aliquots of standard solutions of titanium and zirconium to prepare standards covering the ranges 0.25—1.5% Ti and 0.05—0.4% Zr. Transfer to 50-ml volumetric flasks and dilute to volume.

Excitation and Exposure

Produce and record the spectra according to the following excitation conditions:

Plasma arc:
Current . . . . 230 V; 22 A
Tangential gas flow. . . . 55 liters/min helium
Siphon gas flow. . . . . . . 2 liters/min argon
Aspirating rate. . . . . . . 0.5 ml/min
Atomizer. . . . . Beckman, medium bore
Preburn . . . . 5 sec
Exposure. . . . . 45 sec
Electrode
   upper. . . . . 7-mm orifice graphite
   lower. . . . . 4-mm orifice graphite
   transfer . . . . $3/32$-in. diameter tungsten

Spectrograph:
Grating . . . . 30,000 lines/in.
Emulsion. . . . . SA-1 spectrographic plate
Filter . . . . 100-10% step filter
Spectral region. . . . . : . 2600-3600 A first order
Slit width. . . . . 0.02 mm
Slit length . . . . 2.5 mm

Exposures

Make duplicate exposures of the samples. Single exposures of two or three standards are sufficient after the analytical curves have been established.

Photographic Processing

Process the plate using the following conditions:

Develop . . . . 3 min in D-19 at 68°F
Shortstop . . . . . . 20 sec
Fix . . . . Kodak Rapid Fixer for 2 min
Wash. . . . . 5 min with running water
Dry. . . . . Remove excess water with chamois
                 and dry under hot air for 2 min

Photometry and Calibration

Photometer the 3367.97-A molybdenum line and the 3372.80-A titanium line on the 10% step. Photometer the 3438.23-A zirconium line and adjacent background on the 100% step. Convert the transmittance values to intensities using an

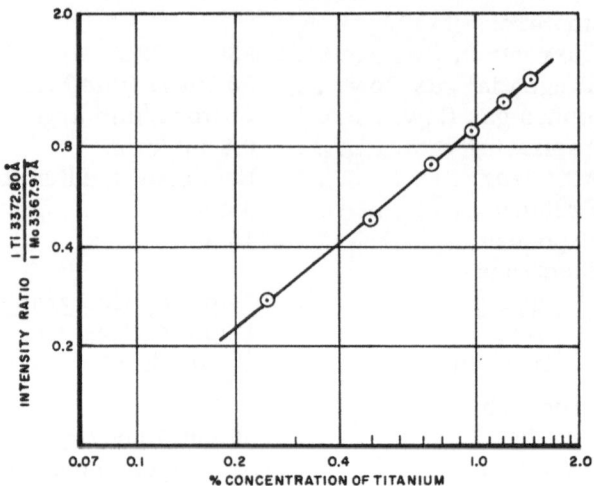

Fig. 1. Analytical curve for titanium in molybdenum.

emulsion calibration curve derived from the two-step method.
Calculate the intensity (*I*) ratios of *I* Ti/*I* Mo and *I* Zr—*I* back-
ground/*I* Mo. Prepare the analytical curves by plotting
intensity ratios versus concentrations for the standards.
Obtain the concentration of the titanium and zirconium in the
sample by relating intensity ratios to the analytical curves.
Figures 1 and 2 show typical analytical curves for titanium
and zirconium, respectively.

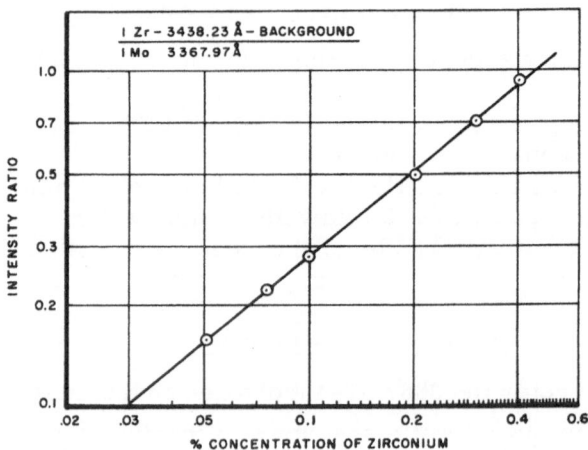

Fig. 2. Analytical curve for zirconium in molybdenum.

TABLE I
Reproducibility

Same Solution	% Ti	% Zr
Sample A	0.47	0.085
	0.46	0.09
	0.45	0.085
	0.46	0.08
	0.45	0.085
Average	0.46	0.085
Relative Standard Deviation	± 2%	± 4.3%
Sample B	1.2	0.305
	1.2	0.3
	1.2	0.29
	1.2	0.305
	1.2	0.29
Average	1.2	0.3
Relative Standard Deviation	± 0%	± 2.7%

TABLE II
Reproducibility

Run	Percent titanium	Percent zirconium
1 a	0.52	0.094
b	0.59	0.105
2 a	0.53	0.098
b	0.55	0.100
3 a	0.57	0.105
b	0.59	0.105
4 a	0.55	0.098
b	0.55	0.105
Average	0.56	0.101
Relative Standard Deviation	± 4.7%	± 4.4%

DISCUSSION

One of the main difficulties involved in the plasma-arc technique is the clogging of the atomizer. This fact dictated the selection of the concentration of the molybdenum solution to be used. A more concentrated solution would give better sensitivity, but would pose more clogging problems. A more dilute solution would decrease the clogging problem, but would require excessive exposure times. Excitation parameters were determined experimentally to provide the sensitivity needed for the determination of zirconium in the selected molybdenum solution. It was found that, with these excitation conditions, a background correction was necessary to provide a straight-line analytical curve for zirconium over the concentration range of interest. There are some fluctuations in the intensity of the spectra, which is probably due to slight differences in the aspirating rate. This is compensated for by using molybdenum as the internal standard.

An attempt was made to use the vacuum cup technique with the same solutions for comparison, but neither titanium nor zirconium were detected at the concentration levels used.

Reproducibility is indicated by the results obtained from five determinations on two different sample solutions and is listed in Table I. The relative standard deviation increases as the concentration decreases. The extremely good precision for titanium on sample B probably can be explained on the basis of the equality of the transmittance values for molybdenum and titanium at this concentration level. To study the variations to be expected in sample preparation, four different portions of the same sample were prepared and analyzed in duplicate. Results are given in Table II. The results indicate that the

TABLE III

Recovery

	Zirconium, ppm		Titanium, %	
Determined in sample	890	890	0.46	0.46
Added	500	1000	0.3	0.5
Expected	1390	1890	0.76	0.96
Found	1350	1925	0.74	0.97
Percent	97	102	97	101

TABLE IV

Comparison of Spectrochemical Analysis with Wet-Chemical Analysis

Sample	Zirconium, %		Titanium, %	
	Spectrochemical	Wet-chemical	Spectrochemical	Wet-chemical
1	0.09	0.11	0.46	0.53
2	0.26	0.25	1.02	1.07
3	0.10	0.12	0.56	0.58

overall reproducibility of the method, including sample preparation, should be within 5%. From data obtained for the standards, variations up to ±15% can be expected for both elements at the lowest level of the concentration range. Precision could be improved at this lower concentration level by the selection of another molybdenum internal standard line whose intensity is closer to the intensity of the analytical lines of titanium and zirconium. The recovery of added titanium and zirconium is indicated in Table III and shows that recovery is within the precision of the method. The accuracy of the method is shown in Table IV by the comparison of spectrochemical and wet-chemical values for three different samples. Five out of six of the values for the spectrochemical method were lower than the wet-chemical values; therefore, a systematic error could be involved.

REFERENCES

1. D. O. Landon, Determination of trace metals in petroleum products using plasma jet, Paper 176, Pittsburgh Conference on Analytical Chemistry and Applied Spectroscopy (1963).
2. M. Margoshes and B. F. Scribner, The plasma jet as a spectroscopic source, Spectrochim. Acta 15: 138 (1959).
3. A. J. Mittelaorf and D. O. Landon, The stabilized plasma jet fluid analyzer, Spex Speaker 8: 1 (1963).
4. L. E. Owen, Stable plasma jet for excitation of solutions, Appl. Spectroscopy 15: 150 (1961).
5. J. T. Roza and J. Stone, Plasma-arc analysis for wear elements in lubricating oil, Paper 128, Pittsburgh Conference on Analytical Chemistry and Applied Spectroscopy (1963).
6. P. A. Serin and K. H. Ashton, Some studies on the use of the plasma jet for aqueous solutions, Paper 7, Ottawa Symposium on Applied Spectroscopy (1963).

The Performance of the Interrupted Discharge in Argon

H. T. Dryer and F. Borile

Applied Research Laboratories
Detroit, Michigan

The effects of the argon discharge upon the performance of the Multi-source discharge are of vital importance in the applications of this analytical technique to control programs. The usual guideposts of the air discharges are not adequate for the argon discharges and require that new or additional studies be made to evaluate the performance of this method.

A study has been made to determine the effects of source parameters and components upon spectral lines and the performance of the analytical method.

The results of this study and the analytical data for several materials will be presented.

Although many workers have used an inert gas atmosphere to improve the performance of arc discharges, it was not until the introduction of the Quantovac, about ten years ago, that practical use was made of interrupted discharges in inert atmospheres. This innovation was required to provide a transparent path for the wavelengths below 2000 Å. The other alternative was the vacuum spark.

The Quantovac was designed primarily for the determination of the elements which have sensitive lines in vacuum ultraviolet, notably, carbon, phosphorus, sulfur, arsenic, and selenium. Unfortunately, the usual guideposts of the air discharges were not adequate for the interrupted discharge in argon and required changes in analytical procedures. A number of "burns" are illustrated in Fig. 1 and indicate the effects of the discharge conditions.

The performance of the Multisource discharge in argon provides many advantages to the analyst. Many elements can now be determined routinely which previously could not be

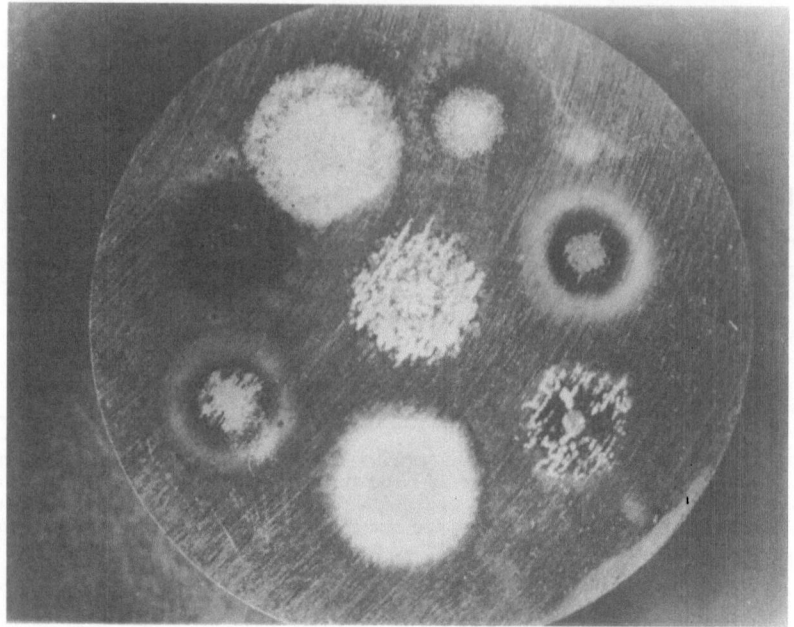

Fig. 1. Typical "burn" pattern as a function of source parameters.

determined except by very specialized techniques. The determination of carbon in ferrous alloys, illustrated in Fig. 2, for 18-8 stainless, is an example of this improvement.

As described several years ago at this meeting [1], this analytical method also provides improved precision and accuracy over that obtained with air discharges for many of the usual elements.

The determination of molybdenum and tungsten at high levels in tool steels has for some time presented a difficult task for spectrochemical analyses. As described by Goldblatt [2], in 1955, for the analysis of alloyed cast iron, and by Hasler [3], for the analysis of tool steels, the oxidation products deposited upon the sample surface produced inconsistent and unreliable results. The Multisource discharge in argon eliminated these effects and greatly improved the performance of the determination of tungsten and molybdenum (Fig. 3).

As described in 1962 at this meeting [4], the analysis of pure

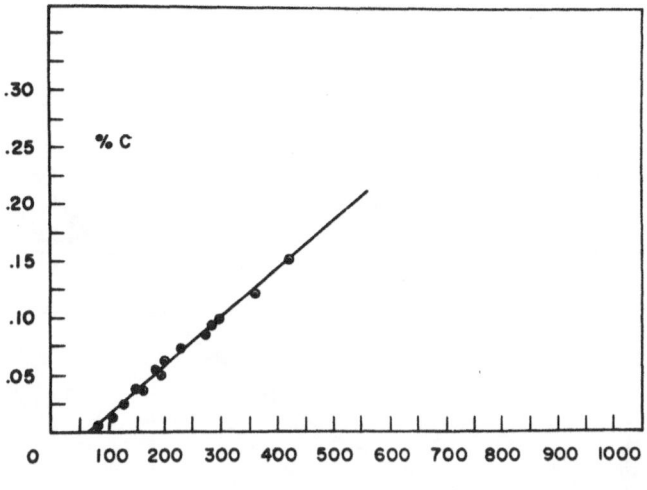

COMPUTER READINGS

Fig. 2. Carbon in 18-8 stainless. Typical performance of the interrupted discharge in argon.

copper samples presents a special problem for the conventional spectrochemical laboratory. Because of the drastic changes in electrical and thermal conductivity, the excitation and control of pure copper samples are difficult and in many instances meaningless. Figures 4 and 5 demonstrate the performance of this discharge for the determination of residual phosphorus and for alloying amounts of nickel.

It is readily apparent that this analytical condition, i.e., the Multisource discharge in argon, provides many improvements over air discharges for the materials described. Many additional elements can be determined, precision and accuracy have been improved, and both trace and alloy concentrations can be determined with a single analytical condition.

Because of these improvements, it was felt that this method might be applied to numerous other materials with similar results. Previous experience with the ferrous alloys and copper base alloys indicated that the transition from the air discharges to the interrupted discharge in argon would require considerable application studies to provide the desired results.

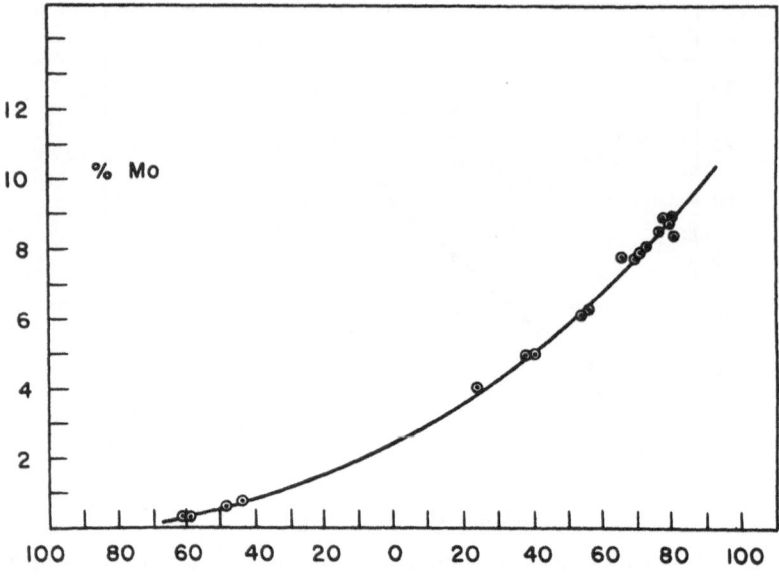

Fig. 3. Molybdenum in tool steel.

Aluminum alloys, because of various interelement effects and wide ranges of alloying elements, require numerous alloy curves in order to provide suitable analytical data. A reduction in the number of alloy curves required would facilitate the various analytical programs of the aluminum industry. Our experience has shown that the Quantovac discharge minimizes the usual interelement effect, suggesting that little difficulty would be encountered with aluminum alloys.

One of the first analytical curves obtained for aluminum alloys is shown in Fig. 6, although not typical of the other curves, it indicates some of the difficulties encountered in the transition from air to argon discharges. Subsequent investigations by photographic methods indicated that this analytical line reverses badly at low concentration levels under these conditions. Another magnesium line was selected and the analytical results are shown in Fig. 7, illustrating a minimum interelement effect.

At the 1964 Pittsburgh Conference, C. Matocha [5] of Alcoa Research described the analysis of aluminum alloys using the

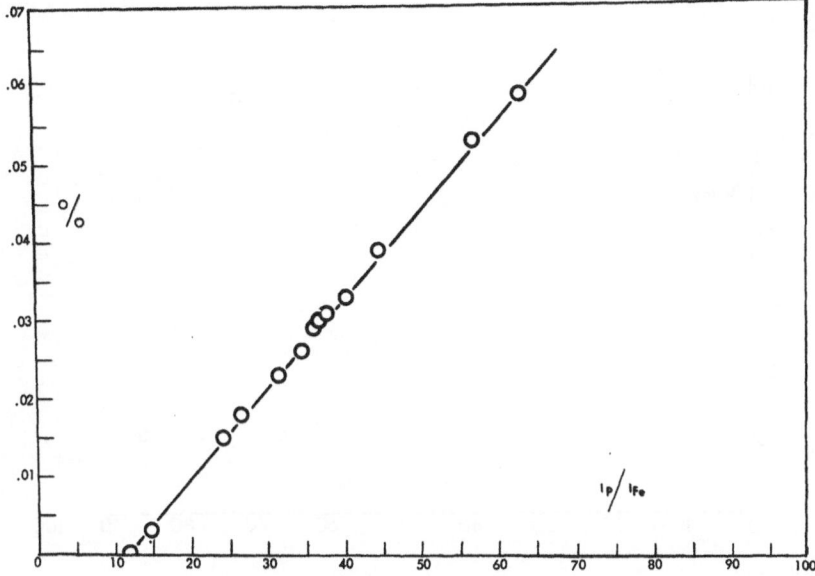

Fig. 4. Phosphorus in copper base alloys.

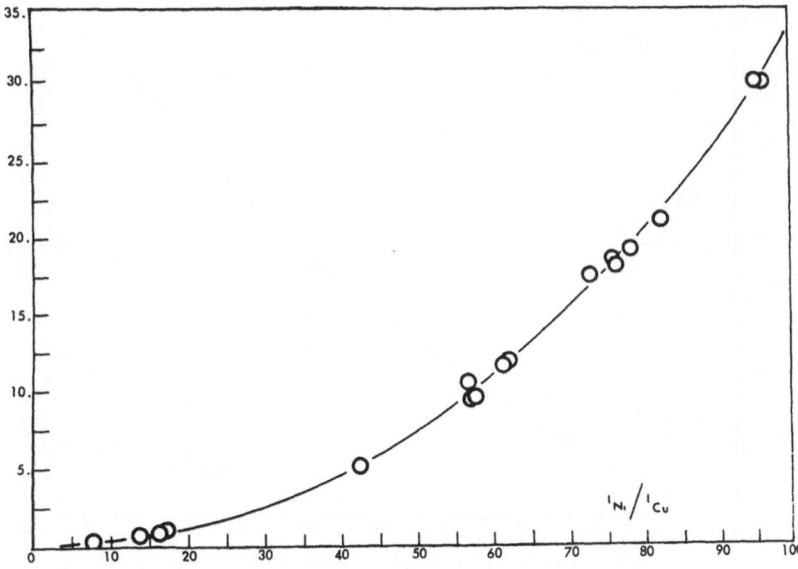

Fig. 5. Nickel in copper base alloys.

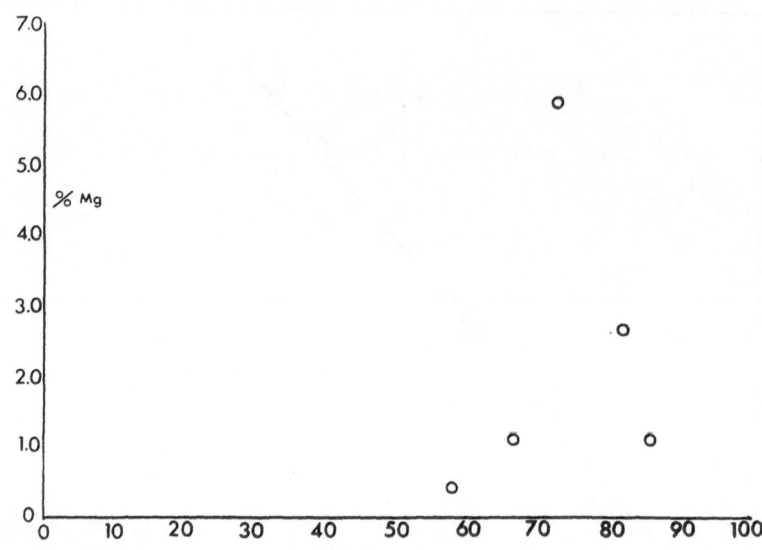

Fig. 6. Preliminary curve for magnesium in aluminium alloy illustrating effect of line reversal in argon.

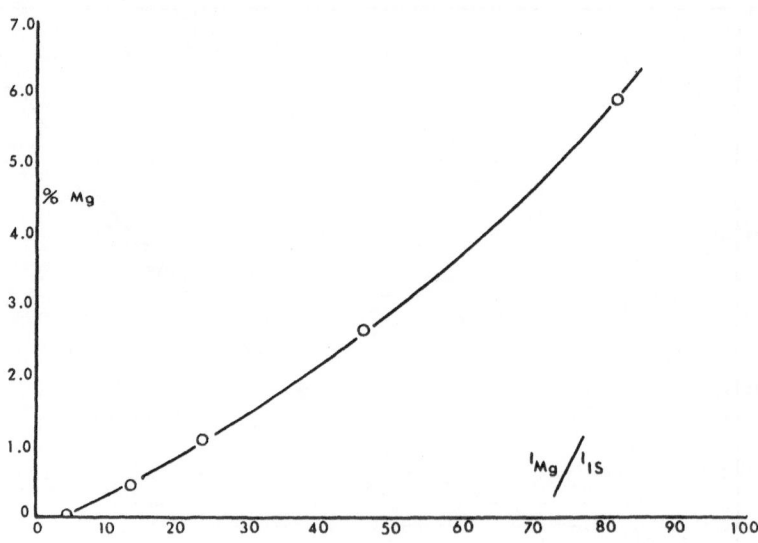

Fig. 7. Final curve for magnesium in aluminum alloy.

high-voltage spark in nitrogen and demonstrated that these analytical conditions also minimized interelement effects.

Arrak [6] of Grumman Research has described the effects of various discharges in nitrogen and also indicated a minimal interelement effect.

Unfortunately, the analytical curves of Matocha and Arrak required the use of ratio percentages which, although relatively easy to use with standard samples, are somewhat difficult to use with unknown samples. In order to use these systems of curves, one must analyze the material for all elements and make the necessary calculations, or use a computer console to facilitate these calculations.

A method of eliminating these calculations has been in limited use for some time. By use of a system of combined internal standards such that the intensity of an analytical line is ratioed to the combined intensities of suitable internal standard lines, the effects of a changing matrix and apparent interelement effects are eliminated.

A typical plot of I_{Zn}/I_{Al} versus concentration is shown in Fig. 8 and indicates the difficulties that generally require alloy curves. These data were obtained using the Quantovac discharge

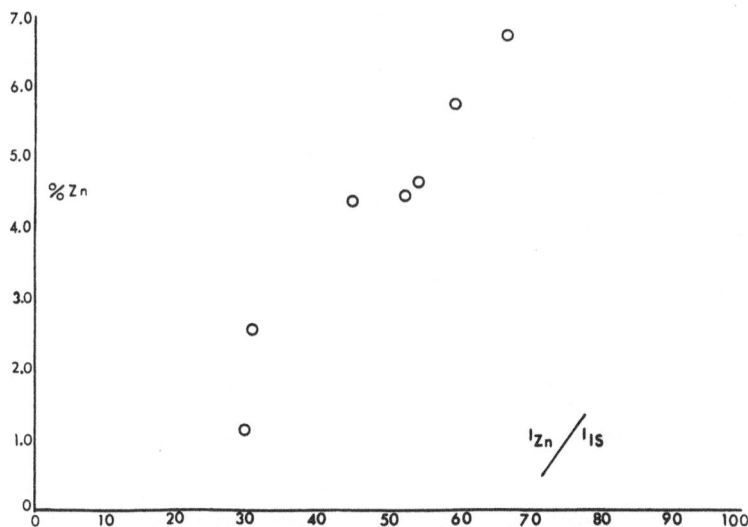

Fig. 8. Zinc in aluminum alloy—7000 series—single internal standard uncorrected.

and, as demonstrated by Matocha, the plot of ratio percentages would provide a usable curve. The same samples were then controlled by the combined internal standard lines using aluminum, zinc, and magnesium as the control elements. A typical plot is shown in Fig. 9 for zinc with no calculation applied.

This analytical procedure, i.e., Quantovac discharge and combined internal standards, has been applied successfully to a number of other materials. Figure 10 illustrates this method for zinc spelter and zinc die cast samples. In Fig. 11, the analytical curve for molybdenum in a high-temperature alloy is shown to illustrate the performance with an alloy which has no real matrix.

Photographic methods were used in a portion of this study to provide the required information for line selection. As shown in Figs. 12 and 13, the Quantovac discharge cannot be compared to typical air discharges because of the drastic changes in line intensities, in the intensity relations of various lines, and in reversal characteristics.

Additional studies were made to determine the effects of source parameters and components upon spectral lines and the

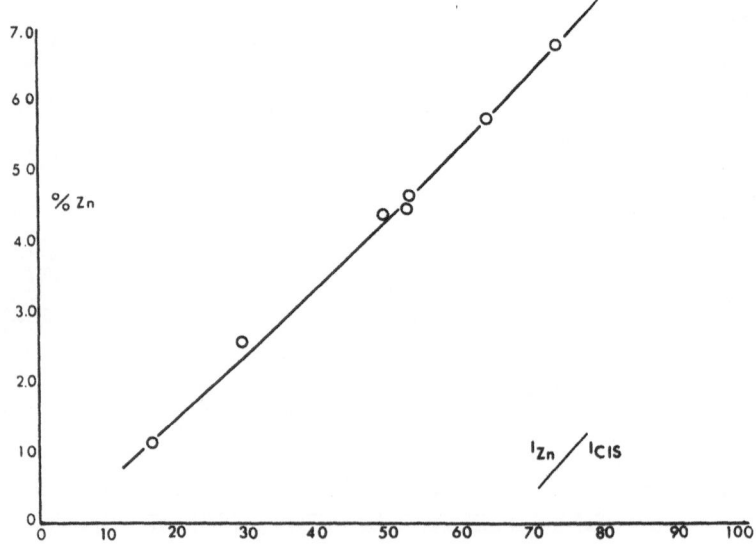

Fig. 9. Zinc in aluminum alloy—7000 series—combined internal standard lines.

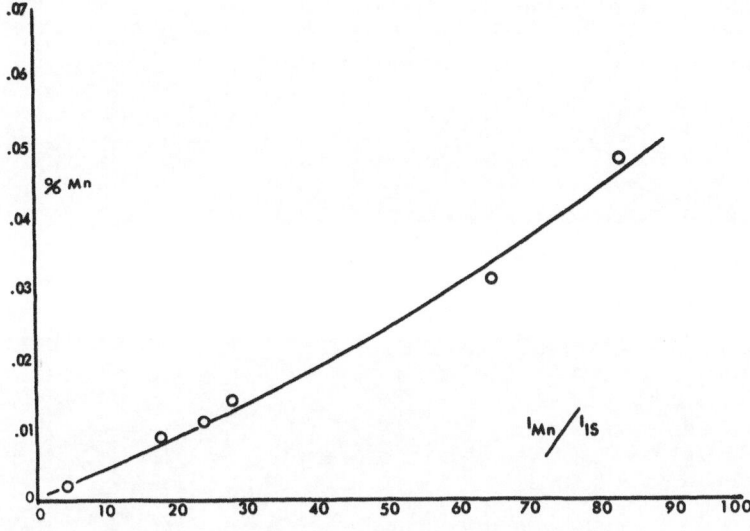

Fig. 10. Manganese in zinc base.

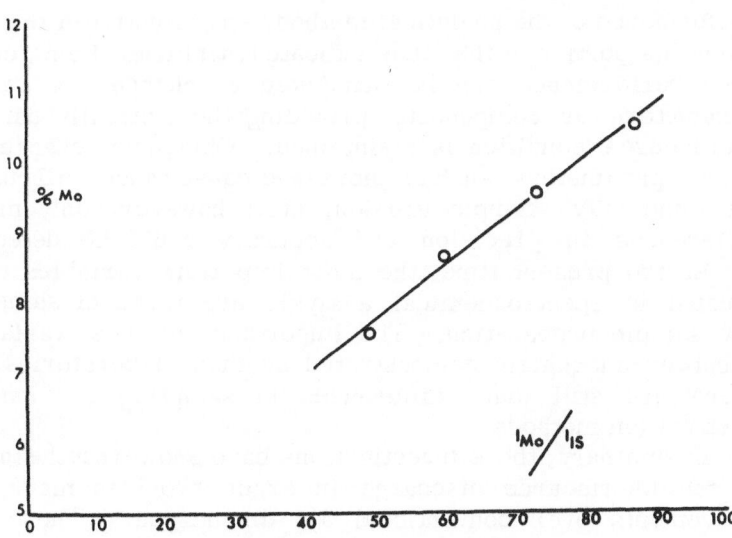

Fig. 11. Molybdenum in high-temperature alloys.

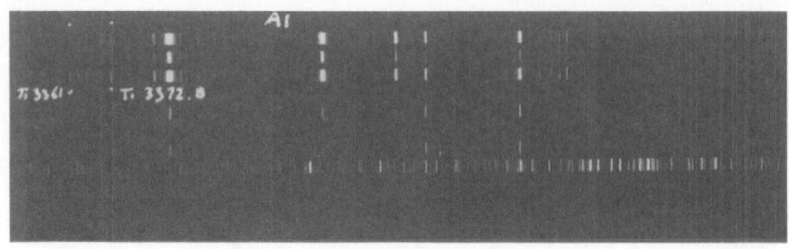

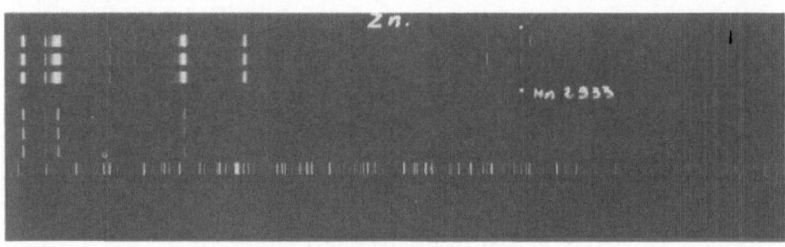

Fig. 12. Spectra of aluminum alloys and zinc base alloys—comparison of argon and air discharges.

performance of the analytical method. An evaluation of the data from this portion of the study indicated that little, if any, effect upon performance can be attributed to changes in source parameters or components, providing the critically damped Multisource condition is maintained. Obviously, changes in source parameters, such as increased capacitance, will change total intensity, sample erosion, etc.; however, only minor differences in precision and accuracy could be detected.

At the present time, the most important variables to be studied in spectrochemical analysis are those of sampling and sample preparation. The importance of these variables has been adequately demonstrated by many laboratories, but there are still many differences in sampling and sample preparation methods.

In summary, these investigations have shown that the inter-rupted Multisource discharge in argon provides many improvements over conventional air discharges. These improvements are (1) improved precision and accuracy, (2) minimal interelement effects, (3) determination of trace and

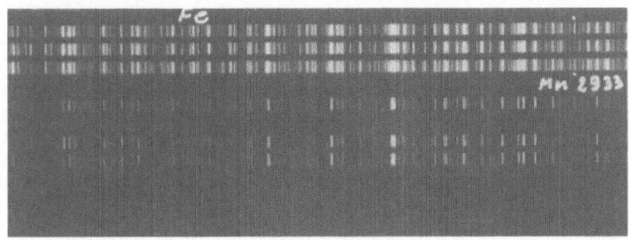

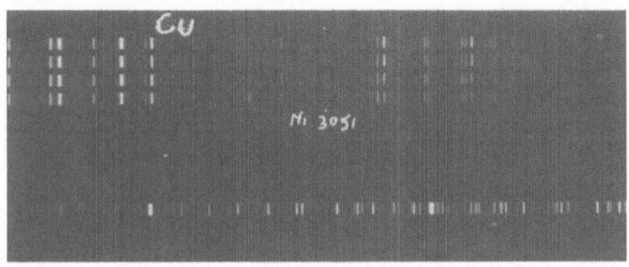

Fig. 13. Spectra of copper and ferrous alloys—comparison of argon and air discharges.

alloying elements by one analytical procedure, (4) routine determination of elements such as copper, phosphorus, sulfur, and selenium, and (5) reduction in number of curves by use of the combined internal standard method. Hopefully, comparable improvements can be made in sampling and sample preparation techniques to utilize fully this method of analysis.

REFERENCES

1. H. T. Dryer and B. R. Boyd, Developments in Applied Spectroscopy, Vol. 2, Plenum Press, New York (1963).
2. R. Bartel and A. Goldblatt, Spectrochim. Acta 9: 227 (1957).
3. M. F. Hasler, "The Direct Reading Analysis of Low Alloy, Tool and Stainless Steel with the Quantometer," Stockholm Lecture (September, 1949).
4. H. T. Dryer and B. R. Boyd, Developments in Applied Spectroscopy, Vol. 2, Plenum Press, New York (1963).
5. C. Matocha, J. Petit, and W. H. Tingle, "Use of Nitrogen Atmosphere to Reduce Matrix Effects in Analysis of Aluminum Alloys Using Point-to-Plane Excitation," Pittsburgh Conference (1964).
6. A. Arrak, "A Study of the Inert Atmosphere Spark in Point-to-Plane Analysis of Steel and Related Alloys," International Conference on Spectroscopy, College Park, Maryland (1962).

Spark Excitation in Inert Atmospheres

Arno Arrak

Grumman Aircraft Engineering Corporation
Bethpage, Long Island, New York

Within the last few years, spark excitation in atmospheres other than air has proven its value in spectrochemical analysis. The beneficial influence of an inert atmosphere is undoubtedly due in part to the elimination of surface oxidation reactions which often strongly influence spectral intensities. Beyond this obvious chemical factor, there are other, more subtle influences due to the presence of the gas which can strongly influence the nature of the spectrum in different atmospheres. Thus, the spectrum of a given sample sparked in helium and in argon will behave quite differently in regard to total and relative line intensities, line form and width, relative strength of ion lines, line to background ratio, tendency of certain lines for self-reversal, reproducibility of intensity ratios in replicate exposures, etc. These observable effects depend upon the ionization and excitation potential of the gas, as well as upon the atomic weight and the collision cross section of the gas atoms with sample vapor. Since the sample vapor enters the spark gap in the form of supersonic vapor jets from the cathode, it is quite evident that the slowing down of these jets by collisions with gas atoms plays a major role in the excitation of spark spectra in different inert atmospheres. An elementary theory of this process will be developed and some of its predictions will be compared with experiments.

In recent years, spectrochemical spark excitation techniques using an inert gas to shield the spark gap have become increasingly important. In this paper, we shall review the reasons for this trend and discuss some of the factors influencing spark excitation that are connected with the choice of the atmosphere.

When the inert atmosphere spark is used the primary function of the shielding gas is to exclude atmospheric oxygen from the spark gap. If oxygen is not excluded from the gap it reacts with the sample surface during the sparking process, often selectively. This selective oxidation is responsible for many of the so-called "matrix effects" which were

Fig. 1. Petrey stand with continuous-flow gas injection attachment for sparking in an inert atmosphere.

formerly attributed to a variety of other causes. Thus, if the spark gap is shielded with nitrogen or argon, it is possible to use common working curves to analyze all steels as well as many high-temperature alloys, using just one universal method of analysis [1]. Such a method was first proposed in a paper given at the Tenth International Colloquium on Spectroscopy, College Park, Maryland, in June, 1962 [2]. To recapitulate briefly, we surrounded the spark gap with the shielding gas by using the arrangement shown in Fig. 1. The sample is placed flush with the top of the chamber and the gas flowing through the system escapes continuously through an opening in the direction of the slit. An incidental byproduct of shielding the gap is a carbon deposit formed on the sample surface during sparking. The appearance of such "black burns," as they are called, depends upon the atmosphere and upon the effectiveness of shielding (Fig. 2). Observation of the

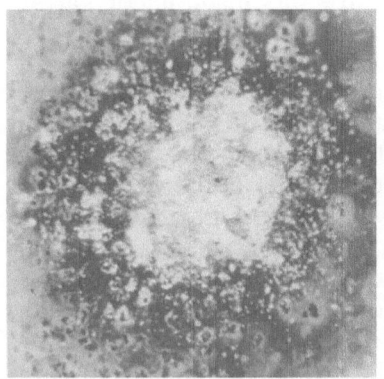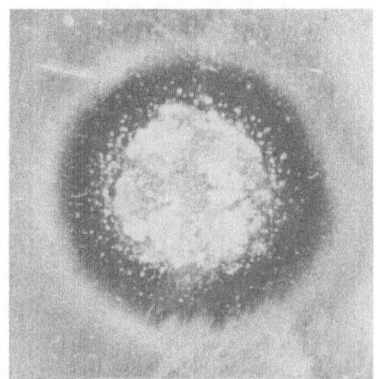

Fig. 2. Appearance of black burns in nitrogen (left) and in argon or helium (right).

TABLE I

Standard	Composition range and type
⊙ BAS SS1 — SS8 ○ NBS 805A, 816A, 1161 — 1164, 1168	Low alloy steels: Fe 94-97, balance Cr, Ni, Mo, V, Mn, Si, Cu, Al, etc.
◊ NBS 836 — 841	Tool steels: Fe 69-79, Cr 2-8, Mo 0.8-8, V 0.6-3, W 2-18, Co 3-12, Mn, Si, etc., balance.
▽ NBS 845 — 850	Stainless steels: Fe 62-83, Cr 3-24, Ni 0.3-25, Mn, Si, Mo, Ti, Nb, W, etc.
△ NBS 1184 — 1187	Stainless and high-temperature alloys: Fe 27-66, Cr 16-21, Ni 9-24, Mo 1.5-6, W 1.4-2.4, also Co, Nb, Mn, Si, etc.
☐ NBS 1188, 1189, 1191, 1192	Nickel base alloys: Fe 1.4-6.6, Cr 15-20, Ni 55-72, Mo 4-7, Ti 2-3, Al 0.7-1.5, Co 11-13, Mn, Si, etc.

deposit after an exposure has been made makes it possible
to judge whether a particular exposure was made under satisfactory shielding conditions or not. When dealing with as wide
a variety of alloys as were included in our study (Table I),
it is somewhat difficult to find a suitable internal standard
element common to all. One way out of this difficulty is to
plot concentration ratios instead of concentrations against
intensity ratios of lines. This procedure is of course permis-
sible only when the "working curve" for the internal standard
line (i.e., plot of relative intensity vs. concentration of internal
standard element, for constant exposure time) is linear and
has the same slope as the working curves for the other elements
to be determined. In our work (which was done photographi-
cally), we had to use internal standard lines that deviated from
this behavior (Fig. 3). To compensate for this effect, an
empirical variable internal standard principle was introduced
that makes allowance for the actual behavior of the internal
standard line used [1,2]. Figure 4 shows typical working curves
obtained with this method. Referring to the key in Table I,
we see that points from low alloy steels, tool steels, and
stainless and high temperature alloys all fit the same curves

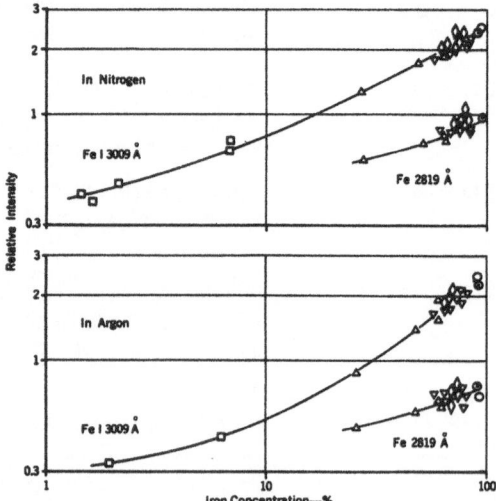

Fig. 3. Relative intensity of Fe I 3009 A and Fe 2819 A as a function of iron concentration in the sample when exposure time is constant. Such "constant time"
working curves cannot be constructed when the same samples are sparked in air.

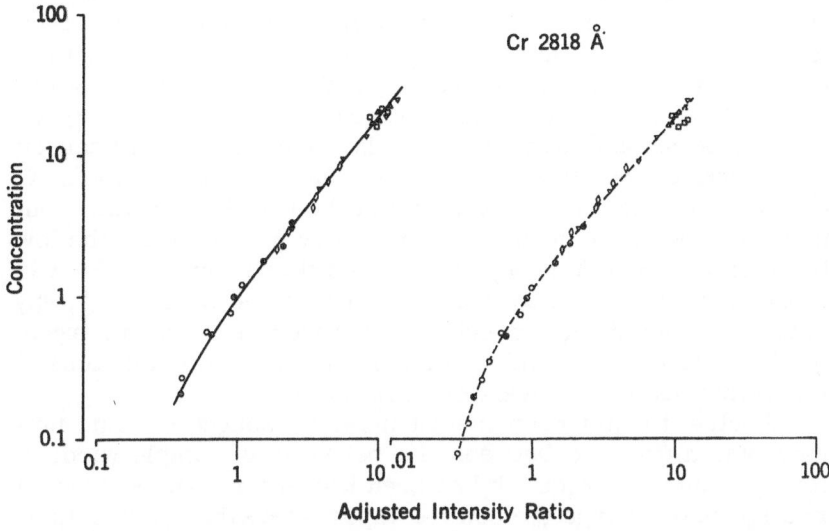

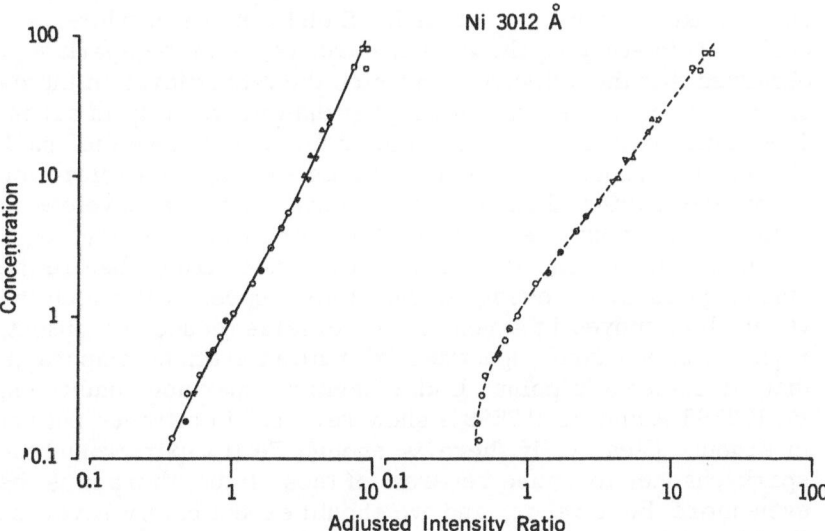

Fig. 4. Working curves for Cr 2818 A and Ni 3012 A obtained by the author's method. [1]. The curve in nitrogen is on the left and the curve in argon on the right. The abscissa represents an adjusted intensity ratio which is corrected for the shape of the "constant time" working curve for the internal standard lines (shown in Fig. 3).

and are intermingled on the graphs. For each element, the curve obtained in nitrogen is on the left and the curve in argon is on the right. Inspection of these graphs shows that the curves for Cr 2818 A are nearly parallel in both gases, while the curve for Ni 3012 A in nitrogen changes slope at higher concentrations and turns up. This behavior is evidently caused by selective self-absorption of Ni 3012 A in nitrogen. In checking the excitation potentials of the two lines, it turns out that the low level of Ni 3012 A is only 0.42 V, while the low level of Cr 2818 A is 4.1 V. This explains why Ni 3012 A is more easily self-absorbed, but still fails to explain why reversal should occur only in nitrogen and not in argon. Similar behavior is typical of other lines from transitions to low levels, even low levels of ionized atoms.

A clue to this phenomenon must be sought in some fundamental difference between the behavior of sample vapor in nitrogen and in argon. It has been known for many years that in a spark discharge the sample vapor enters the spark column in the form of high-speed vapor jets liberated from the cathode at the beginning of each half cycle [3]. After the vapor is liberated, it moves through the gap with a velocity that may approach several kilometers per second. Steinhaus, Crosswhite, and Dieke [4], in studying the motion of iron vapor in the spark gap, observed that the velocity with which the vapor moves in nitrogen may be up to seven times higher than its velocity in argon. The rapid expansion of the vapor in nitrogen allows the spark channel to become surrounded with an envelope of cooler iron vapor before the discharge has ended, and this envelope is then able to cause reversal of lines originating in the spark channel. In argon, this envelope is not formed before the discharge is over, owing to the slower speed with which the vapor cloud moves in argon, and reversal is reduced or absent. Figure 5 shows some spectra taken with a Hartmann diaphragm that illustrate this point. Under identical sparking conditions, Fe II 2598 A and Fe II 2599 A show reversal in nitrogen but not in argon. Clearly, if there is enough Fe II vapor around the spark channel to cause reversal of these lines, there must be even more Fe I vapor, and we should expect easily reversed Fe I lines, such as Fe I 2483 A and Fe I 2488 A, to be even more strongly reversed. Looking at these lines, we see that they are reversed in the arc spectrum as they should be and entirely unreversed in the argon spark exposures. In the nitrogen spark

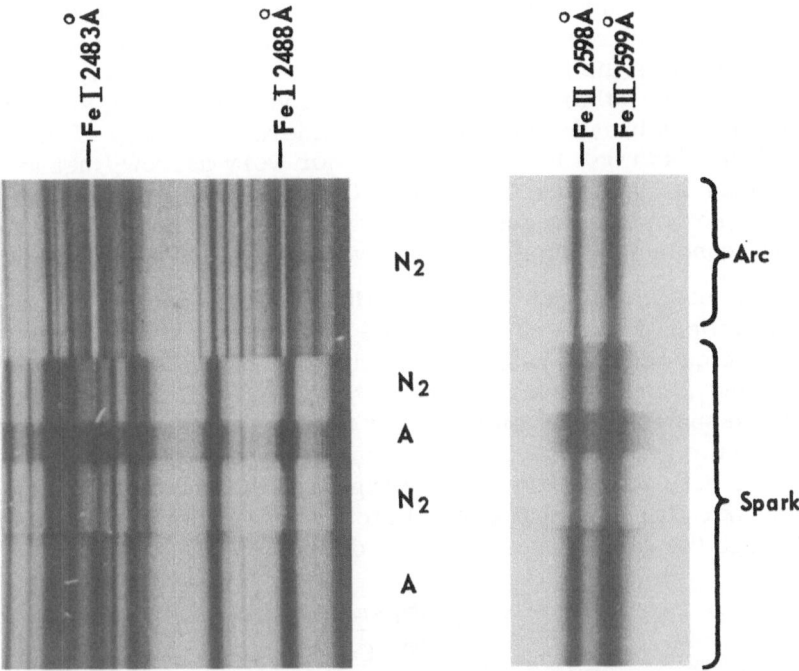

Fig. 5. Comparison of iron spark spectra in nitrogen and argon with the iron arc spectrum. All spectra taken with the Hartmann diaphragm except for the middle spectrum, which is a double exposure obtained by pushing in the fishtail after the three Hartmann steps were used up.

spectrum, they seem to be, at first glance, only weakened and show an odd shift in wavelength. A close examination indicates, however, that they are indeed self-reversed in nitrogen, but the absorption has shifted to lower wavelengths. Apparently we are dealing with a Doppler shift, indicating that the absorbing vapor was moving toward the slit with a velocity of the order of 3 km/sec. If we had not approached this spectrum with the concept of moving high-speed vapor clouds in the spark gap in mind, the apparent shift in wavelength could easily have been interpreted as a Stark shift. It is interesting to speculate on how many Stark shifts reported in the literature are affected by this phenomenon.

Another characteristic of the spark spectra that depends upon the atmosphere is line width. Lines that are subject to pressure broadening tend to be generally somewhat broader in argon than in nitrogen. This may be viewed qualitatively as resulting from resistance or "pressure" of the heavy argon gas that restrains the free expansion of the vapor cloud in argon. If we go in the other direction and use a very light gas, such as helium, this broadening should be considerably reduced. This is in fact observed to be the case, for very narrow lines are obtained by sparking in helium. The discharge in helium behaves somewhat irregularly, however, and it is not yet safe to draw conclusions from the observations available at present.

In order to put these qualitative observations of the interaction of spark vapor with the surrounding atmosphere on a more quantitative basis, the mechanism of collisions between gas and metal atoms must be investigated. If the collisions are assumed to be elastic, and thermal motions of the gas filling the gap are ignored, the mass ratio of the gas and metal atoms determines both the average and the maximum energy loss per collision that a metal atom moving with a high speed through the gas will suffer. If we define

$$a = \frac{(m_1 - m_2)^2}{(m_1 + m_2)^2} \tag{1}$$

where m_1 is the mass of the metal atom and m_2 the mass of a gas atom, we can use a as a parameter that determines the strength of the collisional interaction between metal and gas atoms. Thus, if the kinetic energy of the diffusing metal atom is E and the energy lost in a collision with a gas atom is ΔE, the maximum fractional energy loss suffered by a metal atom in such a collision is

$$\frac{\Delta E_{max}}{E} = 1 - a \tag{2}$$

This is obviously greatest when $a = 0$, i.e., when the masses of the colliding atoms are equal. The average logarithmic energy loss per collision can also be defined in terms of a:

$$\overline{\Delta \ln(E)} = \frac{1 + a \ln(a)}{(1-a)} = \xi \tag{3}$$

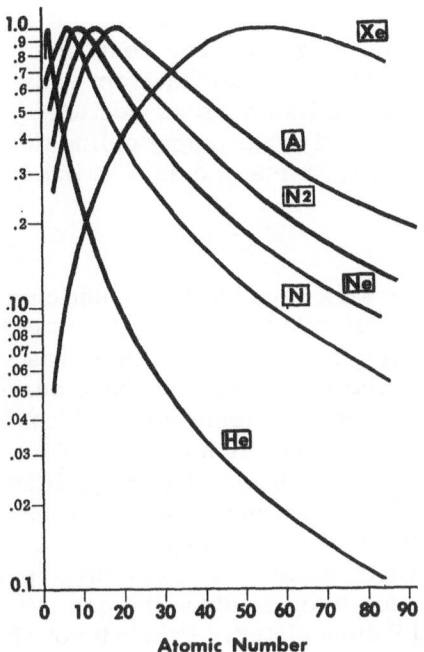

Fig. 6A. Behavior of equation (3) as a function of the atomic number of the diffusing vapor atom for various gases.

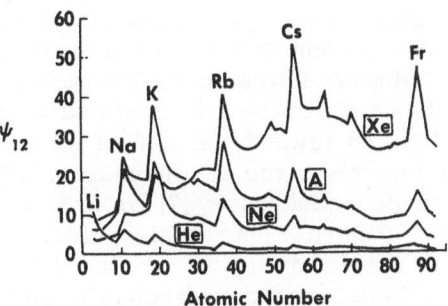

Fig. 6B. Behavior of equation (4) as a function of the atomic number of the diffusing vapor atom for various gases.

The behavior of this quantity as a function of the atomic number of the diffusing atom is shown for various gases in Fig. 6A. More significant than the energy loss of the metal atom per collision is the energy loss per unit path length through the gas. This is obtained from equation (3) by multiplying it with the collision cross section σ_{12} of the gas and metal atoms. We shall accordingly define the "slowing down power" of the gas m_2 for metal atoms m_1 diffusing through it by:

$$\psi_{12} = \sigma_{12} \, \xi \tag{4}$$

Figure 6B shows the trend of this function for a number of inert gases when Pauling's values for atomic diameters [5] and kinetic theory values for the sizes of inert gas atoms are used to estimate σ_{12}. The periodicity of these curves reflects the periodicity of the atomic volume and may not be a prominent feature of the true cross-section curves. While quantitatively unreliable, these graphs should give us qualitative insight into the behavior of metal vapor streams in different atmospheres. The special position of helium is very well brought out in the figures. For nitrogen atmosphere, the curve to be used should refer to dissociated nitrogen, since the gas in the spark gap is nearly completely dissociated. Although not shown, the curve for dissociated nitrogen closely follows the curve for neon in Fig. 6B.

A knowledge of the slowing-down power of a gas for a given metal atom is not sufficient to specify completely the behavior of the vapor streams in the gap, however, for the vapor is simultaneously subjected to an acceleration as well, and its actual motion is determined by the combined action of slowing down by collisions and acceleration due to input of electrical energy to the gap. The input of electrical energy to the gap depends upon the current density in the channel, which is very high at the beginning of each half cycle but then drops steeply [6]. As a result, the acceleration is confined to the beginning of each half cycle, and toward the end of the half cycle slowing down by collisions plays the dominant role. Measurements of jet velocities made by Sukhodrev [7] for a spark in air show this effect clearly. Figure 7 shows his velocity curves for four elements for a spark having 300 μH inductance and 0.125 μF capacitance. Time zero corresponds to spark initiation in this figure, and the gap between the beginning of the discharge

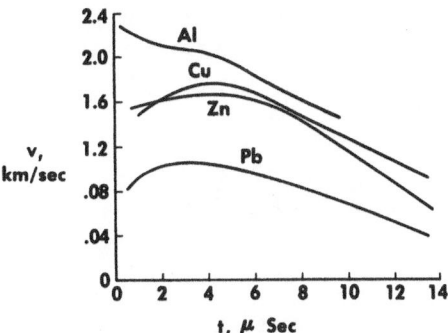

Fig. 7. Velocity of vapor jets from the cathode of a spark in air, after Sukhodrev [7]. Time zero corresponds to spark initiation. Spark constants: $C = 0.125 \, \mu$ F, $L = 300 \, \mu$H.

and beginning of the curve is due to the fact that the vapor is not initially luminous. If we are to describe the motion of the vapor jets with any degree of close approximation, we shall have to set up a differential equation which includes both slowing down and acceleration terms and whose solution will give velocity curves resembling those observed by Sukhodrev and shown in Fig. 7. The slowing-down term will have to be derived from equation (4) and will be independent of spark conditions, while the acceleration term should depend upon spark parameters. A differential equation of this type can be constructed if we assume that the acceleration experienced by the spark vapor diminishes exponentially with time, with a time constant depending upon the spark parameters. The equation becomes linear when written in terms of the variable $u = 1/v$ and then takes the form:

$$\frac{du}{dt} + Pu = Q \qquad (5)$$

where

$$Q = \tfrac{1}{2}\psi_{12}n_2 \qquad (6)$$

is the product of one-half the slowing-down power defined in equation (4) and the average number density n_2 of gas atoms in the channel. The form of the acceleration term P is determined by the hypothesis that when $Q = 0$, then $dv/dt = v_0 ae^{-at}$, where v_0

is the initial velocity (injection velocity) of the vapor, and a depends upon the circuit constants. The solution of equation (5) gives for the velocity (after applying boundary conditions and tranforming variable u to v):

$$\frac{v}{v_0} = \frac{2 + X - e^{-at}}{1 + X(2+X)at + Xe^{-at}} \tag{7}$$

where

$$X = \frac{Qv_0}{a} \tag{8}$$

Since a determines the time scale, the solution cannot be used unless some estimate for a is available. From Sukhodrev's velocity curves it would appear that setting $a = 10^6 (C/L)^{1/4}$ when t is in microseconds will result in an approximately correct time scale for application of equation (7) to his data. Quantitatively, the fit obtained to Sukhodrev's data is not too accurate, owing to the approximations made in deriving the equation and experimental uncertainties. Qualitatively, the predicted velocity curves will reproduce his velocity curves quite well if the parameters a and X are given appropriate values. Because of equation (8) they cannot be independently varied, however, and the constraints imposed on the selection of the applicable solution by equation (8) enable us to make predictions about the effect of changing circuit parameters or physical constant upon the behavior of the vapor streams in

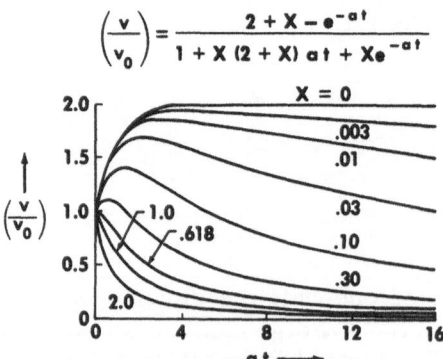

Fig. 8. Behavior of equation (7) for various values of the parameter X defined in equation (8).

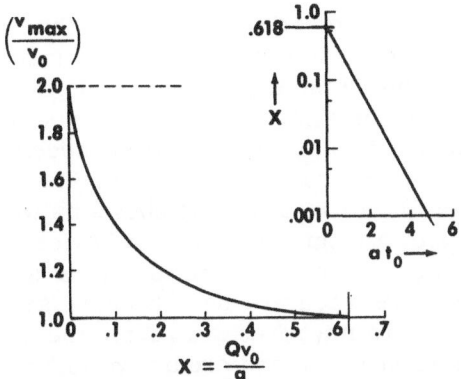

Fig. 9. Dependence of the maximum velocity attained and time when maximum is reached upon the parameter X in equation (7).

the spark gap. The parameter X determines whether or not the jet velocity goes through a maximum such as observed by Sukhodrev. It may be shown that a maximum exits only if $X < 0.618...$. Figure 8 shows the behavior of equation (7) for various values of the parameter X. The ratio of maximum velocity to the injection velocity v_0 as well as the position of the maximum on the time scale are also determined by X. Figure 9 shows these relationships graphically. The limiting forms of equation (7) when P or Q respectively tends toward zero are of interest. Thus, when P in equation (5) tends toward zero, the solution for equation (7) becomes

$$\frac{v}{v_0} = \frac{1}{1 + Q v_0 t} \qquad (9)$$

The solution consistent with our hypothesis of an exponentially diminishing acceleration pulse, when $Q = 0$, is

$$\frac{v}{v_0} = 2 - e^{-at} \qquad (10)$$

This choice leads to a maximum possible velocity just twice the initial (injection) velocity v_0, as may be seen also from Figs. 8 and 9. While these formulas present an over-simplified picture of the behavior of the spark vapor jets in the gap, they do give us some insight into the effect of various physical variables upon the behavior of the vapor. A more accurate description of the motions of the vapor would be obtained by considering the van der Waals' forces between the colliding atoms and by including thermal motions of the atoms in the gap. The situation in the gap is not sufficiently well specified, however, to allow us to apply successfully these more sophisticated concepts to the motions of the vapor. Heuristically, the solution for equation (7) shows us what parameters we should expect to influence the behavior of spark vapor streams, and thus points a way to further systematic experiments. When more accurate data become available, we can hopefully then try to interpret and revise the simple theory to give it a more quantitative and predictive form.

ACKNOWLEDGMENTS

The author is indebted to Frank Hurley of Grumman Spectroscopy Laboratory for construction of the equipment shown in Fig. 1. The work reported in this article was performed under the sponsorship of Grumman Corporate Advanced Development Program, Project No. AD-09-02.

REFERENCES

1. A. Arrak, "A Universal Method for Spectrochemical Analysis of Ferrous Alloys," Spex Speaker 8: 3 (1963).
2. A. Arrak, International Conference on Spectroscopy, College Park, Maryland (June, 1962).
3. H. Kaiser and A. Wallraff, Ann. Phys. 34: 297 (1939).
4. D. Steinhaus, H. M. Crosswhite, and G. H. Dieke, J. Opt. Soc. Am. 43: 257 (1953).
5. L. Pauling, J. Am. Chem. Soc. 69: 542 (1947).
6. L. S. Mandelstam, Spectrochim. Acta 15: 255 (1959).
7. N. K. Sukhodrev, "On Spectral Excitation in a Spark Discharge," Transactions of P. N. Lebedev Physics Inst. 15: Part 3 (Research on Spectroscopy and Luminescence) Moscow (1961).

The Interferometric Control System of the Diffraction Product's Ruling Engine

Edward Leibhardt and John DuBois

Diffraction Products, Inc.
River Forest, Illinois

This paper briefly describes an extensively revised electronic and electromechanical control system based on the system devised by H. W. Babcock [1]. The new system further reduces grating errors by the use of two Michelson interferometers, a piezoelectric crystal motor which positions the diamond in azimuth prior to the ruling of each groove, plus a completely new error-detecting and correction system.

INTRODUCTION

The mechanical ruling engine with its intermittent advance of the grating carriage is not altered by the addition of the control system, nor is the reciprocation of the diamond carriage by its modified Whitworth crank—lever action drive. In fact, if for any reason the control system should fail, the engine continues to rule the grating until it is placed back in operation.

The spacing of the grating is accomplished by change gears between the intermittent rotation of the spiral cam and the globoid worm and gear. A schematic view of the engine is shown in Fig. 1.

CONTROL SYSTEM

The most popular spacing for gratings in the visible region is 600 lines/mm; the closest approach to this with the control system is 610.42 lines/mm, or 6 fringes of the green mercury

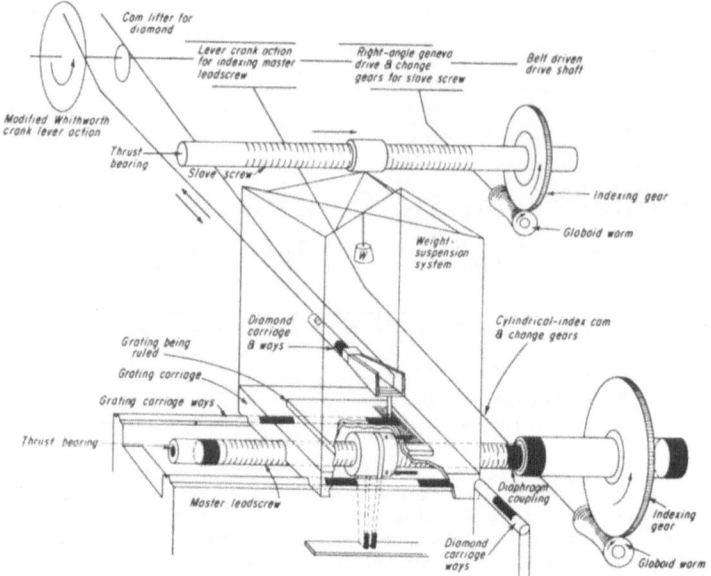

Fig. 1

light 5460.7532 Å using a 198 mercury isotope lamp excited at
2500 Mcps. This spacing is accomplished with the 118—120
[2] change gears which approximate the exact spacing to within
a few places in the fifth decimal. Other spacings are available
using different fringe multiples and also using the blue line
4358.3376 Å, which enables the engine to rule up to 4588.9 lines/
mm.

The control system is based on a Michelson interferometer
mounted on the thrust end of the ruling engine with the movable
plate mounted on the grating carriage. An interference filter
isolates either the green or the blue line of the 198 mercury
lamp and by means of a collimating lens sends a beam of
parallel light to the interferometer. The light from the interferometer is focused by a lens system onto a screen placed in
front of the photomultiplier tube. In the center of this screen is
a small aperture, a fraction of the width of the center fringe at
maximum path difference. To detect the position of the sixth
fringe, which is either appearing or disappearing at the center
of the fringe pattern, the fixed-end mirror of the Michelson
interferometer is modulated. This modulation is accomplished

by means of a piezoelectric crystal motor which vibrates at 70 cps and sweeps the fringe pattern through one-fifth of a fringe.

BAROMETRIC PRESSURE AND ENVIRONMENT COMPENSATION

In order to compensate for spurious modulations of the interferometer output, an interferometric bridge circuit has been employed. One leg of the bridge consists of a Michelson interferometer whose path difference length is fixed, but which is exposed as close as possible to the same environment as the carriage interferometer. The carriage interferometer whose path difference length is determined by the motion of the carriage comprises the opposing leg of the bridge.

An electrical output is derived from the interferometer by slightly modulating each reference path length. The resulting displacement of each fringe pattern is monitored by a stationary photomultiplier. Changes in primary path length due to carriage motion or spurious causes result in phase shift or modulation of the photomultiplier output. Each photomultiplier output is amplified and detected by product-type phase detectors. The phase detector outputs are then analog functions of the fringe pattern displacement. The carriage detector output contains the signal due to desired carriage motion plus spurious modulation of the interferometer due to temperature and barometer changes, etc. The fixed detector output contains only the spurious modulation, which is identical insofar as the optical path environments are identical.

The fixed interferometer detector output must be corrected for a difference in optical path difference length relative to the interferometer with the moving mirror. In general, the fixed interferometer path difference is $m\lambda$, where m is a constant; and the moving interferometer path difference is $n\lambda$, where $n \neq m$ and λ is the wavelength of the line used in the atmosphere surrounding the engine. The correction computer solves an equation of the form $e_{out} = K \cos m[1/n \cos^{-1}(\cos n\lambda)]$ and e_{out} becomes the corrected output of the fixed interferometer detector. The difference of these two detector outputs is derived in a precision DC differential amplifier. This output is an accurate analog function of the carriage motion alone. This bridge

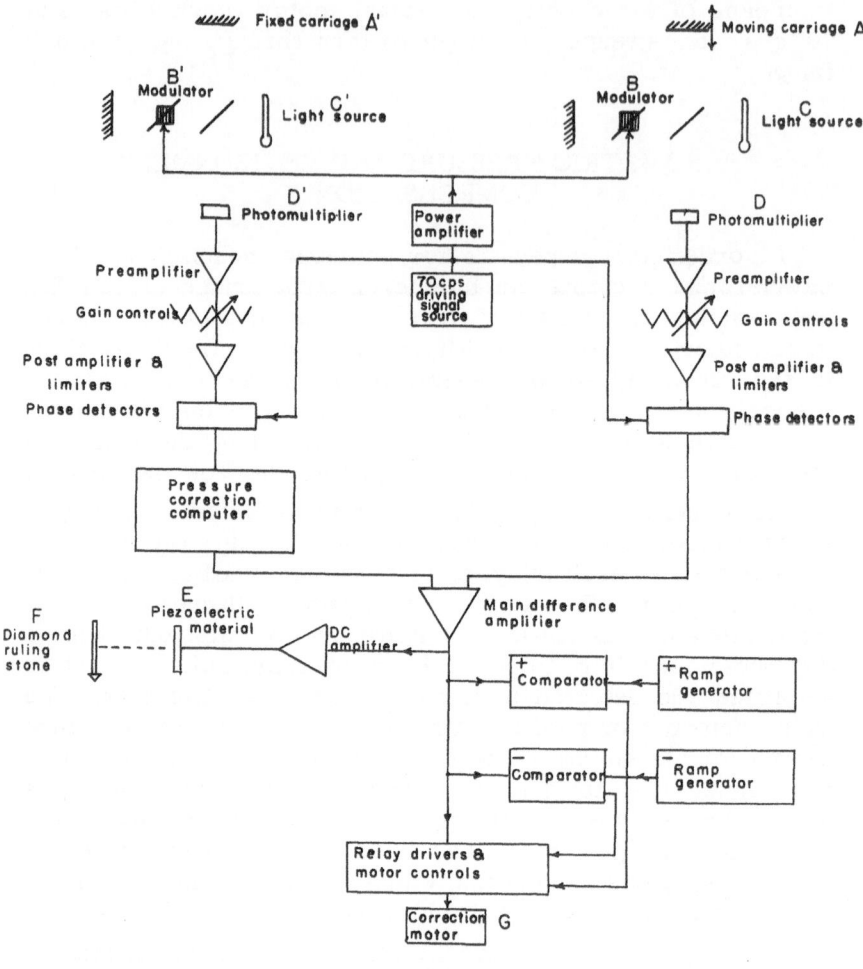

Fig. 2

(difference) output is a periodic, triangular function of carriage motion which goes to zero for carriage displacements of $p\lambda$, $p = 1, 2, 3, \ldots$. Since the desired displacement is $h\lambda$, where h is selected according to the grating spacing and engine constants, nonzero outputs constitute error signals which are used to control a correction motor. This motor causes a small addition to or subtraction from the advance of the grating carriage during the spacing period following the positioning for which an error was detected.

This small addition or subtraction is accomplished by differential gearing attached to the spiral cam mechanism. The amount of correction is determined by the length of time for which this motor runs. The analog error signal must therefore be converted to a contact closure whose length is proportional to the error. This is accomplished by initiating positive and negative linear timing ramps at the beginning of the correction period. These ramps are compared to the error signal by accurate analog comparators. The correction motor, which has already been started clockwise or counter-clockwise according to the polarity of the error signal, is stopped when the positive or negative timing ramp reaches the magnitude of the error signal. The proportionality between the error signal and the length of operation of the correction motor is controlled by adjusting the slope of the timing ramps. In order to provide correction for errors in spacing of each line immediately prior to ruling, the output voltage from the main difference amplifier is used to control the azimuth motion of the diamond. This voltage represents the spacing error corrected for variations in barometric pressure and it is amplified in a DC operational amplifier whose gain is adjusted to obtain the exact correction factor desired. The amplifier output drives a block of Clevite PZT-5 crystal, which in turn moves the diamond perpendicular to the direction of ruling. In this way, the diamond is correctly positioned prior to ruling each line. The operation of the remainder of the control system subtracts each error from successive spacing and avoids a cumulative error beyond the capability of the crystal motor. A complete description of the system, the ruling engine, and the performance of the gratings ruled by the above system will be published sometime in 1965. A block diagram of the system is shown in Fig. 2.

REFERENCES

1. H. W. Babcock, Appl. Optics 1: No. 4 (1962).
2. G. Chrystal, Text of Algebra, Part 2, Dover, chapter 32, pp. 423-434.

Transformation Functions for Photographic Response in Spectrography

J. M. McCrea

Applied Research Laboratory
United States Steel Corporation
Monroeville, Pennsylvania

Because linear plots of photographic response data simplify quantitative analysis by mass or emission spectrography, new forms for treating photographic data have been studied. The probability deviate function, derived from the cumulative normal probability distribution of statistics, has theoretical merit for descriptions of photographic response and can be used as a reference for comparing other transformations, including the similarly shaped Baker—Sampson—Seidel delta function used in spectrochemical analysis. An extensive table of a modified deviate function has been computed for use in transforming transmittance data before plotting. Data transformation with the table and use of graph paper with logarithmic and probability scales are technically equivalent. The shorter direct-plotting method is satisfactory for representing in linear or nearly linear form the response of ion-sensitive plates to both ions and photons.

Whenever photographic detectors are used for quantitative work in emission, mass, or X-ray spectrometry, the relation between the exposure, or magnitude of the stimulus, and the transparency of the resulting image is required for correlation and calculation work. Because easy plotting and simple extrapolation are highly desirable for quantitative spectrometry, the most satisfactory expressions for photographic-plate response are those that result in linear or nearly linear presentations. During research in spark-source mass spectrometry, the author has investigated a number of transformation functions for image transparency as a means of presenting both photon- and ion-exposure data in linear or nearly linear form.

The first quantitative measurements of the relation between exposure and transparency of the photographic image were carried out by Abney [1], who empirically tested a transformation based on the Gaussian or normal probability density function. His data show good agreement with a relation equivalent to:

$$T = e^{-kE^2}$$

where T is the transmittance of the image, E is the exposure, and k is a constant. Levy [2] showed the validity of this relation in the analytically important region of the curves for spectrographic plates, but other workers [3] have generally neglected it as unsatisfactory for describing photographic response characteristics. Although Abney selected the "law of error," or probability density function, from the correct field-statistics, he failed either to note the similarity in shape between the cumulative distribution and one leg of the probability density function or to analyze possible reasons for his success in curve fitting. He thus narrowly missed selecting a function that gives an excellent description of photographic response [4].

Subsequent research in photography followed a circuitous route to the desirable linear representations for quantitative spectrometry. As a result of extensive experience in the alkali industry and in photographic research, Hurter and Driffield [5] introduced a nomenclature and a logarithmic plot presentation which have since been generally used in photographic work. Their term density D, usually denoted "optical density" outside the photographic field to avoid ambiguity, is defined in terms of image transmittance T by the equation:

$$D = -\log T$$

Hurter and Driffield's characteristic plots of optical density against logarithm of exposure have frequently been used in optical spectrography; they were also in the pioneering analytical work with spark-source instruments in mass spectrometry [6,7]. Although such plots appear linear in part of the exposure range, those covering a wide range of exposure are generally sigmoid or s-shaped, as in Fig. 1. The curvature at low exposure is mathematical in origin, because D becomes zero at zero exposure, whereas the logarithm of the exposure tends to minus infinity. Curvature in other regions of the

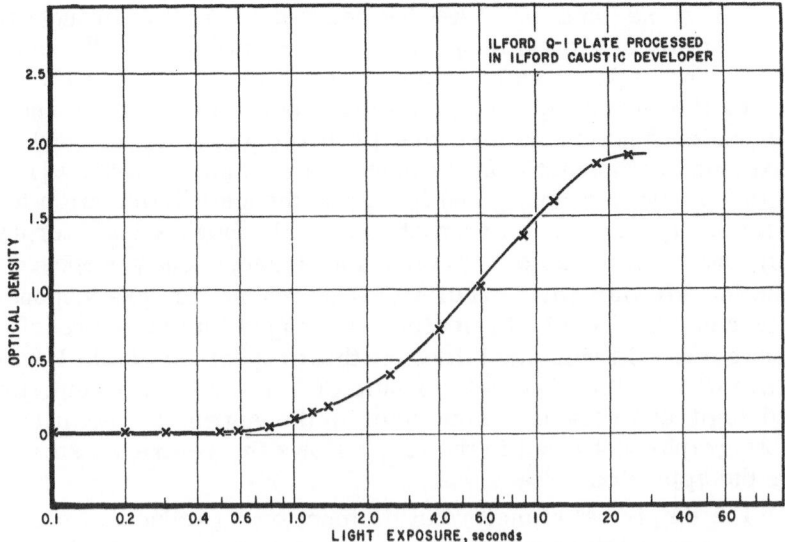

Fig. 1. Optical-density-exposure semilog presentation of response.

characteristic plots results from physical properties of the emulsion [3]. At high exposures, the characteristic curves level off because further exposure causes no additional blackening, although some light still passes through the most heavily exposed region of a plate or film.

As the term optical density became generally accepted in photographic research, many attempts were made to relate it to exposure. Most attempts resulted in ponderous power series quite ineffective for obtaining a linear representation of photographic response. During measurement of stellar temperatures by spectrographic techniques, Sampson and Baker discovered that awkward power series could be simplified by transformation of the optical density terms [8]. Baker discussed a successful transformation function Δ defined by either of the equations:

$$\Delta = \log(10^{D} - 1) \text{ or } \Delta = \log\left(\frac{1}{T} - 1\right)$$

and showed how it could be used for a nearly linear presentation of experimental photographic data [9].

The same function, likewise denoted by Δ, was brought to the attention of emission spectrographers in a talk by W. Seidel which was subsequently reported by Kaiser [10]. Although Kaiser thought Seidel's contribution important enough to name Δ the Seidel function, Seidel was probably aware of the work of Baker and Sampson through a contemporary survey article [11]. Baker has very properly and with great modesty called attention to the early use of Δ in astrophysics [12]. Hughes and Murphy [13], unaware of Kaiser's publication, independently proposed a function similar to Δ. Hull's [14] equations for ion-response data can also be algebraically rearranged into an expression linear in Δ, although he did not use this property to obtain linear calibration plots. The relationship of Δ to the logistic function and logit analysis is summarized in the Appendix. Additional bibliography not cited in the text follows the reference section for the appendix.

The empirical transformation function Δ produces a nearly linear presentation of photographic response data because it is symmetrical about $T = 0.5$ (or 50%) and because it progressively spreads the ordinate scale appropriately as T takes on values increasingly different from 0.5. An infinite number of functions have the same general characteristics as Δ. The probability-deviate function of statistics has the requisite properties, and has been very useful for handling of population data and straightening of s-shaped curves in heredity, waterworks engineering, agronomy, and toxicology. There is a close analogy between photographic plate response and mortality in living populations as a result of dosage of a toxic agent. Photographic exposure may be considered as a series of photon-stimulated catastrophic events to individuals in a population of light-sensitive grains. The experimentally observable variable, transmittance of an emulsion, is decreased cumulatively as a result of an increased number of grains being blackened. Accordingly, cumulative statistical functions should apply to photographic response.

The probability deviate is an excellent reference for comparing other functions useful in describing photographic response. It has been extensively studied and well tabulated. Galton [15] published a preliminary table and induced Sheppard [16] to compute a detailed one. Bliss's [17,18] "probits" are merely probability deviates with 5.0000 added to aid in tabulation. Republished versions of these tables are accessible in statistical compilations [19,20]. The function Δ is not as suitable for reference because it was developed empirically and is mainly used in spectrochemical analysis.

For comparison of various functions, the advantages of uniformity may be gained if symmetrical functions are reduced to a common slope at $T = 0.5$, the point of symmetry. Power series expansions of the probability-deviate function and Δ show that the slopes of the functions at $T = 0.5$ can be matched numerically if values of the probability deviate are multiplied by the factor 0.69303. For spectrochemical work, it is convenient to use a modified probability deviate with numerical values and a slope consistent with those of Δ near $T = 0.5$. Table I gives values of the probability deviate modified by use of a desk calculator in this manner, complete with differences for interpolation. Values in the central portion of the transmittance range are numerically very close to those of Δ [21]. Apart from the need for a conversion factor, Δ is an excellent approximation to the probability-deviate function, a characteristic that accounts for much of the successful use of Δ in spectroscopy. Figure 2 shows that the plot of Δ and the modified probability deviate of Table I follows a straight line of unit slope over a wide range.

Transformations with some of the general properties of the probability deviate can be devised from geometrical relations and trigonometry. If, in the circle of unit radius shown in Fig. 3, T is represented by the area intercepted between the chord and the circumference, then the distance of the chord from the center of the circle exhibits functional behavior of the type specified. The trigonometric expressions $-\tan \pi (T - 0.5)$ and $\cos^{-1} T$ exhibit similar characteristics. A very effective method of comparing the properties of such functions is to adjust for a common slope at $T = 0.5$ and compare plots of the functions against one of them as a standard. The modified probability deviate of Table I is used as a standard in the comparison given in Fig. 4. The transformation based on properties of the circle is obviously not a close approximation to the straight line representing the proability deviate. Neither are the tangent and inverse-cosine functions, which lie on opposite sides of the straight line representing perfect numerical matching. In an effort to discover a trigonometric transformation more closely approximating the probability deviate, the sine—tangent product was tested. The sine factor

TABLE I

Modified Probability Deviate in Terms of Percentage

%	.0	%	0.1	0.2	0.3	0.4	0.5	0.6	0.7	0.8	0.9	1	2	3	4	5	%
			\(Tenths of Percent for Third Column\)									\(Interpolating Differences\)					
			0.9	0.8	0.7	0.6	0.5	0.4	0.3	0.2	0.1						
50	0.0000	50	0.0017	0.0035	0.0052	0.0069	0.0087	0.0104	0.0121	0.0139	0.0157	2	3	5	7	9	49
49	.0174	51	.0191	.0209	.0226	.0243	.0261	.0278	.0295	.0313	.0330	2	3	5	7	9	48
48	.0348	52	.0365	.0383	.0400	.0417	.0435	.0452	.0469	.0487	.0505	2	3	5	7	9	47
47	.0522	53	.0539	.0557	.0574	.0591	.0608	.0627	.0644	.0661	.0678	2	3	5	7	9	46
46	.0696	54	.0714	.0731	.0748	.0766	.0783	.0801	.0818	.0836	.0853	2	4	5	7	9	45
45	0.0871	55	0.0888	0.0906	0.0923	0.0941	0.0958	0.0976	0.0994	0.1011	0.1028	2	4	5	7	9	44
44	.1046	56	.1064	.1081	.1099	.1116	.1135	.1152	.1169	.1187	.1205	2	4	5	7	9	43
43	.1223	57	.1240	.1258	.1275	.1293	.1311	.1329	.1346	.1364	.1381	2	4	5	7	9	42
42	.1399	58	.1414	.1435	.1453	.1470	.1488	.1506	.1523	.1541	.1559	2	4	5	7	9	41
41	.1577	59	.1595	.1613	.1630	.1648	.1666	.1684	.1702	.1720	.1738	2	4	5	7	9	40
40	0.1755	60	0.1773	0.1792	0.1810	0.1828	0.1846	0.1863	0.1882	0.1900	0.1918	2	4	5	7	9	39
39	.1936	61	.1953	.1972	.1990	.2008	.2026	.2044	.2062	.2081	.2099	2	4	5	7	9	38
38	.2117	62	.2135	.2153	.2172	.2190	.2208	.2227	.2245	.2263	.2281	2	4	5	7	9	37
37	.2300	63	.2318	.2337	.2355	.2374	.2392	.2410	.2429	.2447	.2466	2	4	6	7	9	36
36	.2485	64	.2503	.2521	.2540	.2559	.2577	.2595	.2614	.2633	.2652	2	4	6	7	9	35
35	0.2670	65	0.2689	0.2708	0.2726	0.2745	0.2765	0.2783	0.2802	0.2821	0.2839	2	4	6	8	9	34
34	.2859	66	.2878	.2896	.2916	.2934	.2953	.2972	.2991	.3011	.3030	2	4	6	8	10	33
33	.3049	67	.3068	.3087	.3106	.3126	.3145	.3164	.3183	.3203	.3222	2	4	6	8	10	32

%	.0	0.1	0.2	0.3	0.4	0.5	0.6	0.7	0.8	0.9	1	2	3	4	5	%
	.9 (Third)										Interpolating Differences					
				Tenths of Percent for Third Column												
		0.9	0.8	0.7	0.6	0.5	0.4	0.3	0.2	0.1						
				Tenths of Percent for Final Column												
32	.3241	.3261	.3280	.3300	.3319	.3338	.3358	.3378	.3397	.3417	2	4	6	8	10	31
31	.3437	.3456	.3476	.3496	.3515	.3535	.3555	.3575	.3595	.3614	2	4	6	8	10	30
30	0.3634	0.3654	0.3674	0.3694	0.3714	0.3734	0.3754	0.3774	0.3795	0.3815	2	4	6	8	10	29
29	.3835	.3855	.3875	.3896	.3916	.3937	.3957	.3978	.3998	.4019	2	4	6	8	10	28
28	.4039	.4060	.4081	.4101	.4122	.4143	.4164	.4185	.4205	.4226	2	4	6	8	10	27
27	.4247	.4268	.4289	.4310	.4332	.4352	.4374	.4395	.4416	.4438	2	4	6	8	11	26
26	.4458	.4480	.4501	.4523	.4544	.4566	.4588	.4609	.4631	.4652	2	4	7	9	11	25
25	0.4675	0.4696	0.4717	0.4740	0.4762	0.4784	0.4806	0.4828	0.4851	0.4873	2	4	7	9	11	24
24	.4895	.4917	.4940	.4962	.4984	.5007	.5029	.5052	.5075	.5098	2	5	7	9	11	23
23	.5120	.5143	.5166	.5189	.5212	.5235	.5259	.5282	.5305	.5328	2	5	7	9	12	22
22	.5352	.5375	.5399	.5422	.5446	.5469	.5493	.5517	.5541	.5565	2	5	7	9	12	21
21	.5589	.5613	.5637	.5661	.5686	.5710	.5734	.5759	.5783	.5808	3	5	7	10	12	20
20	0.5833	0.5858	0.5883	0.5907	0.5932	0.5957	0.5983	0.6008	0.6033	0.6059	3	5	8	10	12	19
19	.6084	.6110	.6135	.6161	.6187	.6213	.6239	.6265	.6291	.6318	3	5	8	10	13	18
18	.6344	.6370	.6397	.6424	.6450	.6477	.6504	.6531	.6558	.6585	3	6	8	10	13	17
17	.6613	.6640	.6668	.6695	.6723	.6751	.6779	.6807	.6835	.6864	3	6	8	11	14	16
16	.6892	.6921	.6949	.6978	.7007	.7036	.7065	.7095	.7124	.7154	3	6	9	12	15	15
15	0.7183	0.7212	0.7242	0.7273	0.7303	0.7333	0.7364	0.7394	0.7425	0.7456	3	6	9	12	15	14

(cont.)

TABLE I (cont.)

Tenths of Percent for Third Column (value cells) / **Tenths of Percent for Final Column** (tenths labels 0.9–0.1 below)

%	0.0	0.1 (0.9)	0.2 (0.8)	0.3 (0.7)	0.4 (0.6)	0.5 (0.5)	0.6 (0.4)	0.7 (0.3)	0.8 (0.2)	0.9 (0.1)	%
14	.7487	.7518	.7549	.7581	.7613	.7645	.7677	.7709	.7741	.7774	86
13	.7806	.7839	.7872	.7906	.7939	.7972	.8006	.8040	.8074	.8109	87
12	.8143	.8178	.8213	.8249	.8283	.8319	.8355	.8391	.8427	.8463	88
11	.8500	.8538	.8574	.8612	.8650	.8688	.8726	.8764	.8803	.8842	89
10	.8882	.8922	.8961	.9001	.9042	.9083	.9124	.9165	.9207	.9249	90
9	.9292	.9335	.9378	.9422	.9466	.9510	.9555	.9600	.9645	.9691	91
8	.9738	.9784	.9832	.9879	.9928	.9976	1.0026	1.0075	1.0126	1.0177	92
7	1.0228	1.0280	1.0333	1.0385	1.0439	1.0493	1.0548	1.0604	1.0660	1.0717	93
6	1.0775	1.0834	1.0893	1.0953	1.1014	1.1076	1.1139	1.1202	1.1267	1.1333	94
5	1.1400	1.1467	1.1536	1.1606	1.1677	1.1749	1.1822	1.1898	1.1975	1.2053	95
Δ	67	69	70	71	72	74	75	77	78	80	
4	1.2133	1.2214	1.2297	1.2382	1.2469	1.2557	1.2648	1.2741	1.2836	1.2934	96
Δ	81	83	85	87	88	91	93	95	98	101	
3	1.3035	1.3138	1.3244	1.3353	1.3466	1.3583	1.3704	1.3829	1.3958	1.4093	97
Δ	103	106	109	113	117	121	125	129	135	140	
2	1.4233	1.4379	1.4532	1.4693	1.4862	1.5040	1.5228	1.5428	1.5642	1.5873	98
Δ	146	153	161	169	178	188	200	214	231	249	
1	1.6122	1.6395	1.6695	1.7030	1.7410	1.7851	1.8380	1.9043	1.9947	2.1416	99
Δ	273	300	335	380	441	529	663	904	1469	∞	

Interpolating Differences

%	1	2	3	4	5
13	3	6	9	12	16
12	3	7	10	13	17
11	3	7	10	15	18
10	3	8	11	15	19
9	4	8	12	17	20
8	4	9	13	18	22
7	5	10	15	19	24
6	6	11	17	21	29
5	6	12	19	25	31
4					
3	Use specific differences for linear				
2	or higher order				
1	interpolation.				
0					

Sign conventions: When percent refers to proportion of light absorbed, affix negative sign to values below 50.0; when percent refers to transmittance, affix negative sign to values above 50.0.

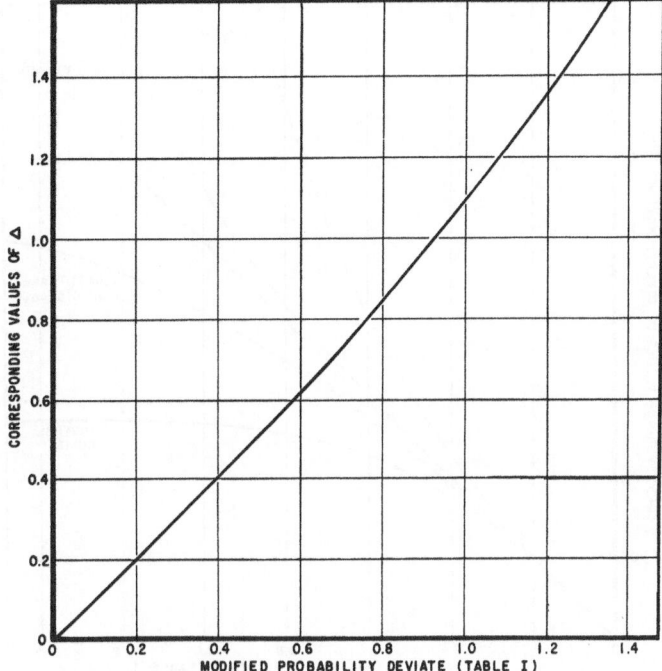

Fig. 2. Comparison of Δ with modified probability deviate.

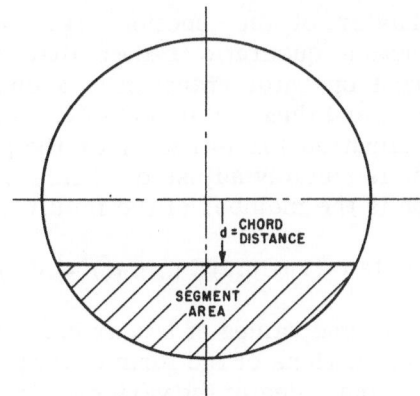

NOTE:
d IS POSITIVE IF SEGMENT IS LESS THAN HALF
THE AREA OF THE CIRCLE, NEGATIVE, IF SEGMENT
IS MORE THAN HALF THE SAME AREA

Fig. 3. Segment-area—chord-distance geometry.

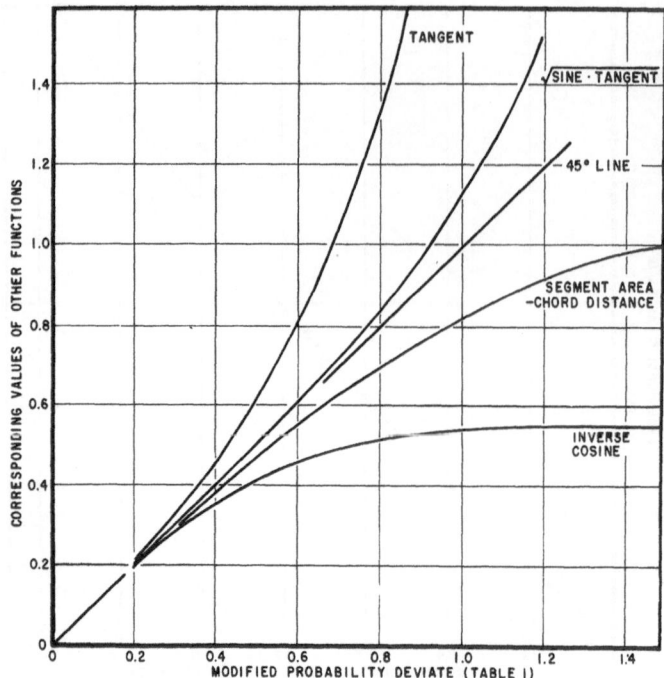

Fig. 4. Comparison of various transformation functions.

improved behavior of the function near $T = 0$ and $T = 1$, but made the function quadratic rather than linear at $T = 0.5$. The square-root operator transforms a quadratic function to linear behavior, and thus $- [\tan \pi (T - 0.5) \sin \pi.(T - 0.5)]^{1/2}$ is a good approximation for either Δ or the probability deviate when the slope is suitably adjusted. This sine—tangent combination function is the member of the family of functions:

$$- [\tan \pi(T - 0.5)]^{\alpha} [\sin \pi(T - 0.5)]^{1-\alpha}$$

for which $\alpha = 0.5$, and values of α near 0.5 give functions which are good approximations of the form of the probability deviate. Figure 4 gives some idea of the variety of transformations that can be empirically employed in attempts to obtain symmetrical linear representations of photographic response data or similar sigmoid curves.

Nonsymmetrical transformations are not as easily compared with the probability deviate, but as a matter of historical

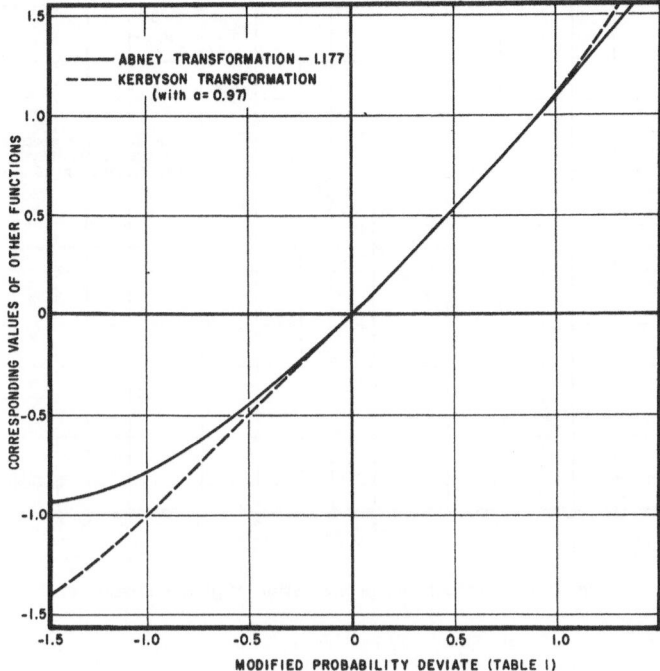

Fig. 5. Comparison of Abney and Kerbyson transformations.

interest Abney's functional relationship has been compared with the probability deviate in Fig. 5. The curvature of this comparison plot is determined by the shape of the cumulative normal distribution and the shape of its derivative, the probability density, between zero and its maximum value. The two functions have similar shapes, and accordingly Abney's transformation is for the most part satisfactory. It fails at vanishing exposure because it remains finite in that region. The inverse cosine and circle—chord transformations fail at both vanishing and infinite exposures for the same reason.

Forms similar to Δ but with additional parameters have been proposed to secure better fit with experimental points in particular cases, and Baker [12] has listed several of these. Kerbyson [22] has recently discussed the nonsymmetrical transformation $\log (1/T - a)$ in connection with a spectral calculating board. Graphically, this transformation is strictly equivalent to the form $\log [(d - T)/T]$ proposed in a report by Crosswhite and Dieke [23] as a simple and adjustable form similar to Δ.

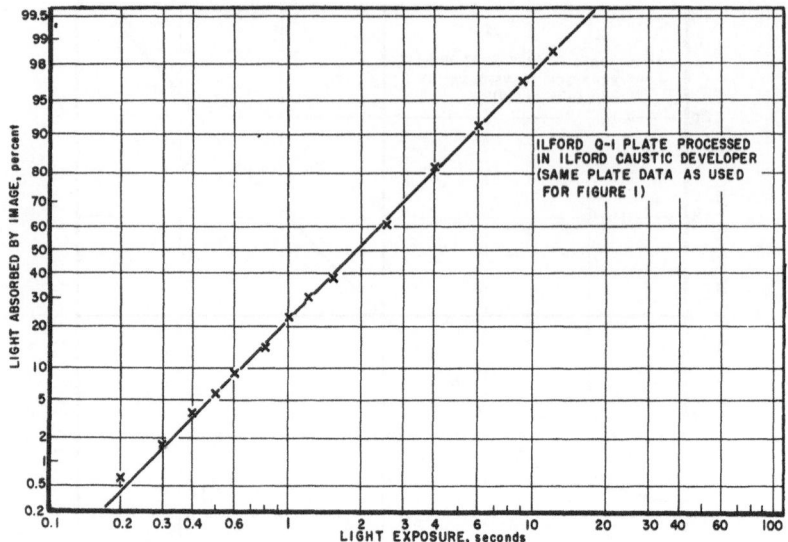

Fig. 6. Log-probability presentation of photon response.

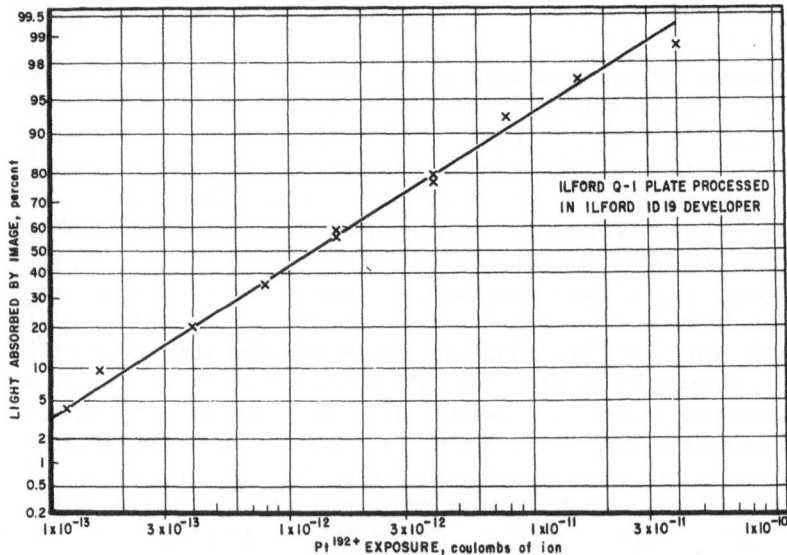

Fig. 7. Log-probability presentation of ion response.

Arrak's [24] neat deduction of the log $(1/T-a)$ transformation from theoretical consideration of the photographic process appeared almost concurrently with a journal citation of the Crosswhite and Dieke formula by Anderson [25]. The expression log $(1/T-a)$ reduces to the symmetrical Baker–Sampson–Seidel Δ for $a=1$ and to the optical density of Hurter and Driffield for $a=0$. Experimental measurements led Kerbyson to conclude that $a=0.9$ gives satisfactory transformations for the type N30 plates that he tested. The log $(1/T-a)$ transformation with $a=0.97$ and adjusted for value and slope at $T=0.5$ is shown in Fig. 5. It has the interesting mathematical characteristic of closely matching Abney's transformation between $T=0.1$ and $T=0.5$ and the probability deviate between $T=0.5$ and $T=0.9$.

With a table or a graph of a transformation function, the conversion of transmittance T to transformed values before plotting is straightforward. The transformed data are then plotted on a linear scale against corresponding exposure values. Conventionally, the scale for the exposure is logarithmic. The resultant plot will be linear or nearly linear if a suitable transformation function is used. For ion-sensitive plates, a probability-deviate transformation based on entries in Table I was successful. The applicability of the normal probability deviate to such data permits direct plotting on special papers without actual transformation of data. Hazen [26] devised graphical scales suitable for obtaining linear plots of normal probability data, and Whipple [27] added the refinement of a logarithmic scale in conjuntion with the probability scale. Probability scales have been extensively used in statistics and biometrics, but they are relatively unknown in emission and mass spectrography. Use of stantard log-probability papers is a strict graphical equivalent of making transformations with Table I for the probability deviate, and will also yield much the same linearization of curves as the use of transformations involving Δ. The technical suitability of log-probability paper for nearly linear presentation of photographic response data and its wide use in other scientific fields are strong reasons for expanded use of such representations in mass and emission spectrography. Data obtained for ion-sensitive plates have shown that log-probability plots such as those in Fig. 6 of photon response and in Fig. 7 of ion response are invariably linear, or very nearly so. The bulk of the data plotted was derived from Ilford Q2 and Q3 plates, although data for several Ilford Q1 and Kodak SWR plates were also included. The photon response characteristics are a valuable control property of ion-sensitive plates, and are useful for evaluating plate-processing conditions and the general suitability of various plates for mass spectrographic work. The log-probability plots for ion and photon data are a great aid in interpreting research results and in obtaining quantitative analytical data from solid mass spectrometers.

APPENDIX—THE LOGISTIC FUNCTION

The general logarithmic form of the function expressing logistic variation of the difference $(y - d)$ with t is given as:

$$\ln \frac{K - (y - d)}{y - d} = \ln C + rt$$

where t is the independent variable, y is the dependent variable, and K, d, r, and C are parameters [A1]. As Reed and Berkson have pointed out, several names have been used to describe special cases of the logistic function that have been applied to the interpretation of research results in different branches of science. The generic name "logistic" originated about 1840 in P. F. Verhulst's studies of human populations [A2]. The usage "autocatalytic function" is prevalent in portions of the chemical literature, and in spectroscopy the names of Baker, Sampson, and Seidel are associated with use of Δ, a term which is easily shown to be implicit in a special case of the logistic function.

The substitutions $K = 1$, $d = 0$, $y = T$, and $t = \ln E$, coupled with conversion to common logarithms, reduces the general logarithmic form of the logistic function to:

$$\log \left(\frac{1}{T} - 1 \right) = \log C + r \log E$$

The term on the left-hand side of the reduced equation is by definition Δ when T is transmittance. Thus if a plot of Δ against the logarithm of exposure is linear, T is necessarily related logistically to the logarithm of exposure. The parallel between Δ and the logistic function fortuitously extends to the history of their discovery and later reintroduction to scientific literature. Some 80 years after Verhulst's work, Pearl and Reed independently rediscovered the application of the logistic function to populations, and the term "Pearl—Reed curve" has often and erroneously been used for the logistic function [A3]. The work of Baker and Sampson preceded Seidel's speech by fourteen years, and the usage "Seidel function" as a name for Δ is likewise erroneous and denies credit to the original publications.

In 1944, Berkson [A4] initiated logit analysis, a system of statistical analysis based on the logistic function, and he has devised special graph papers to aid in this work. These arithmetic—logistic and logarithmic—logistic papers are extensively used in biometrics, and are suitable for the display of photographic response data in nearly linear form without recourse to tabulated values of Δ. These statistical papers are mathematically equivalent to and complement the papers with "Seidel" rulings employed in emission spectrography.

REFERENCES

1. W. de W. Abney, J. Camera Club 3: 93 (1889).
2. S. Levy, J. Opt. Soc. Am. 34: 447 (1944).
3. C. E. K. Mees, The Theory of the Photographic Process, The MacMillan Company, New York (1944), p. 201.
4. J. M. McCrea, submitted to Spectrochim. Acta.
5. F. Hurter and V. C. Driffield, J. Soc. Chem. Ind. 9: 455 (1890).
6. N. B. Hannay and A. J. Ahearn, Anal. Chem. 26: 1056 (1954).
7. R. D. Craig, G. A. Errock, and J. D. Waldron, in: J. D. Waldron, editor, Advances in Mass Spectrometry, Pergamon Press, New York (1959) p. 136.
8. R. A. Sampson, Monthly Notices, Roy. Astron. Soc. 83: 174 (1923); 85: 212 (1925).
9. E. A. Baker, Proc. Roy. Soc. Edin. 45: 161 (1925).
10. H. Kaiser, Spectrochim. Acta 2: 1 (1941).
11. H. Kienle, in: W. Wien and F. Harms, editors, Handbuch der Experimentalphysik (Astrophysik), Vol. 26, Akademische Verlagsgesellschaft, Leipzig (1937), p. 661.
12. E. A. Baker, J. Phot. Sci. 5: 94 (1957).
13. H. K. Hughes and R. W. Murphy, J. Opt. Soc. Am. 39: 501 (1949).
14. C. W. Hull, Mass Spectrometry Conference, ASTM Committee E-14, New Orleans (June 3-8, 1962), Paper No. 72.
15. F. Galton, Natural Inheritance, The MacMillan Company, London (1889), p. 202.
16. W. F. Sheppard, Biometrika 5: 404 (1905).
17. C. I. Bliss, Science 79: 409 (1934).
18. C. I. Bliss, Ann. Appl. Biol. 22: 134 (1935).
19. E. S. Pearson and H. O. Hartley, editors, Biometrika Tables for Statisticians, Cambridge University Press (1954), p. 112.
20. R. A. Fisher and F. Yates, Statistical Tables for Biological, Agricultural, and Medical Research, Hafner Publishing Company, New York (1953), p. 60.
21. D. C. Cordingley, L. J. Lacey, and P. J. Sandiford, Appl. Spectroscopy 8: 138 (1954).
22. J. D. Kerbyson, Spectrochim. Acta 19: 1335 (1963).
23. H. M. Crosswhite and G. H. Dieke, Bumblebee Series Report No. 202, The Johns Hopkins University, Baltimore (September, 1953).

24. A. Arrak, Appl. Spectroscopy 11: 38 (1957).
25. J. W. Anderson, Appl. Spectroscopy 10: 195 (1956).
26. A. Hazen, Tr. Am. Soc. Civ. Eng. 77: 1539 (1914).
27. G. C. Whipple, J. Frank. Inst. 182: 37, 205 (1916).

REFERENCES IN APPENDIX

A1. L. J. Reed and J. Berkson, J. Phys. Chem. 33: 760 (1929).
A2. United Nations-Secretariat, Dept. of Social Affairs, Population Div., Population Studies No. 17, The Determinants and Consequences of Population Trends (ST/SOA/Ser A/17, 1953), U. N. Publication Sales No. 1953 XIII 3, pp. 41-44 and 365.
A3. F. E. Croxton and D. J. Cowden, Applied General Statistics, ed. 2, Prentice-Hall, Inc., Englewood Cliffs (1960), pp. 310-316.
A4. J. Berkson, J. Am. Stat. Assn. 3⁰: 357 (1944).

ADDITIONAL REFERENCES

The following references are not cited in the text in the present article, but are listed here alphabetically for record purposes.

J. W. Anderson, Appl. Spectroscopy 7: 121 (1953).
A. Arrak, Appl. Spectroscopy 12: 171 (1958); 13: 143 (1959).
E. A. Baker, Proc. Roy. Soc. Edin. 47: 34 (1927); 48: 106 (1928). Pubns. Roy. Obs. Edin. 1: 17 (1949).
C. Candler, Appl. Spectroscopy 13: 97, 159 (1959). Spectrochim. Acta 8: 262 (1956).
M. Honerjaeger-Sohm and H. Kaiser, Spectrochim. Acta 2: 396 (1944).
H. Kaiser, Spectrochim. Acta 3: 159 (1948).
H. Kawano, Bull. Chem. Soc. Japan 37: 697 (1964).
S. Kinoshita, Proc. Roy. Soc. London A83: 432 (1910).
J. Noar, Photographic J. 91B: 64, 99 (1951).
R. Schmidt, Rec. Trav. Chim. 67: 737 (1948).
B. Verkerk, J. Opt. Soc. Am. 41: 1017 (1951).

Spectrographic Determination of Rhenium in Molybdenite with the DC Arc*

W. G. Schrenk and Show-jy Ho†

Department of Chemistry, Kansas Agricultural Experiment Station
Manhattan, Kansas

Rhenium in molybdenite can be determined spectrographically with DC arc excitation only by preliminary chemical concentration of the rhenium. The following procedure produced satisfactory results. After grinding, the molybdenite sample was put in solution by treatment with HNO_3 followed by evaporation just to dryness. Five milliliters HCl (0.1 M) was added, the solution was filtered, and then excess NH_4OH added. The solution was filtered again and evaporated to approximately 2 ml.

Rhenium was extracted from the above solution as the quaternary amine. Two milliliters of a tetrabutyl ammonium bromide solution (16 g/100 ml) and 2 ml of NaOH (5M) were added to the sample solution. Five milliliters of 4-methyl-2-pentanone was then added, the solution mixed, allowed to separate, and the organic layer removed. Extraction was repeated three more times and added to the first extract. The 4-methyl-2-pentanone solution was evaporated to dryness, treated with HNO_3 and evaporated to dryness again. The residue was dissolved in 1 ml of 0.1M HCl and 1 ml of 0.1M $CoCl_2$ solution added as internal standard. One-tenth milliliter of the solution thus prepared was placed on a previously prepared electrode and dried.

Spectra were obtained with a DC arc operating at 150 V and 12.5 A. Exposure time was 30 sec on SA-1 plates. D-19 developer was used.

INTRODUCTION

Spectrographic determination of rhenium in molybdenite is complicated by (a) low concentration levels of rhenium and (b) the presence of a number of elements that interfere with the more sensitive rhenium emission lines. It is necessary, therefore, to concentrate rhenium from molybdenite as well as to

---

*Contribution No. 629, Department of Chemistry, Kansas Agricultural Experimental Station.
† Present address: Chemistry Department, University of Oregon, Eugene, Oregon.

reduce the concentration of those elements that interfere with spectrographic determination prior to spectrographic analysis of the molybdenite.

Little information is available in the literature concerning quantitative spectrographic determination of rhenium. Harrison [10] was able to detect 25 ppm in uranium base materials by spectrographic methods. Waring, Worthington, and Weeks [17] reported on rhenium in minerals as part of a procedure involving 51 elements, and Peterson et al. [16] have reported on rhenium in uranite. Borovik and Gudris [3] have examined rhenium lines in the presence of molybdenum and Hurd [12] used spectral methods for qualitative identification.

A number of papers deal with chemical procedures for determining rhenium. Many separation schemes have been proposed. They include precipitations, distillation, a variety of extraction techniques, and ion exchange; all seem lengthy. Colorimetric estimation of rhenium is further complicated by interferences from even traces of molybdenum in the extracted rhenium. For those interested in chemical procedures the following selected references are suggested: 1, 7, 11, 18, and 19.

Since information concerning the determination of rhenium is so limited, the research reported here was conducted to develop a suitable quantitative spectrographic method of analysis.

TABLE I

Possible Interfering Lines That Appear Near the Rhenium Line at 3460.47 A

Element	Wavelength	Relative line intensity (W)
Rhenium	3460.47	1000
Molybdenum	3460.78	25
Manganese	3460.33	60
Chromium	3460.43	40
Copper	3459.43	25
Iron	3459.98	80

EQUIPMENT AND SUPPLIES

A large Bausch and Lomb Littrow spectrograph with quartz optics was used, with a DC arc operated at 150 V and 12.5 A as the excitation source. Spectra were recorded on Eastman SA-1 plates and developed in D-19. The rhenium emission line at 3460.5 A was used for analytical purposes. Line densities were determined with an ARL-Dietert densitometer and all plates included standards.

Potassium perrhenate (K and K Laboratories) was used for preliminary studies and to prepare standards. Reagent-quality cobalt chloride and lithium chloride were used for internal standard and buffer preparation.

PRELIMINARY STUDIES

The sensitivity of the spectrographic procedure was shown to be approximately 15 to 20 ppm with dilutions of potassium perrhenate as the source of rhenium. Samples were prepared for excitation by drying 0.1 ml of the solution on the anode electrode. Exposure time was 30 sec. Thus, the sensitivity of detection required 1.5 to 2μ g of rhenium on the electrode under our combination of conditions.

The rhenium emission line at 3460.5 A seemed most satisfactory for analytical purposes on the basis of sensitivity. There are, however, a number of elements in molybdenite that may directly interfere with rhenium at this wavelength; these are listed in Table I. It is obvious that use of the rhenium line at 3460.5 A depends on chemical separation of rhenium from molybdenum as well as from certain other elements that may occur in molybdenite.

Survey of the literature dealing with separations of rhenium from other elements indicated that a number of techniques have been studied, including distillation, [8,15] precipitations, [9,18] extractions, [7,11,14] and ion exchange [2,4-6]. Attempts to use ion exchange as a quantitative separation technique to prepare the sample for spectrographic analysis were not successful. A variety of exchange resins were tried under a variety of conditions.

After a number of preliminary techniques of solvent extraction were attempted, the one based on the work of Maeck et al.

[14] appeared most promising and was investigated further. The procedure is based on the selective extraction of rhenium as a quaternary amine complex into 4-methyl-2-pentanone. The data of Maeck et al. [14] indicated the extraction to be quite specific in strongly basic solutions. Results of preliminary experiments with known amounts of potassium perrhenate indicated success with quantitative extractions, even at concentration levels in the low ppm range.

EXTRACTION AND SPECTROGRAPHIC PROCEDURES

Following the preliminary studies, these procedures were developed to extract and concentrate rhenium from molybdenite prior to spectrographic analysis:

A weighed sample of molybdenite (0.4 to 1.0 g) was treated with concentrated HNO_3 (5 to 10 ml). The sample solution was heated until almost dry. Ten milliliters of 6 N HCl was added to dissolve the white precipitate. The brownish residue was removed by filtration and the filtrate evaporated almost to dryness. Excess NH_4OH was added and the solution filtered to remove iron, chromium, etc., and the filtrate evaporated to approximately 2 ml. Two milliliters of 0.1 M tetrabutyl ammonium bromide and 2 ml of 5 M NaOH were added. This solution was extracted by four successive treatments with 5 ml of 4-methyl-2-pentanone. The extracts were combined, evaporated to dryness, and taken up in 1 ml of 0.1 M HCl. To this solution 1 ml of internal standard or internal standard and buffer were added. One-tenth milliliter of the solution was then dried on previously prepared electrodes.

TABLE II

Determination of Rhenium in Molybdenite Using Cobalt as an Internal Standard

Sample level	No. of trials	Rhenium/g of sample (ppm)	Range (ppm)	P. E. (ppm)	P. E. (%)
High	5	821	757–880	21.1	2.5
Medium	5	407	354–456	13.6	3.4
Low	5	84.4	67–99	3.6	4.3

TABLE III

Analyses of Molybdenite for Rhenium With Standard Additions of Rhenium

Sample level	ppm added	ppm obtained	ppm sample	
High	465*	1340*	875*	Average = 880 ppm
High	930	1800	870	P. E. = 16.9 ppm
High	1395	2300	895	P. E. = 1.9%
Medium	246	610	364	Average = 368 ppm
Medium	455	825	370	P. E. = 7.1 ppm
Medium	620	990	370	P. E. = 2.0%
Low	38	124	86	Average = 86 ppm
Low	94	186	92	P. E. = 2.7 ppm
Low	186	266	80	P. E. = 3.1%

*All data are averages of duplicate determinations.

Electrodes thus prepared were arced at 150 V and 12.5 A for 30 sec. A sector set at one-fourth open was used to minimize background produced by electrodes. Standard solutions were prepared with known amounts of potassium perrhenate and carried through the extraction procedure. Spectra of standards were placed on all plates.

Cobalt was used as an internal standard. The cobalt emission line at 3453 A was apparently satisfactory for this purpose. Cobalt was selected after investigating several possibilities for the internal standard including manganese, nickel, vanadium, tungsten, and chromium.

RESULTS AND DISCUSSION

Three samples of molybdenite having low, medium, and high concentrations of rhenium were used to determine the usefulness of the procedure. Table II gives the results of analysis of the samples. Probable errors on five single determinations were 2.5, 3.4, and 4.3% for high, medium, and low levels, respectively. Averages were 821, 407, and 84.4 ppm rhenium per gram molybdenite.

TABLE IV

Determination of Rhenium in Molybdenite in the Presence of Lithium Chloride Buffer

Sample level	No. of trials	Rhenium/g of sample (ppm)	Range (ppm)	P. E. (ppm)	P. E. (%)
High	3	890	833−930	22.8	2.56
Medium	3	410	375−440	18.1	4.40
Low	3	79	74−84	2.3	2.91

To further verify these results, the same samples were analyzed with standard additions of rhenium to each sample. Data presented in Table III are average results of duplicate analyses in each case. Probable errors were lower than those obtained without standard additions of rhenium. Average concentrations of rhenium in the samples as obtained by this procedure were 880, 368, and 86 ppm, respectively, in the high, medium, and low concentration samples.

Data presented in Tables II and III were obtained without a spectrographic buffer. Use of lithium chloride as a buffer was

TABLE V

Analyses of Molybdenite for Rhenium With Lithium Chloride Buffer and Standard Additions of Rhenium

Sample level	ppm added	ppm obtained	ppm sample	
High	465*	1350*	885*	Average = 867 ppm
High	930	1740	810	P. E. = 20 ppm
High	1395	2200	805	P. E. = 2.3%
Medium	246	640	394	Average = 395 ppm
Medium	455	815	360	P. E. = 14 ppm
Medium	620	1050	430	P. E. = 3.6%
Low	38	128	90	Average = 86 ppm
Low	94	178	84	P. E. = 3.2 ppm
Low	186	270	84	P. E. = 3.8%

*All data are averages of duplicate determinations.

TABLE VI

Comparison of Spectrographic and Thiocyanate Procedures for Determination of Rhenium in Molybdenite

Sample level	Thiocyanate (ppm)	Spectrographic (ppm)	Deviation (ppm)	%
High	950			
Without buffer		821	−129	−13.6
Standard addition		880	− 70	− 7.4
With buffer		890	− 60	− 6.3
Standard addition + buffer		867	− 83	− 8.7
Medium	442			
Without buffer		407	− 35	− 7.9
Standard addition		368	− 74	−16.7
With buffer		410	− 32	⊢ 7.2
Standard addition + buffer		395	− 47	−10.6
Low	90			
Without buffer		84	− 6	− 6.7
Standard addition		86	− 4	− 4.4
With buffer		79	− 11	−11.1
Standard addition + buffer		86	− 4	− 4.4

investigated. The amount of lithium chloride used produced a final buffer concentration of 0.25 M in the solution placed on the electrodes. Results of these trials are presented in Tables IV and V. Comparisons of probable errors as shown in Table II and IV indicate little advantage from adding buffer. Similar results are shown in Tables III and V. Since the extraction procedure is highly specific, the sample does not contain much extraneous material, which apparently minimizes need for a spectrographic buffer in this system.

The spectrographic procedure also has been compared with the thiocyanate colorimetric procedure [1], which apparently has been widely used. Results are shown in Table VI. Without exception, the spectrographic data are lower than those obtained by the colorimetric technique. The average percentage difference was 9.0, 10.6, and 6.6 for the high, medium, and low level samples, respectively.

The reason for the bias is not known. Work done since that reported here gives similar results. Preliminary investigation of the extraction procedure has produced no helpful information. It may be that bias is due to the thiocyanate procedure, since it depends on efficient extraction to remove interfering ions from the test solution.

ACKNOWLEDGMENT

The molybdenite samples used in this study were kindly furnished by Mr. Fred Ward, United States Geological Survey, Denver, Colorado.

REFERENCES

1. C. J. Rodden, editor, Analytical Chemistry of the Manhattan Project, McGraw-Hill Book Co., New York (1950).
2. W. J. Blaedel, Eugene D. Olsen, and R. F. Buchanan, Sequential Ion Exchange Separation Scheme for the Identification of Metallic Radio Elements, Anal. Chem. 32:1866 (1960).
3. S. A. Borovik and N. M. Gudris, Examination of the Extreme Lines of Rhenium in the Presence of Large Quantities of Molybdenum, J. Appl. Chem. U.S.S.R. 9:937 (1936).
4. A. L. Emanuelli, The Behavior of Molybdates and Perrhenates in Ion Exchange Resins, Anal. fac. quim, Y form, Univ. Chile 10:113 (1958).
5. J. P. Faris, Adsorption of the Elements from Hydrofluoric Acid by Anion Exchange, Anal. Chem. 32:520 (1960).
6. S. A. Fisher and V. W. Meloche, Separation of Perrhenate and Molybdate Ions by Use of Anion Exchange, Anal. Chem. 24:1100 (1952).
7. W. Geilman, Colorimetric Determinations of Rhenium, Z. anorg. allgem. Chem. 208:217 (1932).
8. W. Geilman and F. Weibke, Separation of Molybdenum and Rhenium by Distillations, Z. anorg. allgem. Chem. 199:120 (1931).
9. W. Geilman and F. Weibke, Separation of Molybdenum and Rhenium with 8-hydroxyquinoline, Z. anorg. allgem. Chem. 199:347 (1931).
10. G. R. Harrison, Report A-1031, U. S. Atomic Energy Commission (April, 1944).
11. C. F. Hiskey and V. W. Meloche, Determination of Rhenium in Molybdenite Minerals, Anal. Chem. 12:503 (1940).
12. L. C. Hurd, Determination of Rhenium. I. Qualitative, Ind. Eng. Chem., Anal. Ed. 8:11 (1936).
13. L. C. Hurd and C. F. Hiskey, Determinations of Rhenium in Pyrolusite, Ind. Eng. Chem., Anal. Ed. 10:623 (1938).
14. W. J. Maeck, G. L. Booman, M. E. Kussy, and J. E. Rein, Extractions of the Elements as Quaternary Amine Complexes, Anal. Chem. 33:1775 (1961).
15. O. Mikhailova, S. Pevzner, and N. Archepiva, Separations of Molybdenum and Rhenium, Z. anal. Chem. 91:25 (1932).

16. R. G. Peterson, J. C. Hamilton, and A. T. Myers, An Occurrence of Rhenium Associated with Uranite in Coconino Counts, Arizona, Econ. Geol. 54:254 (1959).
17. C. L. Waring, H. W. Worthington, and A. D. Weeks, Microspectrochemical Analysis of Minerals, Am. Mineralog. 46:1177 (1961).
18. H. H. Willard and G. M. Smith, Determinations of Rhenium by Precipitation with Tetraphenylarsonium Chloride, Ind. Eng. Chem., Anal. Ed. 11:305 (1939).
19. A. F. Williams, Report BR-389, U. S. Atomic Energy Commission (March, 1944).

The Determination of Boron in Metal Particles Using the Copper Fluoride Evaluation Technique

M. E. Waitlevertch, Jr., K. W. Guardipee,

J. E. Paterson, and A. L. Wolfe

Westinghouse Electric Corporation
Research and Development Center
Pittsburgh, Pennsylvania

An emission spectrochemical method has been developed for the determination of boron in metal filings using the basic copper fluoride evolution technique. Filings are obtained from the samples and compared to standards. The method is applicable to the determination of boron in various metallic samples.

INTRODUCTION

A versatile method for the determination of boron in metals was required by our laboratory because of the variety of metallic samples analyzed. The procedure of converting the samples to an oxide cannot always be used because of the various sample compositions and the possible loss of boron. The determination of boron in metal filings is based on a modification of a procedure for the determination of boron and silicon in low-alloy steel by fluoride evolution [1]. The method employs the principle of thermal decomposition of basic copper fluoride (CuOHF) in a relatively low amperage DC arc to provide fluorine, which combines with boron to form a volatile fluoride. The fluoride evolution technique can be used to determine boron in the 10 to 150 ppm concentration range.

PROCEDURE

Materials

ASTM S-3 cup, C-5 counter, S-1 pedestal [2]; boiler cap; boron-free CuOHF;* boron-free graphite-size -48 on 65 mesh.

Sample Charge

20 mg CuOHF, 20 mg filings, scoop of graphite.

Electrode Preparation

The samples are prepared by obtaining filings with particle sizes -100 mesh or smaller. The sample charge is weighed and placed into the electrode in three steps: First, the boron-free CuOHF is added to the electrode and then the sample filings are added. A scoop of boron-free graphite is added on top, and the whole charge is tamped into place and capped.

Standards may be prepared by obtaining fine (-100 mesh or smaller) filings of National Bureau of Standards Boron Steels Nos. 826, 827, 828, 829, and 830, or standards may be prepared from high-purity iron by adding boric acid from standard solutions and dissolving in nitric acid (1:1). The resulting solution is dried, ignited at 400°C to convert to oxide, and ground in a dental amalgamator.

The samples and standards are exposed using parameters in Table I in duplicate on each plate.

TABLE I

Excitation and Exposure Conditions

Excitation	6 A DC
Spectral region	2300 −2900 Å
Filters	2-step (100 −10%) at slit
Slit width	10 μ
Slit length	2 mm
Pre-arc	none
Exposure period	40 sec
Analytical gap	4 mm
Emulsion	S. A. No. 1

*Commercially available: Spex Industries. Prepared from standard laboratory reagents [1].

TABLE II

Effect of Particle Size

File size	Sample	NBS boron value (ppm)	Determined boron value (ppm)
0-cut	NBS No. 826 Cr-Mo (SAE4150)	11	12
2-cut	NBS No. 826 Cr-Mo (SAE4150)	11	13
Bastard-cut	NBS No. 826 Cr-Mo (SAE4150)	11	12
Smooth-cut	NBS No. 826 Cr-Mo (SAE4150)	11	12
Rough-cut	NBS No. 826 Cr-Mo (SAE4150)	11	11

The boron concentrations are determined by plotting the ratio transmittance of boron 2496.8 Å of the standards to a set number versus boron concentration to form the analytical curve. The ratios for the samples are referred to the curve to determine the concentration of boron in the samples.

EXPERIMENTAL

The effect of particle size was studied to determine whether the particle size produces any significant difference in the reproducibility of the method. Table II indicates that there is no detectable difference between particle sizes of -100 mesh or smaller. The bastard-cut file was found to be the simplest tool for obtaining a sufficient reproducible sample quickly. All particle sizes obtained from bastard-cut are -100 mesh or smaller.

The amount of copper fluoride is critical, and at least a 1:1 ratio of basic copper fluoride to sample must be maintained to reproduce burning characteristics and background intensity. Because of the electrode size and maintenance of the 1:1 ratio, the maximum sample size is 50 mg.

Although the method was intended to be applicable to a large variety of metallic samples, it was found to be not applicable to refractory metals. The rate of fluoride evolution was, in most cases, too rapid, and severe wandering of the arc and lost boiler caps usually resulted. The method was applicable for those high-boiling metals that did not form volatile fluorides readily.

TABLE III

Agreement With Chemical Analysis

Type	Sample number	Known boron value (ppm)	Determined boron value (ppm)	Number of determinations
Low-alloy steel D	NBS 1164	50	39	6
Low-alloy steel C	NBS 1163	12	15	10
Low-alloy steel H	NBS 1168	90	93	6
Cr 18.5−Ni 9.5	NBS 443	12	14	1
Cr 20.5−Ni 10	NBS 444	33	40	6
Cr−Ni−Mo	NBS 1186	22	19	2
Waspaloy modified	NBS 1192	15	14	2
Synthetic Fe oxide	—	10	13	2
Synthetic Fe oxide	—	20	19	2
Synthetic Fe oxide	—	100	104	2
Spex Industries standard	G-2	100	130	2
Spex Industires standard	G-3	10	9	2
Carbon steel	RH 1092	65	62	2
Carbon steel	RH 1094	71	68	2
Carbon steel	RH 1095	34	37	2
Carbon steel	RH 1096	48	42	2
Carbon steel	RH 1097	142	115	2

Boron was added to synthetic iron standards as previously discussed; the results in Table III indicate that the evolution characteristics of the boron from the oxides and filings were the same. The synthetic oxide standards can therefore be adapted to the determination of boron in metal filings. Synthetic iron oxide standards can be prepared on the metal or oxide basis.

RESULTS AND DISCUSSION

The reliability of the method for various matrices can be determined from the data presented in Table III.

All of the data from boron analysis in various matrices were divided into two groups for study. Group I consisted of NBS samples only, and Group II consisted of all available samples.   No significant distinction between these groups, either in bias or precision, was observed.

The boron found was assumed to be linearly related to the boron present over the range being studied, and it was further noted that the boron found could be taken as being proportional to the boron present. When the data of group II were analyzed by conventional regression techniques, the variance was observed to increase with the amount present. The ratio of boron found to be present was investigated, and its variance appeared to be independent of the amount present. The estimated slope between boron found and present is given by the relationship:

$$y = kx$$

where y is the boron found and x is the boron present. Therefore, the slope is 0.953, with a confidence interval at the 95% level of 0.90 to 100. In addition, the estimated standard deviation for a single determination is:

$$\sigma = (0.17) \text{ boron present}$$

Thus, there is an apparent downward bias of about 5% on the average in the method and a precision (measured as the standard deviation of a single determination) of about 17%. For Group I, the corresponding results are a slope of 0.924, with a 95% confidence interval of 0.87 to 0.98 and an estimated standard deviation of $\sigma = (0.16)$ boron present. Again, there is a downward bias and an estimated precision of about 17% measured as the standard deviation.

Within the limits of accuracy and precision, the method provides advantages of speed, low cost per determination, and a general method for the determination of boron in miscellaneous metallic samples.

The method for the analysis of boron in metal filings can be adapted for samples of less than 20 mg with the appropriate variations in the copper fluoride to sample ratio, provided the concentration is high enough to produce a detectable boron line.

By use of the maximum sample charge of 50 mg, the method for boron determination can be extended to materials other than the ferrous, such as manganese, chromium, and nickel.

ACKNOWLEDGMENT

The authors would like to acknowledge the assistance of Dr. D. H. Shaffer of the Westinghouse Research Laboratories for the statistical analysis.

REFERENCES

1. J. E. Paterson and W. F. Grimes, Spectrographic determination of boron and silicon in low-alloy steel by fluoride evolution, Anal. Chem. 30: 1900–03 (1958).
2. American Society for Testing and Materials, Methods for Emission Spectro-chemical Analysis, Philadelphia (1964).

The Influence of the Thermal Conductivity of Electrodes on the Spectrochemical Analysis of Small Samples and Trace Concentrations

F. J. Haftka

Union Carbide European Research Associates
Brussels, Belgium

In the spectrochemical analysis of small samples (< 10 mg) and of trace concentrations (< 10 ppm), it is not sufficient only to use sensitive electrical excitation conditions and optically well-adapted spectrographs. Another dominant factor is the support electrode, which influences the evaporation process and codetermines the yield of radiation and the line-to-background ratio for a given sample weight. Apart from the influence of the electrode form and chemical reactions, the evaporation and therefore also the spectrum depend on the thermal properties of the electrode material. This can be demonstrated for various electrode materials such as copper, tungsten, carbon, and graphite, which fact brings us to several general conclusions regarding DC-arc analysis and especially the superiority of carbon.

Support electrodes of the element carbon can have thermal conductivities which differ by one order of magnitude. This leads to a considerable variation in the behavior of the substance to be analyzed in small quantities. A special electrode form for microsamples could be developed which would have many advantages for several spectroscopic techniques. Intensity time curves for substances with quite different boiling points enabled us to elaborate the best conditions for an analysis with high sensitivity for traces in the nanogram range and trace elements in the ppm range. Corresponding oscillographic observations of the electrical parameters helped in studying the evaporation and in ameliorating the constancy of the excitation and the reproducibility of the results.

New developments, for example, the zone-melting technique of Pfann [1], have now made it possible to produce very pure materials. It is therefore necessary to determine the degree of purity with the help of trace analysis. The problem can be solved (1) with mass spectroscopy, (2) with γ-ray spectroscopy, and (3) with emission spectroscopy, especially with DC-arc excitation, which is generally the most sensitive spectroscopic method. Unfortunately, the method of the constant-temperature arc of Addink [2] cannot be used due to lack of sensitivity and also because the method of total evaporation does not help in solving the problem [3]. Furthermore, the sample itself cannot be used as an electrode in the case of DC-arc excitation. One of the spectroscopic reasons for this is that the evaporation of the sample is influenced not only by the specific conductivity but also by the form and the heat capacity of the electrode. So, for example, metals with a high conductivity can exclude the formation of an interrupted arc and carbon electrodes can change an overdamped multisource discharge into a rectifier discharge. The spectra produced depend mainly on the size of the sample and on the place from which the discharge starts. These effects are not only important for the spectrum of the matrix but also for the lines of the impurities. This can be easily proved with copper electrodes of different diameter.

Under these circumstances, the transformation of the samples, mainly metals, into powder form was necessary, even when taking into account the possibility of the contamination of the sample by the grinding tools. It was possible [4] to pulverize metals like aluminum, indium, and gallium with particle size $< 50 \mu$, whereby the effect of particle size can be practically eliminated.

The use of powders makes it necessary to utilize another material as a supporting electrode. One possibility is to work with graphite electrodes, but since 1955 we have had very good results with electrodes of lower heat conductivity, which is in agreement with the experience of Mellichamp and co-workers [5–7]. During the development of the analytical procedures at Union Carbide European Research Associates in Brussels, we also tested supporting electrodes of copper and tungsten in order to have a wide range of properties.

It has been found, however, that tungsten is a bad supporting electrode, because the lines of the matrix elements, iron, aluminum, and arsenic, for example, are very weak compared with those of tungsten. Copper is much better as a supporting electrode, showing with iron powder, for example, line intensities comparable to those obtained using graphite electrodes.

For powders like aluminum or arsenic with low melting points, graphite is superior by one order of magnitude, if one compares the intensities of the lines of the matrix with the intensities of the lines of the trace elements.

If one uses the element carbon as supporting electrode, one has practically an ideal material to influence the evaporation in different ways. One can then choose between graphite electrodes with high electrical and heat conductivity, and carbon electrodes with low conductivity, whereby the difference can be as much as one order of magnitude. Furthermore it is possible, by changing the shape, to influence the temperature of the evaporation zone.

In the development of new methods for trace analysis, the following factors were considered:

1. An amount of sample of the order of 1 mg should be sufficient.
2. Well-exposed spectra should be obtained if (1) is fulfilled and a normal grating spectrometer is used.

TABLE I

Metals and Their Corresponding Oxides
with Melting and Boiling Points Used
for the Experiments

Metal	Melting point, °C	Boiling point, °C
As		615 subl.
Bi	271	1420
Al	660	2057
Cu	1083	2336
Fe	1535	3000
W	3370	5900
As_2O_3	193 subl.	
Bi_2O_3	820	1890
Al_2O_3	2015	3500
CuO	1026 dec.	
Fe_2O_3	1565	
WO_3	1473	
C	3652 subl.	4200

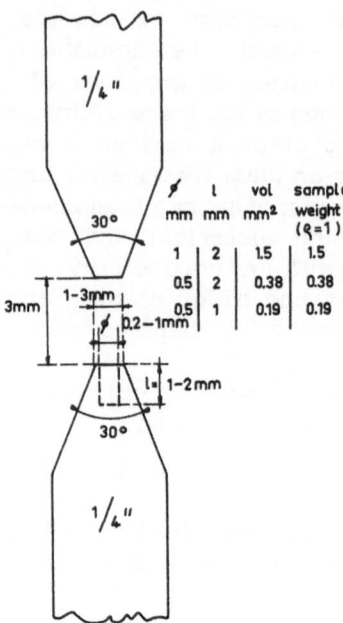

Fig. 1. Electrode combination for microsamples in powder form.

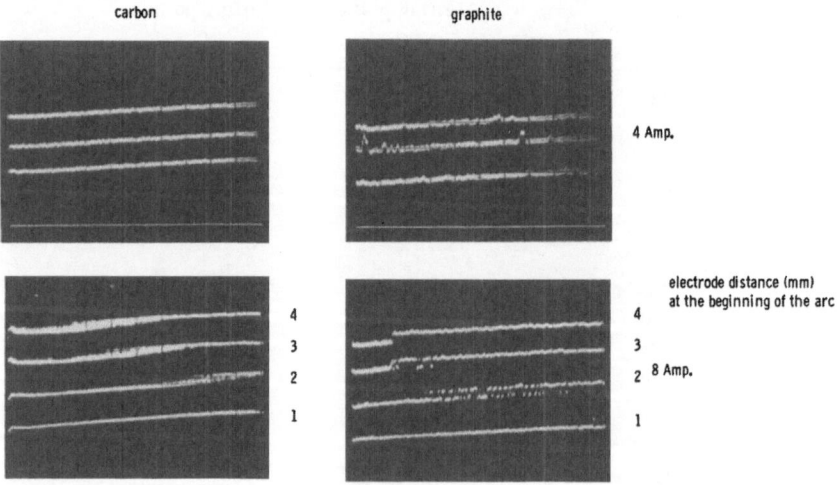

Fig. 2. Voltage-time oscillogram of carbon and graphite samples (without sample).

3. A particularly high signal-to-noise ratio, which means a low detection limit, should be obtained.

For a systematic study, some metals and their oxides (Table I) with widely different melting and boiling points were chosen.

In the course of the development of the shape of the electrode, we aimed to have as high a radiation efficiency as possible; in other words, the substance should ideally evaporate so as to give a plasma concentration sufficient for maximum radiation.   Figure 1 shows the shape of the electrodes used, the main part being the positively charged crater electrode. In principle, the arc should produce a ring-like spot on the supporting electrode, so that the vapor of the sample has to pass from the hole through the plasma and does not stream along the discharge channel on its outer sides.

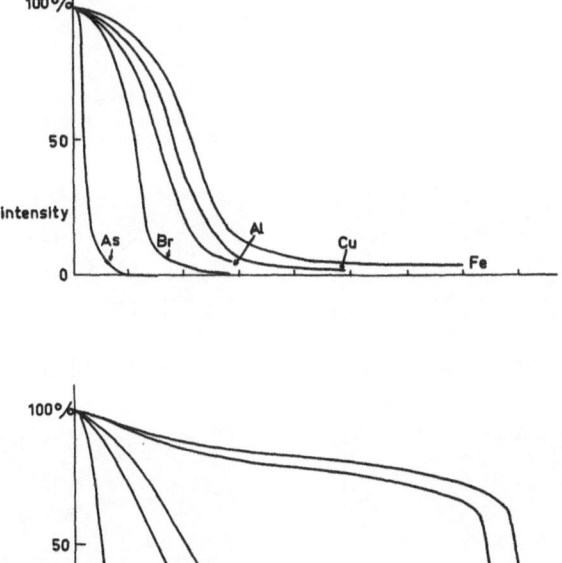

Fig. 3. Behavior of the line intensity of different matrix elements in a 4 and 8 A DC arc (schematic).

With electrodes of this shape and containing no sample, we first studied the current—voltage curves as a function of the time with the help of a double-trace oscillograph (Fig. 2). We obtained recorded curves for graphite and carbon electrodes which established the well-known visual impression that carbon electrodes burn much more smoothly. However, we observed for both types of electrode a change of the arc between two curves with a voltage difference of about 10 V. The meaning of this observation has now been investigated and will be discussed in a future paper. It can definitely be said that such unstable behavior of the arc is by no means to be excluded if analyses are carried out.

Harvey [8] was able to show that, under certain conditions, similar spectra can be obtained with different amounts of

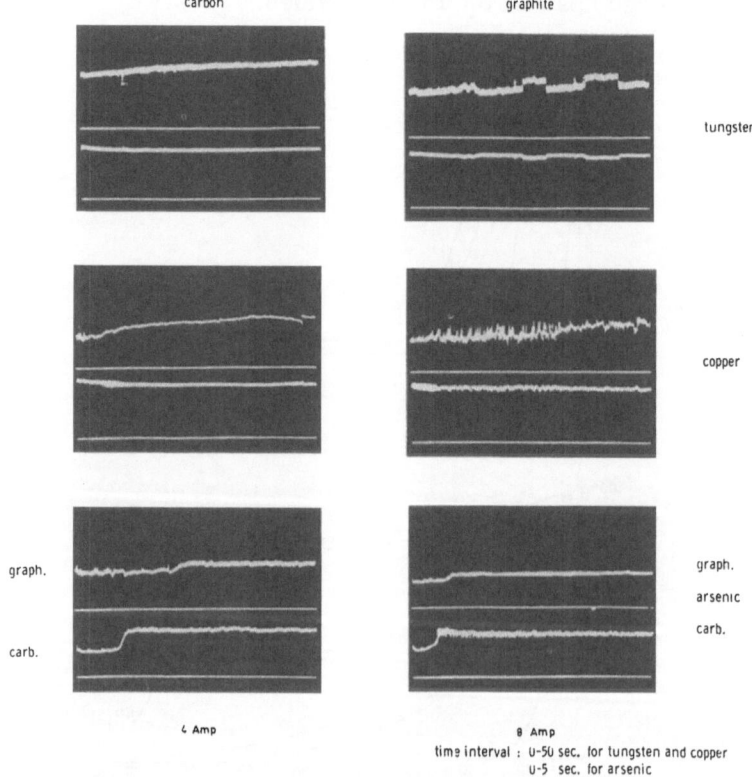

Fig. 4. Voltage-current-time oscillograms of carbon and graphite electrodes with different samples (4A).

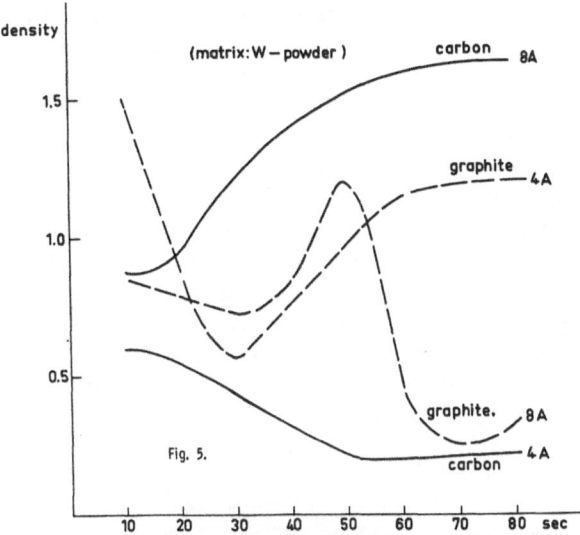

Fig. 5. Variation of the intensity of the tungsten line with time using graphite and carbon electrodes.

sample. Our work, however, was directed toward the goal of obtaining the most intense spectrum with a given amount of sample.

We started with the assumption that an intense spectrum of the matrix substance is normally also connected with a high intensity of the lines of the trace elements, ignoring the effects of selective evaporation.

At first, the behavior of the matrix spectra of selected metals as a function of time, current, and conductivity was measured. The results are summarized in Fig. 3 and are as expected. An element with a low vapor pressure evaporates much slower than an element with a high vapor pressure. The times of evaporation for different elements at 4 A can vary from 1 to 10, if one compares arsenic with tungsten. Higher currents displace the curves to notably shorter time values. The same tendency shows up if electrodes of low conductivity are used, as can be deduced from the oscillograms (Fig. 4). An increase of the voltage always means that the substance is steadily consumed, and the arc then becomes a normal carbon arc.

From these oscillograms and others which have been made on a series of different substances, it follows that the samples in carbon electrodes show a stable and smooth evaporation in nearly all cases. Although the oscillograms are very useful and informative concerning the behavior of the matrix element, they do not allow any conclusions concerning the trace elements. This is especially true in the case of tungsten, because the difference between vapor pressure of the matrix and of the trace elements is very great.

The intensity—time spectrum of tungsten (Fig. 5) shows a completely different behavior from those of the other elements, which decrease in intensity with increasing time. It is surprising that the tungsten spectra with carbon electrodes and 4 A are ten times weaker during the first 10 sec than are those with graphite electrodes. Since the temperature of the evaporation zone of the carbon electrodes is generally higher,

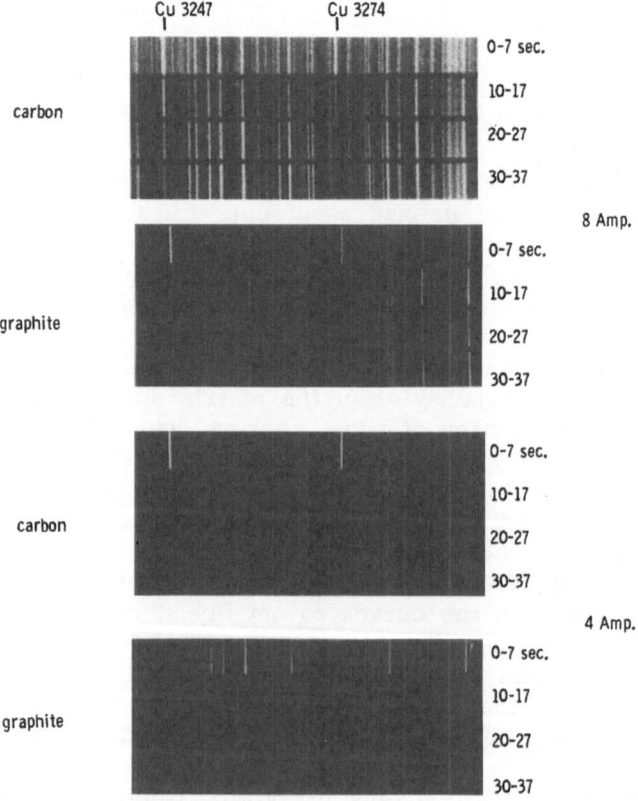

Fig. 6. Intensity time spectra of tungsten on carbon and graphite electrodes.

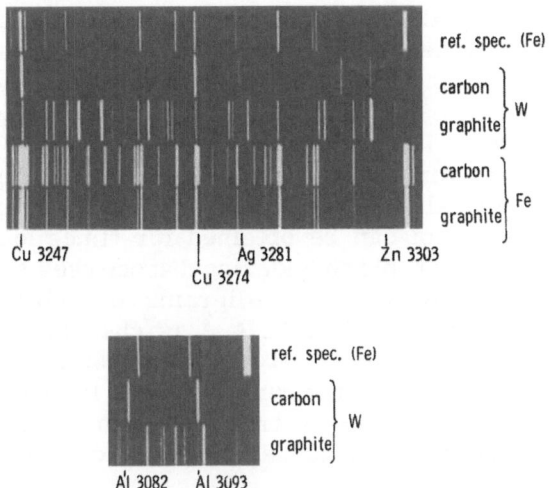

Fig. 7. Influence of the grade of the supporting electrode on the analysis of tungsten and iron powder.

further studies will have to be made before an explanation can be given. On the other hand, the behavior of the lines of the trace elements seems to be normal, because a selective evaporation due to the great differences of vapor pressure can be expected (Fig. 6). It must be mentioned that there is a remarkable difference between carbon and graphite electrodes, since carbon electrodes show—independent of the current and of the intensity of the tungsten spectrum—a highly increased intensity of nearly all trace impurities. This can clearly be seen from the results shown in Fig. 7. The spectra have been photographed during the first 10 sec. An overall view of the spectra with carbon electrodes shows the surprising fact that the matrix spectrum is more or less completely absent. However, this result was admittedly a chance occurrence, since it follows from the combination of thermal data of the substance and of the supporting electrode.

Although this example has shown that no direct conclusions from the behavior of the matrix spectrum can be drawn concerning the line intensities of the traces, in other cases it is still possible. (See copper in iron in Fig. 7.) Again, it should be mentioned that only minute amounts of material were investigated. For larger amounts of sample, the problems are much more complicated.

To determine the optimum conditions for high sensitivity, moving-plate studies were carried out. The substances were mixed with carbon powder of highest purity in the ratio 1:1 and burned at 4 or 8 A using graphite or carbon electrodes as supporting materials.

A few characteristic examples are shown in Fig. 8. With iron powder and all other elements of higher boiling points, an intense spectrum can be obtained for times longer than 1 min. This is remarkable considering that one uses a supporting electrode and only a few milligrams of substance. The behavior of the copper line 3274 Å is characteristic for all other common impurities in iron. The best signal-to-noise ratio occurs in the first spectrum, whereas later the intensity decreases greatly. This is true, in general, for carbon as well as for graphite electrodes. Furthermore, it can be said

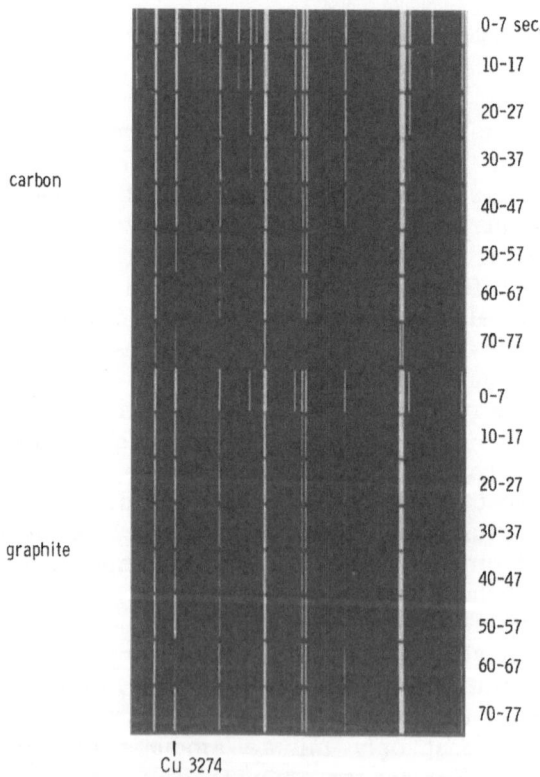

Fig. 8. Moving-plate spectrum of iron.

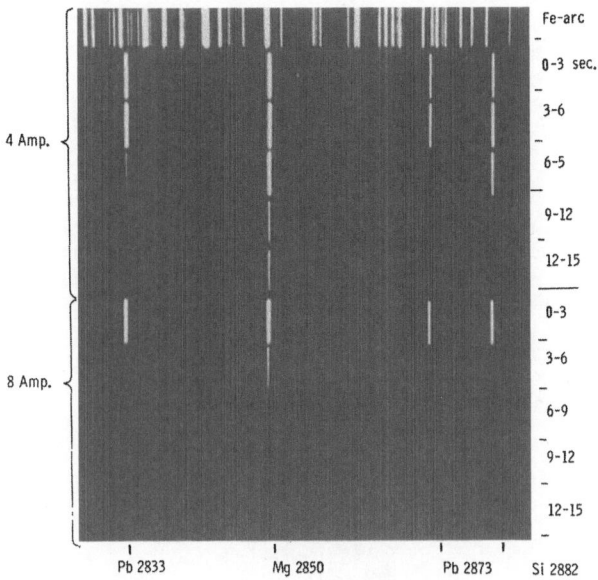

Fig. 9. Time-dependent spectra of tellurium with 4 and 8 A DC–arc excitation.

that carbon electrodes show a steadier and more reproducible behavior of the line intensities than do graphite electrodes. This can clearly be seen from the copper line.

For elements with lower boiling points, it is understandable that the spectra show a much stronger decrease of the intensities for the trace elements also. This has been found valid for other elements which have not been discussed so far. Figure 9 shows the results for tellurium and Fig. 10 the results for selenium during the extremely short time range of the first 8 sec. Especially in the case where selenium is used as matrix element, one can see that the trace elements, which can be detected in another matrix in the same concentration for much longer times, occur only for 1 sec or even less if higher currents are used.

A comparison of the spectra, when either carbon or graphite is used for the supporting electrodes, allows the following conclusions:

1. Carbon electrodes have a tendency to lead to stronger spectra of the matrix elements and especially of the trace elements within short time periods.

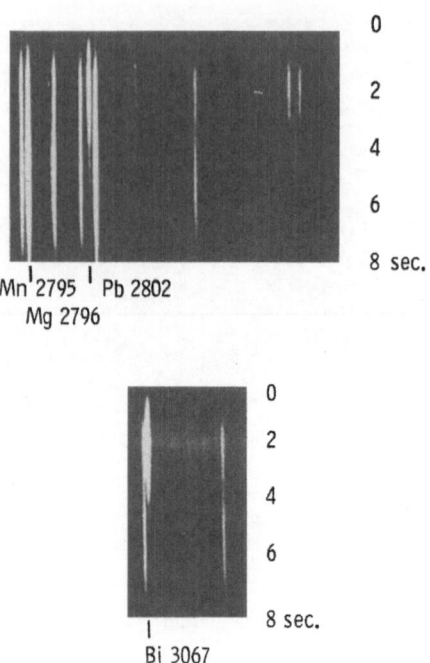

Fig. 10. Moving-plate spectrum of selenium.

2. Carbon electrodes show a stronger but smoother decrease of the intensities than do graphite electrodes.
3. With increasing current and decreasing boiling point of the matrix element, the effect of the nature of the supporting electrode on the matrix spectrum diminishes. Currents higher than 8 A are undesirable for sample amounts of 1 mg because the evaporation becomes too great.

Concerning the first point, the results for a series of elements under different conditions, using either graphite or carbon electrodes, are summarized in Table II. The intensity ratios for the matrix elements with carbon and graphite electrodes are given for the exposure time of up to 10 sec and for the case of nearly complete evaporation.

While for metals—tungsten excluded—carbon electrodes produce a stronger matrix spectrum and show an advantage over the graphite electrodes, no advantage is obtained with the

TABLE II

Rounded Intensity Ratios for the Spectrum Lines of the Matrix Elements Using Carbon and Graphite Supporting Electrodes

	4 A Exposure time		8 A Exposure time	
	0-10 sec	To evaporation	0-10 sec	To evaporation
W	1:10	1:8	1:1	8:1
Fe	2:1	1:1	5:1	2:1
Cu	4:1	2:1	2:1	2:1
Al	2:1	2:1	2:1	2:1
Bi	1:1	1:1	1:1	1:1
As	1:1	1:1	1:1	1:1
WO_3	1:10		1:1	
Fe_2O_3	2:1		1:1	
CuO	2:1		1:2	
Al_2O_3	1:2		1:2	
Bi_2O_3	1:2		1:2	
As_2O_3	1:1		1:1	

use of carbon electrodes in the case of oxides. If we compare the intensities of the traces, the carbon electrodes provide better results with metals and oxides, with a few exceptions. One can usually find a doubling of the line-to-background ratio; some examples are given in Figs. 11 and 12, in which the spectra were made with 10 or even only 2 sec exposure time.

Our results allow the conclusion that trace analyses with sample amouts in the milligram range can be carried out with the help of calibration curves, giving a reproducibility of up to 2%. The sensitivity obtained is, in favorable cases, some hundredths of ppm without enrichment and application of the

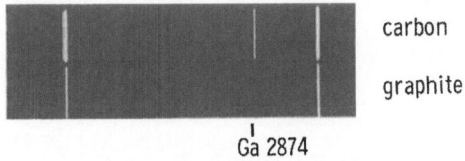

carbon

graphite

Ga 2874

Fig. 11. Comparison of the sensitivity of the Ga line 2874 in the spectrum of aluminum on carbon and graphite electrodes (4A, 10 sec).

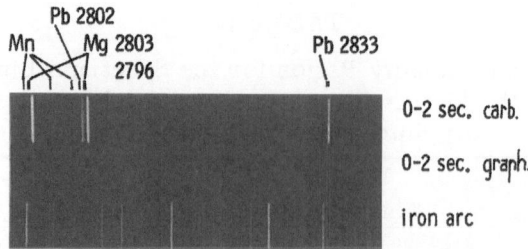

Fig. 12. Comparison of the sensitivity of a short time exposure with arsenic on carbon and graphite electrodes (8A).

cathode-layer method. The use of carbon electrodes generally results in an increase of the line-to-background ratio by a factor 2 to 10, which leads to a remarkable improvement of the detection limit. Small quantities down to 10^{-11} g can often be detected with supporting electrodes of carbon, whereas for graphite electrodes the limits of detection are one order of magnitude higher. Only short exposure times. are required.

This work, showing the superiority of support electrodes with low heat conductivity for the analysis of extremely small samples and trace concentrations, may lead to an increased application of emission spectroscopy in microanalysis.

REFERENCES

1. W. G. Pfann, Trans. AIME 194: 747 (1952).
2. N. W. Addink et al., Spec. Acta 7: 45 (1955).
3. M. Slavin, Ind. Eng. Chem., Anal. Ed. 10: 407 (1938).
4. F. J. Haftka, Coll. Spec. Int., XI, Belgrad (1963).
5. J. W. Mellichamp and R. K. Buder, Appl. Spec. 17: 57 (1963).
6. J. W. Mellichamp and J. J. Finnegan, Appl. Spec. 11: 158 (1957).
7. J. W. Mellichamp and J. J. Finnegan, Appl. Spec. 13: 126 (1959).
8. C. E. Harvey, November Meeting of Spec. N.Y. (1959).